Physical and Technical Aspects of Fire and Arson Investigation

Physical and Technical Aspects of Fire and Arson Investigation

By

JOHN R. CARROLL, P.E., B.S.C.E.

Director, National Association of Fire Investigators
International Association of Arson Investigators
National Fire Protection Association
National Society of Professional Engineers
American Society of Civil Engineers
President, Forensic Engineers, Inc.
Minneapolis, Minnesota

CHARLES C THOMAS · PUBLISHER
Springfield · Illinois · U.S.A.

Published and Distributed Throughout the World by
CHARLES C THOMAS • PUBLISHER
Bannerstone House
301-327 East Lawrence Avenue, Springfield, Illinois, U.S.A.

© *1979, by* CHARLES C THOMAS • PUBLISHER
ISBN 0-398-03785-X
Library of Congress Catalog Card Number: 78-1611

*With THOMAS BOOKS careful attention is given to all details of
manufacturing and design. It is the Publisher's desire to present books that
are satisfactory as to their physical qualities and artistic possibilities and
appropriate for their particular use. THOMAS BOOKS will be true to those
laws of quality that assure a good name and good will.*

Library of Congress Cataloging in Publication Data

Carroll, John Richard, 1919-
 Physical and technical aspects of fire and arson
investigation.

 Bibliography: p.
 Includes index.
 1. Fire investigation. 2. Arson investigation.
I. Title.
TH9180.C37   364.12   78-1611
ISBN 0-398-03785-X

Printed in the United States of America
C-1

To "them"—they know who they are

Preface

F IRE INVESTIGATORS, fire marshals, policemen and firemen will find this book useful in the investigation of fires. It will also be of interest to insurance investigators, adjusters, engineers and attorneys who are involved in litigation.

The main objective of any fire investigation is to determine the cause of the fire and eliminate all other possible causes. Once this is done, the next objective may be to develop a case for litigation. In civil code, only a preponderance of the evidence is required for the jury's decision. The legal requirements of proof in a criminal case are more strict because the liberty of the defendant is often at stake. This contrasts with a civil case in which the outcome of the trial is generally limited to an exchange of money for damages. Where the liberty of the defendant is involved, the court requires that "proof beyond a reasonable doubt" be provided by the prosecution.

In either case, criminal or civil, if it is possible to show that another cause of the fire could have been the true cause, then the prosecution or the plaintiff has failed to remove "reasonable doubt" or has failed to provide a "preponderance of the evidence."

This is particularly true in an arson case, where the case is usually based upon circumstantial evidence. In an arson case, it is usually easily shown that other causes could have started the fire.

The elimination of other possible causes is the main subject to be dealt with in this book. In eliminating other possible causes, it is probable that the true cause of the fire may be discovered. This could save an innocent person from prosecution and could guarantee prosecution of the guilty.

This book is based upon years of practical experience in the field of fire investigation. Many of the techniques presented are based upon original methods developed in determining the cause of fires.

vii

Contents

Physical and Technical
Aspects of
Fire and Arson Investigation

CHAPTER I

Introduction

COSTS OF FIRES

O N OCTOBER 6, 1975, "Dear Abby" devoted her entire column to the subject of fire prevention–and with good reason. According to the statistics of the National Fire Protection Association, more than 11,700 lives were lost in 1974 as a result of nearly 3 million fires.

Americans were appalled by the loss of life during the Vietnam conflict. The total loss of United States military personnel from 1961 to 1972 was 45,925. In that same period, deaths resulting from fires in the United States totalled 143,550. These statistics, from the Department of Defense and the National Fire Protection Association, dramatically point out the need for a more conscientious and detailed study of the origin and cause of fires which result in these tragedies. Through this vital step the loss of lives can be reduced.

In addition to the tragic death toll, the cost in dollars is in excess of $3 billion a year. Local and state government efforts to solve this problem resulted in the creation of the National Commission on Fire Prevention and Control. This commission has developed a *National Uniform Fire Reporting System* of standardized fire reporting. The purpose of this system is to gain a clearer picture of fire incidents.

CAUSES AND TRENDS

The current statistical studies by the National Fire Protection Association (NFPA) are based upon approximately 2,000 reporting fire departments from the 50 states. The data from these sources are extrapolated to a national average. The figures in Table I are estimates intended to show the relative order of magnitude of fire causes and annual trends. The NFPA cautions that the figures do not show the relative safety in the use of the various

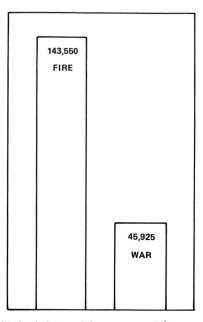

Figure 1. Deaths—United States Fires versus Vietnam War. Comparisons of deaths in United States military personnel (Army, Navy, Coast Guard, Marine Corps and Air Force) resulting from actions by hostile forces in Vietnam, 1961 through 1972, and deaths from United States fires for the same period. (Statistics from the Department of Defense and the National Fire Protection Association.) From *America Burning,* National Commission on Fire Prevention and Control, 1975.

types of materials, devices, fuels or services involved and should not be used for that purpose.

The breakdown of fires by number and percentage according to cause is shown in Table I. *Unknown Fires* are shown to exceed 12 percent of all fires and *Incendiary Fires* approximately 10 percent. Assuming that 50 percent of *Unknown Fires* may be incendiary, the total number of *Incendiary Fires* may be as high as 16 percent. Some authorities estimate this total to be as high as 25 percent.

Incendiary fires are clearly a major factor in the total fire problem. This demands that qualified fire investigators be provided so that the numbers in this category can be reduced.

TABLE I
NUMBER OF FIRES CLASSIFIED BY CAUSE (1,000s)*

Classification	%	1971	%	1972	%	1973	%	1974	%	1975
H and C equipment	16	157.7	15	155.2	15	165.8	13	160.0	13	165.6
Smoking-related	12	118.4	10	109.7	11	115.2	10	121.6	11	137.8
Electrical	16	160.9	16	162.6	16	170.7	13	165.0	12	150.5
Trash burning	3	34.4	3	36.0	3	35.2	13	177.0	12	155.5
Flammable liquids	7	64.9	6	65.2	6	67.3	4	56.1	5	61.9
Open flames and sparks	7	74.1	7	71.9	6	70.0	6	77.5	7	85.5
Lightning	2	22.2	2	22.7	2	21.6	1	16.6	1	14.2
Children and fire	7	70.4	7	69.2	7	70.8	5	59.6	5	64.2
Exposure	2	23.2	2	25.4	2	25.2	3	44.2	3	34.1
Incendiary and suspicious	7	72.1	8	84.2	9	94.3	9	114.4	11	144.1
Spontaneous ignition	2	15.7	1	15.1	1	14.9	1	11.0	0.9	11.0
Gas fires and explosions	1	8.2	1	8.7	1	9.6	1	11.9	0.8	9.5
Fireworks, explosions	0.5	4.4	1	4.2	1	4.3	1	4.2	0.3	3.9
Miscellaneous known causes	0.5	3.8	6	65.9	6	70.5	7	91.7	7	89.3
Unknown causes	17	166.2	15	154.2	14	150.5	13	159.2	11	137.3
Total	100	996.6	100	1,050.2	100	1,085.9	100	1,270.0	100	1,264.4

* From National Fire Protection Association.

This book is intended to assist the fire investigator, whether he is employed as a fireman, fire chief, insurance adjuster, consulting engineer, attorney, police officer or police investigator. It can be a valuable tool in the determination of the origin and cause of fires.

Fire service personnel will find this book useful in completing the Uniform Fire Reporting form regarding cause of ignition and the type of material first ignited in the fire.

Classification of Fires

The estimates of the National Fire Protection Association are based upon the following classifications (see Table I):

Heating and Cooking Equipment

The greatest percentage of fires in this category are the result of defective or misused equipment. Subcategories include combustibles near heaters and stoves; chimneys and flues; and hot ashes and coals.

Smoking-related

This refers to cigarettes, pipes, tobaccos or matches used in the process of lighting smoking materials.

Electrical

This category has two major subdivisions: (1) Wiring distribution equipment and (2) motors and appliances. The former is responsible for twice the number of fires as the latter. This category does not include fires originating in heating and cooking equipment.

Trash Burning

This includes rubbish and waste material fires whose ignition source is unknown.

Flammable Liquids

This category includes all flammable liquids, excluding fires originating in heating and cooking equipment.

Open Flames and Sparks

An open flame is described as one exposed to the ambient

atmosphere, such as candles, fireplaces, chimneys, propane torches and acetylene torches used in welding and cutting. Sparks include friction sparks, sparks from machinery, overheated bearings or belts.

Lightning

Lightning is unique in that it is the only cause of fire regarded as natural by the National Fire Protection Association. Floods, earthquakes, tornadoes, winds, rain or other natural occurrences are not considered in this category.

Children and Fire

This category excludes incendiary fires by minors.

Exposure Fires

This category includes fires started in combustible objects exposed to radiant heat or sparks from an adjacent fire.

Incendiary and Suspicious Fires

This classification includes those fires set by man, including arson and fires which are suspected of being arson.

Spontaneous Ignition

Spontaneous ignition includes fires which have started through chemical causation. This category includes fires started by the formation of pyrophoric carbon, agricultural, chemical and oil fires where the primary cause is not outside ignition as found in exposure fires.

Gas Fires and Explosions

This classification includes fires which have resulted from an explosion but does not include fires originating in heating and cooking equipment.

Fireworks and Explosives

These are fires which resulted from the use of pyrotechnics and/or commercial explosives, including blasting agents.

Miscellaneous Causes

These are causes other than those listed and are too numerous to classify separately.

Unknown Causes

These are the target of the fire investigator, as well as the goal of anyone involved in the fire service. The purpose of determining the cause of a fire is to establish better control over these causes so that they may be eliminated.

It is interesting to note that in 1974, the category of *Unknown Causes*, which constituted 13 percent of the total number of building fires, accounted for 38 percent of the dollar loss (more than $3 billion total loss).

Incendiary fires, which account for 9 percent of the total number of building fires, represent 17 percent of the dollar loss. According to insurance figures, fires with the largest dollar loss comprised less than 0.02 percent of all fires; the total amount lost in these fires, however, comprised 14 percent of the dollar loss for the year.

Unfortunately, the conviction record for arson fraud fires has not kept pace with the increase in the number of arson fraud fires. Arson cases upon which convictions are obtained usually involve the emotionally unstable students who burn down a high school or church in rebellion against established authority.

Several reasons for the lack of conviction of arsonists are becoming apparent. One of the reasons is the sophisticated methods by which professional arsonists start fires. Many of the methods are a result of military training received in World War II, Korea, and Vietnam. The hippie movement in the 1960s and 1970s resulted in the publication of a flood of underground material such as *The Arsonist's Cookbook,* which is openly sold in bookstores.

FIRE REPORTING

NFPA Standard 901, Uniform Coding for Fire Protection

To identify a problem, statistical evidence must be available which will isolate the various aspects of the problem. To accumulate meaningful data, it is necessary to have a wide source of information reported in a logical, concise and uniform manner. Recognizing this problem, the National Fire Protection Association established a committee on fire reporting.

The result of this committee's efforts to date has been the

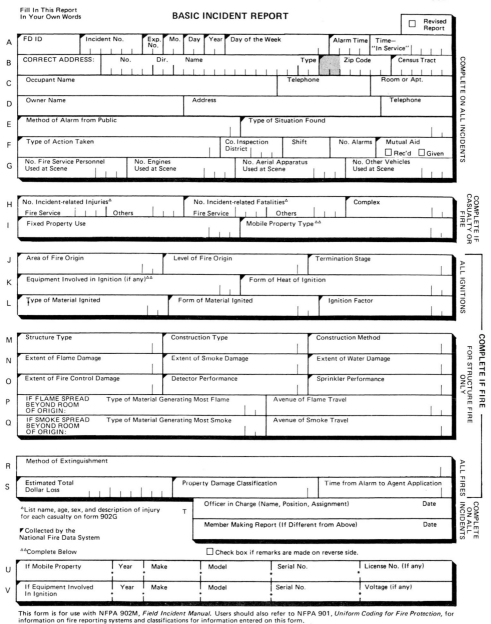

_____ **Fire Department** 902F

Fill In This Report
In Your Own Words

BASIC INCIDENT REPORT

☐ Revised Report

A FD ID | Incident No. | Exp. No. | Mo. | Day | Year | Day of the Week | Alarm Time | Time— "In Service"

B CORRECT ADDRESS: | No. | Dir. | Name | Type | Zip Code | Census Tract

C Occupant Name | Telephone | Room or Apt.

D Owner Name | Address | Telephone

E Method of Alarm from Public | Type of Situation Found

F Type of Action Taken | Co. Inspection District | Shift | No. Alarms | Mutual Aid ☐ Rec'd ☐ Given

G No. Fire Service Personnel Used at Scene | No. Engines Used at Scene | No. Aerial Apparatus Used at Scene | No. Other Vehicles Used at Scene

COMPLETE ON ALL INCIDENTS

H No. Incident-related Injuries△ Fire Service Others | No. Incident-related Fatalities△ Fire Service Others | Complex

I Fixed Property Use | Mobile Property Type △△

COMPLETE IF CASUALTY OR FIRE

J Area of Fire Origin | Level of Fire Origin | Termination Stage

K Equipment Involved in Ignition (if any)△△ | Form of Heat of Ignition

L Type of Material Ignited | Form of Material Ignited | Ignition Factor

ALL IGNITIONS

M Structure Type | Construction Type | Construction Method

N Extent of Flame Damage | Extent of Smoke Damage | Extent of Water Damage

O Extent of Fire Control Damage | Detector Performance | Sprinkler Performance

P IF FLAME SPREAD BEYOND ROOM OF ORIGIN: | Type of Material Generating Most Flame | Avenue of Flame Travel

Q IF SMOKE SPREAD BEYOND ROOM OF ORIGIN: | Type of Material Generating Most Smoke | Avenue of Smoke Travel

COMPLETE IF FIRE FOR STRUCTURE FIRE ONLY

R Method of Extinguishment

S Estimated Total Dollar Loss | Property Damage Classification | Time from Alarm to Agent Application

COMPLETE ON ALL FIRES ALL FIRES

△List name, age, sex, and description of injury for each casualty on form 902G

T Officer in Charge (Name, Position, Assignment) | Date

☛Collected by the National Fire Data System

Member Making Report (If Different from Above) | Date

△△Complete Below | ☐ Check box if remarks are made on reverse side.

COMPLETE ON ALL INCIDENTS

U If Mobile Property | Year | Make | Model | Serial No. | License No. (If any)

V If Equipment Involved In Ignition | Year | Make | Model | Serial No. | Voltage (if any)

This form is for use with NFPA 902M, *Field Incident Manual.* Users should also refer to NFPA 901, *Uniform Coding for Fire Protection,* for information on fire reporting systems and classifications for information entered on this form.

100M-12-76-FP

Printed in U.S.A.

Figure 2. Basic Incident Report. This form is for use with NFPA 902M, *Field Incident Manual.* Users should also refer to NFPA 901, *Uniform Coding for Fire Protection,* for information on fire reporting systems and classifications for information entered on this form. Courtesy of the National Fire Protection Association.

NFPA Standard 901, Uniform Coding For Fire Protection. This is available from the National Fire Protection Association, 470 Atlantic Avenue, Boston, Massachusetts 02210.

The standard consists of more than 210 pages of alphanumeric coding which establishes a uniform definition of terms used to describe property, contents, construction method and use of the property prior to a fire, including obstacles to fire rescue and fire fighting. Postfire information includes the origin and cause of the fire, flame-and-smoke spread avenues and the extent of char. The purpose of this standard is to provide guidance to the fire service in completing the Basic Incident Report, NFPA Form 902F. A copy of this form is shown in Figure 2.

NFPA Standard 902M—Fire Reporting Field Incident Manual

In addition to NFPA Standard 901, Standard 902M is necessary in developing 902F. Standard 902M will assist the fire investigator in the form required in completing this report. The most significant section of the form and Standard 902M is *General Application, Form Completion:*

> "The Basic Incident Report and the Basic Casualty Report should be in the words of the person completing the report and should give the details necessary for someone not at the incident to understand exactly what happened. The symbol N/A should be used in any data space that is Not Applicable. If information cannot be determined, the abbreviation UNDET can be used to indicate Undetermined. All data spaces in each applicable block should be completed."

The manual deals with *Lines* beginning with *Lines J* and *K* data. *Line K* includes "identifying the type of equipment that failed or permitted ignition." "The form of heat, spark, arc . . ." is identified in *Line K*. The investigator should enter the most appropriate word (heating or cooking equipment, air conditioning, refrigeration equipment, electrical distribution equipment, appliances and special equipment, processing equipment, service and maintenance equipment or heat from exposure fire). Line K data continues for several pages until finally, if indications are not found there, the investigator should refer to NFPA 901, Chapter G, or the problem should be explained under *Remarks*.

Line V describes the type of equipment involved in ignition, including the year, make, model, serial number and voltage of the equipment.

Line L concerns the type of material such as gas, flammable or combustible liquids, volatile solids, chemicals, natural products, wood, paper, fabric, textiles, furs or oil products. It also includes the form of the material first ignited such as wood shingles and the structural component.

Lines M through *Q* refer to construction details and fire spread. When in doubt, refer to NFPA 901 for clarification.

Basic Casualty Report, NFPA Form 902G, which is concerned with injuries or deaths of persons as a result of fire incidents, is also covered in 902M.

The National Commission on Prevention and Control has indicated that the *Uniform Fire Reporting System* will be adopted for national use. Therefore, familiarity with the form and content is a necessity for a fire investigator.

Use of Fire Reports

The information from fire reports becomes part of statistical studies such as those tabulated by the NFPA and NFPCA. These statistics are for various uses such as market research to determine the need for smoke detectors, fire alarm systems, sprinkler systems or other fire protection products and sociological studies regarding the frequency and loss from various causes of fire. They are also helpful in determining the safety of appliances and devices in common use, as well as the effectiveness of safety systems designed to prevent explosions or fires.

Publicizing these statistics serves to alert the general public to hazardous materials or equipment and may result in recall campaigns. In this way, dangerous items may gradually be eliminated from use.

FIRE INVESTIGATOR TRAINING

The present method of training fire and arson investigators consists primarily of on-the-job training and attendance at five-day seminars sponsored by various states, the International Association of Arson Investigators or private institutions.

According to the fourteenth edition of the *Fire Protection*

Handbook published by the NFPA, formal training in investigative procedures is conducted in very few departments. Lecture methods are used due to lack of texts or published material on fire investigation techniques.

On-the-job training is a function of the expertise of the trainer. It consists of the more experienced man training the younger, less experienced investigator. The new investigator serves an apprenticeship under the master craftsman and learns by working with the expert. On-the-job training is supplemented by attendance at seminars (see Chapter 2).

FIRE RECONSTRUCTION

The investigation of every fire, regardless of size, is mandatory if accidental fires are to be eliminated and a determination made as to how they can be prevented. The state of the art of fire investigation is greatly hindered by the reluctance of responsible parties such as fire marshals and insurance companies to investigate small fires or fires involving small claims. This is an unfortunate paradox since the smaller the fire, the easier it is to determine the origin and cause and the better the chance for a successful determination of the cause of the fire.

For example, let us trace a fire back to its inception. Assume that you have been called to investigate a fire where total destruction has occurred. The remains of the building are located in the basement following total consumption of all structural members of the building (see Fig. 3).

To determine the origin and cause of the fire, you would have to depend upon eyewitnesses, as there is no physical way of determining how this fire started. To determine the cause means that you would have to eliminate all but one possibility, starting with electrical and gas appliances, children playing with matches, spontaneous ignition, lightning, etc. The task is enormous, and the results are usually frustrating.

Assume that in this same fire, instead of total destruction, the fire consumed both first and second stories but did not destroy the structural integrity of the building. The floors and walls are still intact, and the burn pattern is plainly evident.

In this case, by measuring the depth of char and by studying the

Figure 3. Due to lack of water, rural fires are permitted to burn themselves out. As a result, the remains consist of the noncombustible walls of the foundation, plaster and metal objects which fall into the basement. Total destruction of this type presents a formidable challenge to the investigator.

burn pattern and determining how the fire was extinguished, along with other pertinent facts, the fire can be reconstructed. The reason it can be reconstructed is that the physical evidence (walls, floors and ceiling) has not been destroyed by the fire.

The chances of success in this case are greatly improved. By checking char depths in the room of origin, the *low point* (lowest point of burning) can be found. This is where the fire probably started (Fig. 4).

The next step is to determine the cause of the fire. In this same example, let us assume that the origin was the kitchen and the low point was the top of the gas range. Further assume that a kettle was found on the front burner of the range with charred contents and that the control knob regulating flow to the burner was in the full *on* position. Assume further that the housewife stated that she heard a muffled explosion when she was outside mowing the lawn and called the fire department immediately.

The conclusion that the fire investigator would draw is that the pot of stew boiled over, extinguishing the pilot light and flame for

Figure 4. This fire was extinguished before the burn pattern had been destroyed. The *low spot* is between the left side of the building and the window. The fire started on the porch, burned upward to destroy the porch roof, burned inward, spreading throughout the building, but was extinguished before total destruction.

that burner. However, gas continued to flow through the burner unit until it found a source of ignition. The resultant minor explosion ignited combustible material above the range, causing the house to burn.

Let us examine still another set of circumstances. This time we will assume that the housewife was in the living room and heard an explosion from the kitchen. She immediately called the fire department, who extinguished the fire before it had spread from the kitchen.

We have immediately confined the origin of the fire to the kitchen. In a few moments, the fire investigator can follow the burn pattern down the "V" in the wall directly to the top of the range. A few moments with the housewife would reveal that the pot had contained stew which probably boiled over, and the entire deduction process is completed as in the previous problem.

Now to compare the three fires. In the first case, the entire residence was destroyed; assume a $30,000 loss. In the second case, the interior of the building was destroyed; assume a $15,000 loss. In the third case, the interior of the kitchen was destroyed; assume a $2,000 loss. It is obvious that the fire investigator can immediately determine the cause of the third fire with a minimum expenditure of time and money. The second case would take a little longer. In the first case, regardless of the number of dollars spent, the investigator may never discover the cause of the fire.

INVESTIGATION COSTS

It is readily apparent that more could be learned about the cause of the fire per dollar value on smaller fires. In smaller fires, there is no problem eliminating other causes. In the case of the small kitchen fire, the origin of the fire was immediately detected and did not require extensive investigation. Since no other source of ignition or energy was involved, there was no need to eliminate other possible causes. This saved the investigator considerable time.

Unfortunately, economic factors prevent insurance companies from investigating small fires. Economic pressures on fire investigation facilities or governmental agencies require that they investigate the larger fires. No time is left to investigate the small ones. The experience and knowledge gained in investigating the smaller fires would lead to a more efficient and highly trained fire investigator.

Every fire, regardless of size, should be investigated to determine the cause so that the detection of technological failure of controls can be discovered at an early stage and before repetitious failures of faulty components cause a rash of fires. The manufacturing industry, particularly the firms providing gas controls for furnaces, heating appliances and water heaters, does not have recall campaigns similar to those instituted by the federal government in the automobile industry. Unless the manufacturers of gas controls begin policing themselves, it is inevitable that such a government control will evolve.

Any fire investigator who has the chance to investigate a small fire should take advantage of the opportunity. Investigation of small fires will produce results which will instill confidence in the successful investigator. There is no better training ground than a small fire.

ARSON INVESTIGATOR OR FIRE INVESTIGATOR

One problem facing the arson investigator is the motive of the arsonist. This is not a concern to the civil investigator who is only concerned with the cause. Once an arson investigator has determined that arson has been committed, his problem is to determine who started the fire and why.

The arson investigator is, by necessity, an individual accustomed to working with and questioning people. The arson investigator may not be accustomed to the scientific approach to problems. He may be more comfortable working with people than with ideas or mechanical devices. Therefore, the two functions should be separated so that the arson investigator can concentrate on *who* started the fire and *why* it was started, leaving the question of *how* the fire started to the trained fire investigators.

A *fire investigator* should be assigned the responsibility for determining the cause of every fire, regardless of size. Once arson has been established as the cause of the fire, an *arson investigator* should be called in to determine the answers to the question of who and why. This will free the arson investigator of the burdensome task of investigating accidental fires. He can then devote all of his time to arson cases instead of only 20 percent of his time, as is the present case.

The role of the fire investigator would be to determine the cause of every fire. If arson is the suspected cause, he will continue his investigation until he has eliminated all other possible causes. This would greatly enhance the role of the arson investigator. It would also bring the focus onto the true causes of accidental fires so that they would be eliminated by improved design of mechanical controls. This would also result in lowering the incidence rate of accidental fires. Insurance companies, the general public and the government would all benefit.

CHAPTER 2

Requirements of a Fire Investigator

THE FIRE INVESTIGATOR must be intelligent and physically fit, with a natural curiosity and deep analytical ability. He must be keenly observant and completely objective in his investigations. With his high moral and ethical background, he will adhere to the canon of the fire investigator's code of ethics which reads as follows:

I will bear in mind always that I am a truth-seeker, not a case-maker; that it is more important to protect the innocent than to convict the guilty.

The investigator's work is highly technical and requires on-the-job training as well as formal classroom training, seminars and self study. His training will not stop as long as fires continue to destroy persons and property. As modern technology develops mechanisms which fail and cause fires, the fire investigator is required to develop scientific techniques to determine the cause associated with the latest technological developments.

The course of a fire investigator's inquiries will require the use of tools ranging from a pick and shovel to highly sophisticated laboratory equipment such as scanning electron microscopes or atomic absorption spectrophotometers.

QUALIFICATIONS

The qualifications of a fire investigator (which includes the arson investigator) are based upon the nature of his work. His work may involve the disposition of court cases resulting in decisions which may affect the lives and properties of persons or large corporations. In a civil case, the outcome of a multimillion-dollar lawsuit may hinge upon the testimony of a fire investigator. In a criminal case, the life and liberty of a suspect is in jeopardy. Regardless of the purpose, the outcome will be influenced by the ethical, physical and educational qualifications of the fire investigator.

Ethical Qualifications

Regardless of the reason for making his investigation, the fire investigator must have a high moral and ethical background. He should be diligent in seeking the truth though the truth may not be helpful to his client or enhance his position in his organization. The fact that revealing the truth does not enhance his prestige, please his supervisors or save his company enormous sums of money should not deter the fire investigator from the impartial performance of his task. The investigator should bear in mind at all times that his work may ultimately end up as testimony in court. With this goal in mind, he should conduct himself at all times in a manner befitting the nature of his responsibility.

Since a high moral and ethical background is a result of family, religious and educational training, the fire investigator must possess the necessary qualities before becoming a fire investigator. These qualities cannot be acquired through on-the-job training, by attending seminars or by reading books. These qualities are engrained from childhood and would only be enhanced by association with persons of equally high moral and ethical backgrounds.

Physical Qualifications

Fire investigation requires the meticulous examination of piles of debris in search of clues and evidence of the origin and cause of the fire. The work entails walking through burned-out buildings and climbing burned-out stairs to the upper stories of fire-weakened structures. The debris may include heavy floor joists, sections of walls, furniture and fixtures.

Since evidence of the origin and cause of the fire usually lies beneath several feet of overlying debris, the use of mechanical equipment such as bulldozers or cranes would be a useful aid. However, their use might destroy the underlying evidence. Therefore, the fire investigator is seldom permitted this luxury. Because most debris is removed by hand, the fire investigator must be physically capable of heavy manual lifting.

The fire investigator should have keen eyesight, with or without glasses, and excellent color perception. Although the color-blind individual can function as an investigator, color perception plays an important role in the differentiation of burn patterns. Color blindness would be a handicap in this respect.

Educational Qualifications

At present, no courses are offered by any universities leading directly to a degree of "Fire and/or Arson Investigator." The majority of college courses and degrees are in the fire protection engineering or fire service area. As a result, most fire investigators are trained on the job.

The background requirements for a fire investigator at the present time demand little more than a high school education. About 50 percent of the agencies hiring fire investigators require fire fighting experience.

It should be pointed out that the defense in a civil or criminal fire case will generally hire an "expert" to investigate and determine the cause of a fire. This expert will most often be a college graduate, with a degree in physics or the sciences, who specializes in fire investigation. In the eyes of the court, this man would be considered an "expert witness." He usually has *no* fire fighting experience.

Since the credibility of an expert witness is based upon his education, training and background, it follows that the more education the fire investigator has, the more favorable his testimony will appear in the eyes of the jury. The lack of previous fire experience on the part of the expert witness does not detract from his credibility.

TRAINING

The shortage of trained fire investigators and the poor conviction record of arson cases has contributed to the action of the federal government in forming a National Fire Academy. This academy was established under the Fire Prevention and Control Act of 1974. Part of the curricula of the academy will be directed toward training in the science of fire and arson investigation. Inquiries regarding admittance to the academy can be made at the National Fire Prevention and Control Administration, Washington, D.C. This academy will fulfill a long-felt need in the fire investigation field.

University Courses

Several major universities offer courses which are helpful to a fire investigator. The majority of colleges will have courses useful to firemen, including instruction in fire behavior. None of the universities teach a course leading directly to a degree in fire investigation. Most fire-related courses are either designed for firemen or fire protection engineers.

The Illinois Institute of Technology has several excellent courses which are of great value to the fire investigator. Colleges offering similar courses include the University of Maryland, Rutgers University, Purdue University, University of Illinois, University of Miami, University of Oakland, University of California at Berkeley and others. Inquiring at your local university or local adult extension center may reveal courses not given wide publicity. By contacting your local fire department and State Fire Marshal divisions, courses given by these government entities may be tailored to meet your needs. Additional courses may be organized by contacting other interested investigators and joining in a request for seminars or classes.

On-the-Job Training

The majority of the fire investigator's education will come from on-the-job training, working with more experienced investigators. This method has a distinct advantage in giving the novice first-hand experience. There is no substitute for this experience. This type of training eliminates many questions and prevents the beginner from performing the same mistakes his mentor made as a novice. Working shoulder to shoulder with an experienced investigator is one of the quickest methods for gaining expertise in this field. This experience, coupled with other training, will quickly develop the potential of the novice investigator.

The disadvantage to this method is that the novice will learn no more than his instructor is capable of teaching. Training will be limited to the training by one individual, and all the inherent prejudices, biases and deficiencies of the master will be passed on to the student.

This lack of broad experience can be overcome by self-training, communication with other investigators, attendance at seminars and the use of in-house training devices, such as FIFI.

FIFI

As an adjunct to on-the-job training, the National Fire Protection Association has written two programs for the education and training of fire investigators. These are the *Fire Information Field Investigation Units A* and *B,* commonly referred to as *FIFI.*

FIFI consists of a set of color slides, tape cassettes, workbooks and a well thought-out Hexagonal Elimination Process (HEP) (see Fig. 5). The FIFI units are developed to be used either as a course taught by one of the more experienced fire investigators or as a self-taught course.

Included in FIFI is the *Hexagonal Elimination Process.* HEP is a systematized method for determining the important circumstances regarding the origin and cause of a fire. These are broken down into numbered descriptions which are used to fill out printed forms as the fire investigation is made. It is a very efficient checklist and a welcome tool in the technique of fire investigation. While FIFI has not been adopted as a universal teaching aid by NFPCA, it or something similar will probably be adopted so that a uniform teaching system will be used by all fire investigation agencies. This will ensure that the methods taught through this process will be more effective and result in better and more uniform fire investigation and reporting.

UNIT A. Unit A consists of eighty 2 by 2 inch color slides, one cassette tape, one instructor's guide and ten participant workbooks. This program is a training device for developing the power of observation of the student. It will teach the student to be more aware of what he observes while at the fire scene. It is primarily directed to firemen who are on the scene and have a better opportunity to observe the fire as it develops. The information conveyed to the fire investigator will lead to more rapid conclusions regarding the origin and cause of the fire. The stress in Unit A is on "looking for the unusual"–things out of place or the abundance of unexpected substances or circumstances.

UNIT B. The objective of Unit B is to improve the effectiveness of the student in determining the origin and cause of fires.

The primary purpose is to teach the student to recognize when a fire requires further investigation. The assumption is that the student is already familiar with fire behavior. A detailed checklist in logical sequence is provided. By following this checklist, the student develops the ability to determine the point of origin, the cause of the fire and whether the fire was of incendiary origin.

The fire fighter, using this unit as a guide, will then minimize water damage to vital evidence and will also maintain furniture and other material arrangements in the room so that the origin of the fire will be preserved.

HEXAGONAL ELIMINATION PROCESS. The HEP chart has been developed from the National Fire Protection Bulletin, NFPA 901, *Uniform Coding for Fire Protection,* 1976, through the efforts of the NFPA and the National Bureau of Standards. It is a systematized approach to the determination of the origin and cause of a fire. It is based upon the following chapters in NFPA Standard 901:

Chapter	*HEP Chart Section*
E	Area of origin
F	Equipment involved in ignition
G	Form of heat of ignition
H	Type of material ignited
I	Form of material ignited
J	Ignition factor

The HEP chart assists the fire investigator in answering questions of where the fire started and the circumstances involved in the start of the fire. The chart is used as a checklist starting at the *Area of Origin* and proceeding clockwise to *Form of Heat of Ignition.* Then proceed to *Type of Material Ignited* and then counterclockwise. These charts are available from the NFPA and are useful devices to ensure that nothing is overlooked during the investigative process.

Seminars

Most of the State Fire Marshal divisions participate in an annual seminar open to anyone interested in the program. The seminars are attended by a heterogenous group of fire investigators.

"Hexagonal Elimination Process" (H.E.P.) Chart

Area of Origin

0. Means of Egress
1. Assembly, Sales Areas
2. Function Areas
3. Function Areas
4. Storage Areas
5. Service Facilities
6. Service & Equipment Areas
7. Structural Areas
8. Transportation, Vehicle Areas
9. Other Areas

Type of Material Ignited

1. Gas
2. Flammable or Combustible Liquid
3. Volatile Solid, Chemical
4. Plastic
5. Natural Product
6. Wood, Paper
7. Fabric, Textile, Fur
8. Material Compounded With Oil
9. Other Type of Material Ignited

Equipment Involved In Ignition

1. Heating Systems
2. Cooking Equipment
3. Air Conditioning, Refrigeration Equipment
4. Electrical Distribution Equipment
5. Appliances & Equipment
6. Special Equipment
7. Processing Equipment
8. Service & Maintenance Equipment
9. Other Object, Exposure Fire

Form of Material Ignited

1. Structural Component or Finish
2. Furniture
3. Soft Goods, Wearing Apparel
4. Adornment, Recreational Material
5. Supplies or Stock
6. Power Transfer Equipment or Fuel
7. General Form
8. Special Form
9. Other Form of Material Ignited

Ignition Factor

1. Incendiary
2. Suspicious
3. Misuse of Heat of Ignition
4. Misuse of Material Ignited
5. Mechanical Failure or Malfunction
6. Design, Construction, or Installation Deficiency
7. Operational Deficiency
8. Natural Condition
9. Other Ignition Factor

Form of Heat of Ignition

1. Heat from Fuel-Fired or Fuel-Powered Object
2. Heat from Electrical Equipment Arcing or Overloaded
3. Heat from Smoking Materials
4. Heat from Open Flame or Spark
5. Heat from Hot Object
6. Heat from Explosives, Fireworks
7. Heat from Natural Source
8. Heat Spreading from Exposure
9. Other Form of Heat

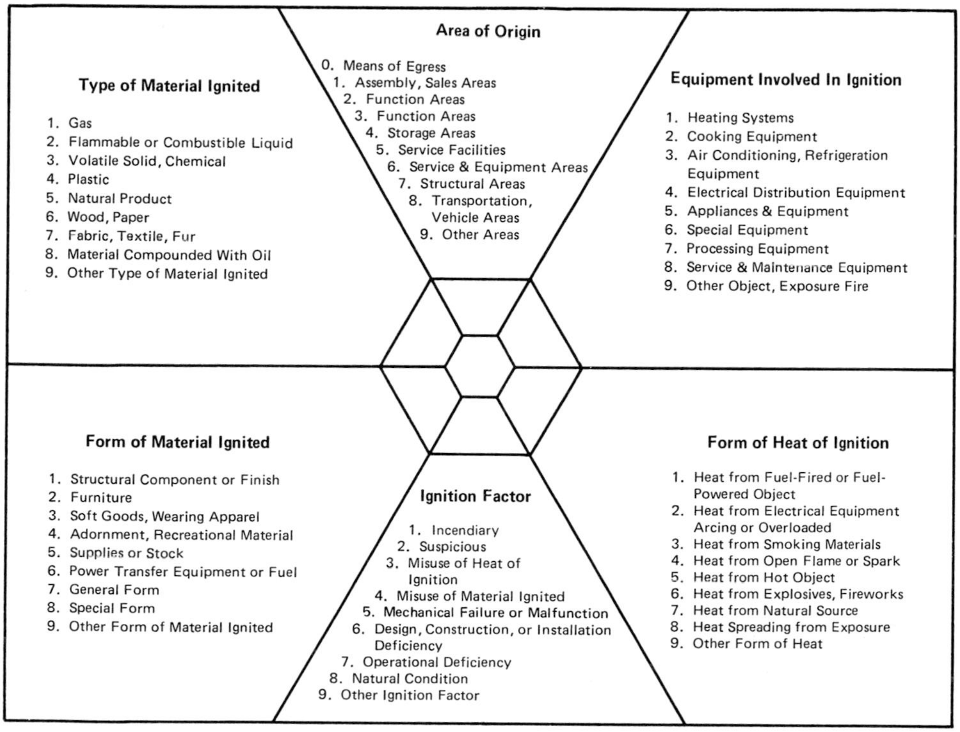

Figure 5. "Hexagonal Elimination Process" (H.E.P.) Chart. From NFPA Publication No. SPP-42, *Fire Incident Data Coding Guide,* 1976. Courtesy of National Fire Protection Association, Boston, Massachusetts.

They range from the complete novice to investigators with decades of experience. The seminars usually run for a five-day week and are basically a lecture-type presentation with slides, movies and practical demonstrations. The quality of speakers varies considerably, but on the whole the speakers are authoritative, experienced and have a message for someone in the audience. Therein lies the problem with seminars.

As the seminars draw a diversified crowd, the program has to be diversified. Lectures suitable for a beginner will be found on the same program with a sophisticated presentation of a method for determining the difference in the atomic spectrophotometer differentiation of brands of gasoline. It is impossible to please everyone and to find courses to satisfy everyone's needs. As a result, all seminars must compromise as to the tone and depth of their presentation.

If the fire investigator has the opportunity to attend a seminar, he should seek out other investigators with similar experience, discuss his needs with them, and gain knowledge from the exchange of ideas through active participation in informal sessions after the formal presentations have been made.

Your local fire marshal can tell you when the next seminar in your area will be held and how you can attend. One of the most successful seminars is the annual International Association of Arson Investigators' meeting. Information regarding this seminar can be obtained by writing the International Association of Arson Investigators, 970 Pacquin Drive, Marlboro, Massachusetts 01752.

TOOLS OF THE TRADE

The fire investigator performs a most unusual role. He investigates fires which have destroyed property (and sometimes life) in an attempt to determine how and why they occurred. Once his opinion is formed regarding the origin and cause of the fire, he must then convince his superior, or a jury, of the reasonableness, logic and credibility of his opinion. All of the equipment that he has at his disposal is designed to fulfill this objective.

As a fire investigator sits on the witness stand, his voice will convey to the jury a mental image of his observations at the fire scene. This is the basis of his opinion regarding the origin and cause of the fire. Photographs and other evidence can be used to refresh his memory and to display the scene at the time of his investigation. These are the building blocks used by the fire investigator to construct his case.

It has been said in court many times that "photographs speak for themselves." However, to the trained eye of a fire investigator, many things appear in a photograph that the investigator saw at the scene but that the jury is incapable of interpreting. The significance of the objects shown in the photographs is often lost to the jury unless explained by the fire investigator.

All of the tools available to the fire investigator are predicated toward the ultimate goal of presenting his opinion to a jury. The tools he uses are either extensions for his memory or physical confirmation of his sensory perception.

Note Taking

The goal of any investigation is the presentation of an opinion, whether it be to your supervisor or in court. Any tools which will enhance your ability to give a clear and lucid description of what you did, saw, smelled or felt at the time of your investigation are a good investment. Time taken to draw a sketch, take a photograph, jot down a note or dictate a memo is time well spent.

Handwritten notes are and have been the fire investigator's main method of recording details and "memory joggers." The axiom regarding photographs that "you can't take too many" also applies to note taking. Days, weeks, months or even years may pass between the time of your investigation and the time your opinion is presented to your supervisor or in court. The average fire investigator will spend from 10 to 40 hours per week investigating a fire. This can total up to approximately 50 to 200 fires per year. It is not surprising to see a fire investigator thumbing through notes on the witness stand two years later trying to recall whether he examined an electrical fuse box to see whether pennies had been placed behind the fuse.

The opposing attorney is certain to ask whether you performed a particular examination, inspection or measurement. If your notes indicate whether you did or did not, then it is impossible for him to confuse your recollection. Having good notes eliminates answers such as "I always do and, therefore, I must have," or "I don't remember." If the notes are written down in chronological order, it is a greater stimulant to your memory and helps recall details.

The selection of notebooks and writing implements is a matter of individual taste. Any form is admissible in court as long as it is the established method of the fire investigator and can be shown that it is his "standard practice."

One reason that the fire investigator does not take adequate notes is because, in the process of the fire investigation, he will soon get his hands dirty. Thus the fire investigator is reluctant to pick up pad and pencil to write down what he did and saw as the work progresses. He is more interested in determining the cause of the fire than in recording the steps taken. A notebook and pencil are burdensome to carry with all the other necessary paraphernalia. One tool which can eliminate the inconvenience of the note pad and pencil is the tape recorder.

Many portable tape recorders are available which are battery operated with rechargeable batteries. These provide a convenient and rapid method for recording observations, measurements and reactions to the inspection of a fire scene. Most batteries provide eight hours of continuous recording, which is more than enough to get through a normal work day. Portable hand-held tape recorders are available which use standard cassettes. The standard cassette provides a convenient method for handling tape. It can be unloaded and loaded in seconds, even with heavy gloves.

After completing the investigation, the fire investigator can have the dictated notes transcribed on standard 8½ by 11 inch stationery which fits into a loose-leaf binder or file. There is no question about legibility or concern with stains from dirty hands or gloves, snow or rain.

The process of using a tape recorder with a transcribed record of the investigation is acceptable in court. The fire investigator should read the transcription before erasing the original tape to ensure that it was correctly typed. Critical measurements or other numerical data should be double-checked for typographical errors.

After the investigation is completed, a summary sheet of the investigation should be made and a brief summary report inserted into the file. This should be done while the case is still fresh in the fire investigator's mind. It serves two purposes: it firms his opinion on how the fire started and reveals any weaknesses in his investigation.

When preparing the file report, time sheets for the daily times spent on the investigation can be used to prepare a chronological summary of the important dates. Included in the file should be sketches and scale drawings showing pertinent measurements, openings, room arrangements and sources of energy available. The sketch should show whether the doors and windows were open or closed, the condition of circulating fans, whether the air conditioner or furnace was operating and whether the circulating fan on the furnace was on or off. It can be used as a reference sheet when appearing in court for the broad picture, with the actual field notes as backup material for detailed questions. In terms of convenience, the accuracy and amount of detail resulting from the use of the tape recorder makes it one of the most important tools of the fire investigator. Its importance is exceeded only by the camera.

The Camera

The camera is the most important tool that the fire investigator can own. It enables him to preserve evidence which can be preserved in no other way. For example, the camera can record the condition of a burned-out wall so that it can be shown in court as evidence. It would be physically impossible to bring the wall into court. Conversely, it is usually impossible to bring a jury to the scene of the fire to examine a wall.

It is often impossible to preserve all of the physical evidence as it was found by the fire investigator. For example, when debris is removed, a photographic sequence of photographs, showing the various layers as they are removed, can be taken. This sequence will show in graphic detail the condition and composition of the debris. No amount of verbal record could surpass the clarity and accuracy of the photographs to portray the evidence as the investigator uncovered it.

The choice of cameras lies solely within the limits of the budget of the fire investigator. It is imperative that each photograph taken be given a high probability of success. The camera equipment selected by the fire investigator should be reliable, well maintained and compatible with his photographic skills. He should be trained in the use of the camera so that he may obtain the best possible results. The investigator should thoroughly familiarize himself with the equipment, its capabilities and its limitations. The reason is obvious. Once a photograph has been taken, the scene is altered and can never be restored to its original condition. The opportunity of rephotographing is lost forever.

The choice of film is an economic function. In today's courts, either color or black-and-white photographs are acceptable. Where the photograph reflects merely the depth of char, black-and-white film will show more detail and give better resolution and rendition than color film. When a "low spot" is revealed by the traditional "V" burn pattern on a painted surface, then color photographs are better suited to show the separation between the painted and burned surfaces. On serious cases, the use of both black and white and color photographs is warranted. By using both types of film, all of the definition and detail of the scene will be preserved.

It is good practice to limit yourself to two films—one black and white and one color film—which will yield usable 8″ by 10″ prints. The key to good photography and results is simplicity. This will be further explored in the chapter on forensic photography.

The Sniffer

The "Sniffer" is the nickname of a gas detection device used primarily to detect the presence of hydrocarbons. Technically, it is known as a "general purpose, combustible gas detector." They are very reliable and simple devices to use. "Sniffers" are manufactured by the Mine Safety Appliance Company, Bacharach Instrument Company and others. The Bacharach Model G® "Sniffer" detects hydrocarbons from 0 to 100 percent of the lower explosive limit range. Bacharach also makes a Super Snooper® which is capable of detecting hydrocarbons at very low densities. Included in this group by Bacharach is the JW Model SS-P Super-Sensitive Indicator®, which can detect hydrocarbons in the 0 to 1,000 parts per million (PPM) range.

A "Sniffer" is used to detect the presence of hydrocarbons in arson cases or suspicious fires. Should the circumstances regarding the fire indicate the possibility of arson, a "Sniffer" is used to determine whether a flammable hydrocarbon was used as an accelerant. However, experienced arson investigators have found that the human nose can detect the presence of hydrocarbons at about the same level as an ordinary "Sniffer." This will be covered in greater detail in Chapter 5.

Hand Tools

After one or two investigations, an investigator will learn the problems encountered in reaching evidence of the cause of a fire.

His first problem will be to clear away the debris. The hand tools for this purpose would include a saw for cutting structural members. Both a hand saw for cutting wood and a hacksaw for cutting metal are desirable. A pick and shovel will be useful for removing and uncovering piles of debris. A flat shovel can be used for cleaning floors. Once the major debris has been removed, smaller tools such as a mason's trowel or hand rake will be convenient for lifting layers of debris without destroying the evidence below each layer.

Once the floor has been uncovered, brooms and small brushes can be used for cleaning and sweeping the heavier and lighter materials from evidence. Once evidence has been discovered, devices for handling the material such as pliers and tweezers will be found useful.

For removal of electrical or mechanical controls, such as those found on water heaters or furnaces, an assortment of wrenches, screwdrivers, pry bars, hammers and other tools will be indispensable. When examining electrical circuitry, a small portable multimeter will be useful for checking the continuity of electrical circuits, circuit breakers and fuses.

A depth gauge, such as is used for measuring tread depth on tires, is a useful device for measuring char depth on burned structural members. The device is used for measuring the depth by pressing the measurement bar below the surface of the charred wood and reading the depth of penetration.

Equipment needed for the preservation of evidence would include empty unused paint containers for storing volatile fluids which could have been used as accelerants, or materials thought to contain them. These containers should be of the pressurized lid-type to prevent evaporation of the volatile fluids.

An ample supply of manila tags and wires for attaching them should be available to mark and identify evidence as it is discovered. Assorted sizes of polyethylene self-sealing bags can be used for storing small pieces of inert evidence. The envelopes can be identified by inserting a tag within the bag, or an adhesive tag can be attached to the envelope with a written description.

After completion of the initial investigation, the fire investigator needs tools to take measurements of the rooms and building. This would include a fifty- or one hundred-foot tape measure with a self-hooking end and an additional ten- or twenty-five-foot self-retracting tape. A carpenter's hand level will be useful for checking the slope of floors and to see whether the walls and doors are out of plumb. It can also be used for determining the pitch of roofs.

Lights

Due to the nature of burned wood, the illumination in a burned-out building is exceedingly dim. Most of the light entering will be absorbed by the dark surfaces of the burned-out building. Artificial lighting is a must. If the structure's electrical system has been destroyed by fire, it may be necessary to run extension cords from nearby buildings. The fire investigator should also have extension cords of adequate size to carry sufficient floodlights for proper illumination of the scene. Under no conditions should the fire investigator attempt to work using hand-held flashlights. While these sources of illumination are adequate for a preliminary inspection, they should only be used when no other source is available. The importance of adequate lighting on a fire scene cannot be overemphasized.

Portable floodlights should be available to the investigator so that an entire room can be examined with minimum readjustment of the floodlights. By being able to see an entire room at one time, a rapid comparison of the burn pattern can be made visually while standing on one side of the room. The results are well worth the effort it takes to set up the additional lights.

If the investigator is photographing the scene on black-and-white film, ordinary tungsten lights are suitable as long as the illumination level is uniform. The illumination from the floodlights can be used for the light source rather than using flash photography.

If the photographer is using color film, the illumination should be provided by appropriate photoflood lamps to give a color temperature matching the film used. This will be covered in greater detail in Chapter 13. The object in using adequate lighting is twofold. It enables the fire investigator to see the room as an entity rather than one small portion at a time. This gives a better view of the burn pattern and its relation to the fire. The second objective of using adequate lighting is its use as a source of illumination for photography.

Microscope

A binocular optical microscope with a magnification power of 6 to 100 will be found very useful. It is indispensable for the examination of opaque objects such as fuses, electrical wires, control switches, gas controls, regulators and other items of evidence. It is a great aid in the identification of various metals.

It is not practical to bring a microscope into court for the jury to see what you observed in a microscopic examination. To overcome this problem, many optical microscopes are equipped so that *photomicrographs* (pictures of what is seen under the microscope) can be taken. Various methods of taking photomicrographs are explained in further detail in Chapter 13.

Miscellaneous Equipment

In addition to the previously mentioned tools, the fire investigator has some unusual equipment requirements. These include molding material for casting footprints, tire marks or the striations made by burglar tools. There are several sources for these materials. One such source is Sirchie Laboratories, Moorestown, New Jersey 08057. Other supply houses are available, including a plastic molding material distributor, the Duplicast Corporation, P. O. Box 1373, Sonoma, California 95476. Duplicast makes a casting medium for copying fragile materials. It has a physical characteristic that is very useful in casting: after it solidifies, it is flexible enough to be removed from even the most intricate holes.

Clothing

The tools required of a fire investigator are based upon the peculiar nature of fire. The clothing worn by a fire investigator is based on the same predication. A hard hat is mandatory due to the hazard of falling structural material. In addition, a pair of rubber boots is necessary for wading through water. These boots should have a steel instep to avoid the penetration of nails or other sharp objects. A good, snagproof protective garment will prevent clothing from becoming soiled. The outer garments should be easily washable. A good pair of rugged gloves will round out the ensemble.

The Investigator's Library

The cause of fires involves so many branches of science that the investigator cannot operate for long without accumulating an extensive library. A bibliography for the nucleus of his library should include the following books:

America Burning. Washington, D.C., National Commission on Fire Prevention and Control, 1975.

American Society of Heating, Refrigerating and Air Conditioning Engineers: *Handbook of Fundamentals.* Menasha, Wisconsin, Banta, 1972.

American Gas Association: *Gas Engineers Handbook.* New York, Industrial Press, 1965.

Bates, Edward B.: *Elements of Fire and Arson Investigation.* Santa Cruz, California, Davis Publishing Co., Inc., 1975.

Baumeister, Theodore and Marks, Lionel S.: *Standard Handbook for Mechanical Engineers,* 7th ed. New York, McGraw, 1958.

Battle, Brendan and Weston, Paul B.: *Arson.* New York, Arco, 1970.

Brady, James E. and Humiston, Gerard: *General Chemistry: Principles and Structures.* New York, Wiley, 1975.

Clifford, Earl A.: *Practical Guide to LP Gas Utilization.* Duluth, Minnesota, Harbrace, 1969.

Crocker, Sabin and King, Reno C.: *Piping Handbook.* New York, McGraw, 1967.

Directory of Fire Research in the United States, 7th ed. Washington, D.C., National Research Council, National Academy of Sciences, 1975.

Eastman Kodak Company: *Fire and Arson Photography,* Catalog #M67.

Eschbach, Ovid W. and Souders, Mott: *Handbook of Engineering Fundamentals,* 3rd ed. New York, Wiley, 1975.

Federal Bureau of Investigation: *Handbook of Forensic Science.* Washington, D.C., U.S. Government Printing Office, 1975.

Fire Protection Handbook, 14th ed. Boston, National Fire Protection Association, 1976.

Fitzgerald, A. E. and Higginbotham, David E.: *Electrical and Electronics Engineering Fundamentals.* New York, McGraw, 1964.

Gas Appliance Service Manual. New York, American Gas Association, 1970.

Gas Heating Controls Service Manual. New York, American Gas Association, 1969.

Huron, Benjamin S.: *Elements of Arson Investigation.* New York, Dun Donnelley Publishing Corp., 1976.

Kennedy, John: *Fire-Arson Explosion Investigation.* Chicago, Investigations Institute, 1977.

Kirk, Paul L.: *Fire Investigation.* New York, Wiley, 1969.

Marrero, T. R. and Mason, E. A.: *Gaseous Diffusion Co-efficients.* Washington, D.C., American Chemical Society and American Institute of Physics for the National Bureau of Standards, 1972.

Meyer, Eugene: *Chemistry of Hazardous Materials.* Englewood Cliffs, New Jersey, P-H, 1977.

National Bureau of Standards Technical Notes. Washington, D.C., National Bureau of Standards, 1977.

Philo, Harry M., Robb, Dean A. and Goodman, Richard M.: *Lawyer's Desk Reference,* 5th ed. Rochester, New York, Lawyer's Co-operative and San Francisco, California, Bancroft-Whitney, 1975.

Resnick, Robert and Halliday, David: *Physics.* New York, Wiley, 1966.

Technicians Manual. New York, National Oil Fuel Institute, 1970.

Walls, H. J.: *Forensic Science.* New York, Praeger, 1974.

Watt, John H. and Summers, Wilford I.: *NFPA Handbook of the National Electrical Code,* 4th ed. New York, McGraw, 1975.

The Way Things Work, Volumes 1 & 2. New York, S&S, 1971.

In addition to his personal library, the investigator should have access to his departmental library and the public library. He should become familiar with the *Thomas Register* and its invaluable list of manufacturers of products. This makes it easy to contact the manufacturer of a product when the identification of a particular piece of equipment is necessary.

In addition to the formalized publication of books, a vast wealth of literature is available through many governmental agencies. These agencies are engaged in fire research and periodically publish reports of tests performed on full-scale building fires or other types of tests.

The most active government agencies include the following.

United States Forest Service

The Forest Service conducts many tests pertaining to wood and wood products at the Forest Products Laboratory in Madison, Wisconsin.

Department of Commerce

Under the NFPCA, the Department of Commerce is establishing a National Fire Academy to engage in training, research and development which will be of great interest to the fire investigator.

The National Bureau of Standards, at the Fire Research Center, has been engaged in fire research pertaining to building technology since the early 1900s.

The Consumer Product Safety Commission, which operates under the Department of Commerce, is responsible for the Hazardous Substances Act and the Flammable Fabrics Act.

This commission publishes numerous reports regarding tests conducted under these two important acts. The Consumer Product Safety Commission's Toll-free Hotline (800-638-2666) offers information about products (except for food, drugs, automobiles and boats) that are unsafe or potentially unsafe.

Department of the Interior

The Bureau of Mines, through various research centers, has conducted extensive tests on explosions and spontaneous combustion of various substances. It is also responsible for the development of gas detectors and establishing burning rates of substances.

Department of Transportation

The Department of Transportation (DOT) is responsible for all modes of transportation, including airlines, highway trucks, passenger automobiles, ships and rail transportation. DOT establishes standards, conducts research and regulates the transportation of hazardous materials.

Department of the Treasury

The Alcohol, Tobacco and Firearms Division of the Treasury (ATF) regulates interstate transportation, sale and storage of explosives and has the authority to investigate accidents and criminal acts involving explosives. Any local law enforcement agency can request the ATF laboratory in Washington, D.C., to test suspected arson-related materials. Laboratory personnel will test materials and also provide expert testimony in court.

Other Sources of Research Material

The National Academy of Sciences, the National Academy of Engineering, the National Research Council and the National Science Foundation are sources of fire research material.

In addition to these, many agencies within the federal government have published papers on fire investigation and techniques. The accumulated bibliography of reports of fire research is published biannually in the *Directory of Fire Research in the United States* by the National Academy of Sciences, Washington, D.C. The bibliography is prepared by the Committee on Fire Research, Division of Engineering, National Research Council. It includes fire research performed currently by the federal government and through grants to private and industrial laboratories, universities, colleges and private corporations.

The Laboratory

Each fire investigator should determine the laboratory facilities available locally and the services provided. A local laboratory is usually familiar with sources for tests which they themselves do not perform. The investigator should acquaint himself with the personnel and facilities available at each laboratory. This is time well spent and will provide the fire investigator with a better insight into problems facing the laboratory technician in analyzing debris brought to him. The investigator should work out arrangements with the laboratory regarding the type of sample containers to be used and the quantities of samples required for each test to be performed.

Testing Laboratories

Testing laboratories are those which provide standardized tests devised by such organizations as the American National Standards Institute (ANSI), American Society of Testing and Materials (ASTM) and others. These testing laboratories perform standard tests such as the melting point of a substance, ignition temperatures, flammable and explosive limits, and the identification tests regarding the composition of substances. Testing laboratories include independent testing laboratories, research laboratories and governmental agency laboratories.

INDEPENDENT TESTING LABORATORIES. Independent testing laboratories are used by manufacturers to test their products and gain laboratory approval for the products. The fact that a product is listed or unlisted by a laboratory, such as Underwriters Laboratories, is no reflection or guarantee that the product could not or did not malfunction.

The independent laboratories take sample units from a manufacturer's product line. These random samples are subjected to careful examination and a set of standardized laboratory tests. They are usually set up by the interested manufacturers and laboratories to determine and establish a standard satisfactory to both parties involved. The tests determine whether the sample of the product meets these standards and performs its function reliably and consistently for a reasonable product life. The samples are taken on a random basis, and tests apply only to the samples taken. They cannot possibly apply to untested products.

When these specifications are met, the testing laboratory "approves" and "lists" the equipment. It should be borne in mind, however, that a company which does not submit its product for testing can have better quality control than a competitive company which does send its product in for testing.

It should also be obvious to the fire investigator that in order to make a product completely fail-safe, the cost of inspection would bankrupt any company. Therefore, regardless of who or what the company is or manufactures, there will always be defective products, based upon the law of averages.

RESEARCH LABORATORIES. Research laboratories are those

engaged in basic or specialized research to discover certain aspects (physical, chemical or electrical) of substances or equipment. The more common of these are the government laboratories, such as the National Bureau of Standards, the Federal Bureau of Investigation, the Forest Products Research and various city and state crime laboratories.

The function of these laboratories is to perform standard and nonstandard tests and to answer questions. These also include private research laboratories which are available to perform standard and nonstandard tests at the request of the general public.

Exotic Analytical Instruments

Progress in the analysis of substances has developed analytical instruments unheard of ten years ago. The analytical instrument industry is finding it difficult to keep pace with the increase in technology required for the identification of chemical substances.

In addition to optical microscopes, many new electronic instruments and methods are available to aid the investigator in the identification of hydrocarbons or other substances found at the fire scene. A whole new industry has grown up around the identification of chemical compounds. Leading the field is the science of chromatography. Chromatography is a method by which the separation of a solution into its components is made by adsorption of its compounds so that they are separated rapidly and accurately. Once they are separated, identification of the compounds can be made by other methods.

Gas Chromatography

In gas chromatography, the fire investigator will obtain a solid, liquid or even a gas sample which he needs identified. This can be taken to a laboratory for identification on a gas chromatograph. The sample, either gas or liquid, is run through a tube at varying temperatures with a carrier gas, depending upon the type of material being tested. The components of the sample are separated and a time study made at the rate at which components emit from the downstream end of the gas chromatograph. The concentration of the various compounds is reflected on a time graph as the components are emitted.

At the time of emission from the tube, the compound is detected by a thermoelectric sensor which records its presence graphically. A visual comparison of the chromatograph (a chart produced by the sample) with the chromatograph produced by a known substance will indicate the similarities between the two charts. The gas chromatograph has become commonplace in the crime laboratories of the nation. Most commercial laboratories will have at least one gas chromatograph available.

The gas chromatograph is a useful tool, but it is no better than its operator and his ability to interpret the graphs resulting from its use. The important factor involved in gas chromatography, or any other tool, is the proof of the fact that the sample was not contaminated in its procurement, transportation or insertion into the gas chromatograph. It is also imperative that the gas chromatograph does not contaminate the sample.

The investigator should become familiar with the operator of the gas chromatograph, his capabilities and limitations. If the operator is experienced in conducting hydrocarbon detection tests, then the investigator can place his faith in the expertise of the operator. It will be the operator's testimony that will establish the credibility of the tests, not the testimony of the fire investigator.

A warning should be given to the fire investigator at this point. A gas chromatograph, or any other exotic instrument used to detect hydrocarbons, can only do that. It will detect the presence of hydrocarbons, but it cannot reveal the source of the hydrocarbons. It must be remembered that many substances, such as rugs, upholstery, and varnish, contain hydrocarbons. Therefore, when a sample of rug or carpeting is taken to determine whether it contains an accelerant, a test sample (or *control sample*) should be taken of the same material not involved in or near the origin of the fire. The control sample should be subjected to the same tests as the suspect material, and the two results should be compared. If these two gas chromatographs produce identical charts, it indicates that the hydrocarbons are a result of the manufacturing process and not from an arsonist's accelerant.

Thin-layer Chromatography

One method used for identification of compounds is called *thin-layer chromatography*. It is much less expensive than gas chromatography and gives similar results. In actual use, a small drop of the sample or suspect solution in liquid form is deposited on a glass coated with an adsorbent. It is then exposed to a solvent which travels across the plate, carrying the solution with it. The rate at which the compound moves is a function of its chemical relationship to the adsorbent, so that the compounds are separated according to their different solubility rates.

Once the chemical reaction has discontinued, the plate is then dried and exposed to a reagent which reveals the components to the naked eye. These are identified using comparison techniques such as infrared spectrophotometry, ultraviolet spectrophotometry and/or atomic absorption spectrophotometry.

Infrared Spectrophotometer

The infrared spectrophotometer is used in conjunction with, or in preference to, the gas chromatograph. It is useful in identifying compounds once they have been separated on a gas chromatograph. Infrared spectrophotometry (I.R.) can also be used to classify a compound according to type by comparing it with an *Atlas Reference Spectra*. The Atlas Reference Spectra is a listing with photographs and charts of various specific compounds which aid in identification. Usually it is desirable to run the sample through a gas chromatograph to separate the compound and then expose it to an infrared spectrophotometer for identification. This method is preferred where positive identification of individual compounds is required. It can be used for characterizing polymers, fibers, paints and other organic substances as well as for confirmation of gas chromatographs and thin-layer chromatographic fractions.

Ultraviolet Spectrograph

Ultraviolet spectrophotometry (U.V.) is similar to infrared spectrophotometry except that it uses the opposite end of the visible light spectrum to identify the compounds. The measurements are rapidly carried out and are very conclusive, although not as specific as I.R. While it is possible to classify the compound, its specific identity is not as readily found by U.V.

Atomic Absorption Spectrograph

One of the most exotic tools available in the laboratory is the atomic absorption spectrophotometer. This tool burns a small sample of the substance, and, through analysis of the spectrograph of the flame, a simple, rapid and reliable detection of trace elements can be made. The greatest deterrent to its use is its initial investment cost, and therefore, it is only found in the larger, more extensive laboratories.

Scanning Electron Microscope

Quite often the magnification of optical microscopes is insufficient to identify substances. When this occurs, the laboratory can use a scanning electron microscope for greater magnification. These are electronic devices capable of giving extremely fine pictures which are actually shadow photographs of substances magnified up to 65,000 times. Photographs can be taken with the scanning electron microscope. Unlike the optical microscope, the resultant image is in black and white only. The scanning electron microscope operates on a shadow optical principle so that no color can be detected. This is similar to an X-ray film where only shadow photographs (black and white) are obtained. However, not being limited by optical principles, no depth of field or focus problem is encountered with the scanning electron microscope. The photographs give very sharp, distinctive resolutions of the substance being examined.

Any laboratory equipped with a scanning electron microscope will also have in its library a set of the *Particle Atlas,* Second Edition. This four-volume set of books contains all of the necessary principles and techniques for obtaining and identifying photomicrographs of various particles, including products of combustion. In addition, X-ray analysis graphs showing the chemical composition of the same particles are provided.

X-Rays

X-rays are short-wavelength, electromagnetic radiations. They can pass through objects which are opaque in ordinary light. By passing the X-ray through an opaque object onto photographic film, the variation in density of the object will create a photograph or shadow image on the film. The fire investigator will find X-rays useful in examining the interior of containers without destroying the device. It is also useful for determining the contents of a melted mass of debris without running the risk of destroying any evidence within.

For example, if a "set" has been used to start a fire by an arsonist and the debris has melted and encased some suspicious-looking wires, X-rays will reveal what is connected to the wires within the mass of resolidified material.

X-rays are also useful in examining the charred remains of a fire victim. This will determine whether the body contains a bullet or suspicious wounds where surface evidence of such inflictions was destroyed by the charring effects of the fire.

Questioning

The investigation of a fire will reveal much physical evidence which will lead to the origin and cause of a fire. Invariably, however, it will be necessary to talk to eyewitnesses such as the owner, workmen, firemen or other persons familiar with the building or room before, during and after the fire. Therefore, the investigator must be able to communicate effectively and accurately to gain a mental picture of the building, its contents and the activities of individuals in relation to the building before, during and after the fire. The techniques involved encompass two features of communication—interviewing and interrogation.

Interviewing

An interview is a meeting with a person or persons in which information is obtained regarding the fire. Interviews would normally be conducted with firemen who saw the fire so that their opinion regarding its origin and circumstances surrounding its progress can be obtained.

The owner and employees of the business, or residents of the home who were the last to leave the scene before the fire, should also be interviewed. An interview can be held at a location mutually agreeable to both parties, conducted on an informal person-to-person basis. Witnesses have a tendency to become tense and on their guard when a note pad or tape recorder is used. Should the investigator feel that the presence of these items might deter the witness, they should be dispensed with during the interview. After the interview, the notes can be written down or dictated to cover the highlights of the conversation.

In the information-gathering phase of the investigation, the insurance agent and adjuster for the insured should also be contacted concerning legal aspects of the case. Once the facts of the case have been well established through eyewitnesses and supported and confirmed by physical evidence found at the scene of the fire, the investigator should have established a firm conviction as to whether arson has been committed. By then he will have a list of possible suspects to interrogate. Before interrogation begins, all possible accidental causes should be eliminated and a prima facie case established that arson was committed. It is at this time that the civil fire investigator should notify the proper authorities that evidence of arson has been found, and the arson investigator can perform the interrogation.

Interrogation

Interrogation differs from interviewing in the thrust of the questions, based upon the fact that the person being questioned is suspected of having committed a crime and being dishonest in his answers. At times, the interviewing of an adverse or hostile witness who "doesn't want to get involved" will take on all of the aspects and characteristics of an interrogation. Normally, techniques and methods of interrogation are reserved for persons suspected of being involved in a crime.

If a person has knowledge of a criminal act, he will try to hide the facts concerning his involvement. The fire investigator will have evidence of the true facts, and in discussing the case, discrepancies will appear as the suspect tries to hide his involvement. The purpose of the interrogation is to clear up the discrepancies between the story as told by the suspect and the facts as known by the investigator.

Outside Experts

During the investigation of a fire the investigator may need to retain outside experts to eliminate or confirm other possible causes of a fire. It may be necessary to call in several experts to eliminate or confirm some causes. For example, in a typical residential fire it may be necessary to retain an electrical expert to check the wiring, a gas expert to test the furnace and chemists dealing with odorants to check the presence or absence of odorants in the gas supply. Metallurgists may be required to determine the corrosive elements in a gas piping system.

After all possible causes except one have been eliminated (often with the help of outside experts), the fire investigator can confirm his opinion with the aid of the outside experts. This is particularly true in accidental fires in which the cause is malfunction of a control or another accidental cause. The appropriate expert, basing his opinion on sound physical evidence, can give an opinion in court, stated with "reasonable scientific certainty," that will carry considerable weight to the jury. The investigator should become familiar with the local experts available and know their qualifications, specialties and educational background (see Chapter 15).

Sometimes a second opinion may be necessary before criminal or civil action is taken when the evidence is not overwhelming, or if the fire investigator has a nagging doubt regarding the cause of the fire. Quite often an outside fire investigator can assess the situation from an entirely different viewpoint and can assist the fire investigator in reaching a conclusion and forming an opinion.

Professional Organizations

One of the characteristics of a profession is the establishment of an organization for persons with similar interests which polices its own members. Every fire investigator should join an organization and actively participate in its goals. Organizations available to the fire investigator and worthy of his support include, but are not limited to, the following:

Federal Fire Council
National Bureau of Standards
Washington, D.C. 20234

Fire Department Instructors Conference
P.O. Box 1089
Chicago, Illinois 60690

Fire Marshal's Association of North America
60 Battery March Street
Boston, Massachusetts 02210

International Association of Arson Investigators
97 Pacquin Drive
Marlboro, Massachusetts 01752

International Association of Fire Chiefs
1725 "K" Street NW
Washington, D.C. 20006

International Association of Fire Fighters
1750 New York Avenue Northwest
Washington, D.C. 20006

International Fire Photographers Association
558 West DeKoven Street
Chicago, Illinois 60607

International Society of Fire Service Instructors
Box 382
College Park, Maryland 20740

International Fire Service Training Association
Oklahoma State University
Stillwater, Oklahoma 74074

Mill Mutual Fire Prevention Bureau
2 North Riverside Plaza
Chicago, Illinois 60606

National Association of Fire Investigators
53 West Jackson Boulevard
Chicago, Illinois 60604

All of these organizations have a contribution to make, and it behooves the fire investigator to seek out those organizations which reflect his philosophy and give them his full support.

CHAPTER 3

The Physics of Fire

CHEMICAL COMPOSITION

FIRES AND EXPLOSIONS are phenomena involving chemical and physical laws from chemistry, physics, mechanics, thermodynamics, combustion and, at times, electromagnetism. Therefore, to understand how a fire burns and the course it takes, it is necessary for the fire investigator to have a working knowledge of the chemical and physical principles involved.

Chemistry is the science of composition, structure, properties and reactions of matter. Physics is the science of matter, energy and their interactions.

Atoms

The basic building block of the universe is the atom. It is so small it cannot be seen under even the most powerful optical microscope. We know of its existence by inference and what might be called, in the parlance of the fire investigator, "circumstantial evidence." (Most scientific knowledge is based upon "circumstantial evidence.")

The diameter of most atoms is about 10^{-4} centimeters. In nuclear physics, the atom has been broken down into smaller units such as electrons, neutrons and protons, but these particles are the building blocks of the atoms themselves and are beyond the scope of this discussion. We are not concerned with the energy released in splitting an atom. As of this date, no arsonist has resorted to the use of atomic or nuclear energy as an accelerant.

Elements

An *element* consists of one or more identical atoms containing electrons and a nucleus. Electrons have the same relationship to a nucleus as the relationship of the planets to our sun. In fact, it can be said that the sun is our nucleus and the earth is an electron spinning around the nucleus. In this concept, our planetary system becomes an atom of gigantic proportions.

Our universe, however, is usually considered as being all in the same elliptical plane. The atom is considered to be three-dimensional, with electrons spinning not only around the equator of the nucleus but also around the poles of the nucleus at various distances from the nucleus.

The electrons which revolve around the nucleus of an atom vary in the ease of their removal from orbit. The inner electrons are very difficult to remove, while the outer electrons are easily removed. The ease with which these are removed is measured by *conductivity* and *resistivity*. The easier the removal of electrons–the more free electrons there are–the better *conductor* the element is.

All matter is composed of elements which combine in many ways to form different substances. These substances in the combined state are called molecules and are actually combined groups of atoms. The molecules will differ in their physical or chemical properties, depending upon the kind of elements involved and the strength of their bond–their "glue." This is a function of the outer electrons revolving around the nucleus.

Molecules

The next step up the ladder of structure is the molecule. A molecule is a combination of atoms bonded by the sharing of electrons, similar to the way in which the moon is bound to the earth by the gravitational attraction of one to the other. Molecules can vary in size from small combinations of two simple elements to complex molecules containing 10^{23} atoms or more.

Compounds

A compound is a mixture of two or more different types of molecules. Ordinary chemical reactions involve changes in molecules by rearrangement of the electrons with a reaction which can either absorb heat (mechanical energy) or give off heat (mechanical work). It is this aspect of chemistry which interests the fire investigator.

Chemical Reactions

Chemical reactions can be slow or rapid, resulting in a gradual change, such as the weathering of roofing through the slow oxidation of roofing material and its contact with air and the elements, or a rapid change, such as the burning of a piece of paper or the explosion of a flammable gas, or even more rapidly as the detonation of a high explosive. The fire investigator is interested in the more rapid reactions, with some exceptions. These exceptions involve the slow *pyrolysis,* or destruction of wood by heat.

A reaction where heat is given off is called *exothermic.* This is typified by the heating of a substance through release of energy by the combining of the atoms of two compounds. An *endothermic* reaction is one which is instigated through the adsorption of heat with a cooling effect upon the two compounds involved. These reactions are usually not associated with the cause of a fire. They do come into consideration by the fire investigator when an endothermic reaction may cause a cooling effect by adsorbing heat during a fire, thus slowing its progress.

An example of endothermic reaction is the conversion of water to steam. It takes an enormous amount of energy to convert water to steam, and the temperature of the water is not raised in the process. This is one reason why water is such an effective medium for extinguishing fires.

A third factor involved in chemical reactions is that of a *catalyst.* A catalyst is a substance which does not enter directly into a reaction between two compounds but has a profound effect upon the *rate* of the reaction of the compounds. A catalyst will accelerate certain reactions at a much faster rate than would normally occur were it not for its presence.

A catalyst might be compared to greasing the skids on a sled running on dry pavement. The sled could be moved across dry pavement by pulling the sled by several men. By using grease, the men could pull it much faster, or one man could pull it at the same speed as several men. In a like manner, a catalyst will speed up the reaction of two compounds or enable a lesser amount of compound to do the same work that a larger amount of the same compound would perform without the aid of the catalyst.

PHYSICAL PROPERTIES OF COMPOUNDS

Compounds take various forms and shapes and are commonly in the form of gases, liquids and solids. The shape a substance takes is the function of the internal activity of the molecules. If they are loose and running at random in a disconnected form, they will take the shape of the container in which they are held. They will exert pressure against the walls of the container and will assume a homogeneous mixture as a gas. If a gas or vapor is condensed, the molecules have less room to move and become involved with each other so that the distance between molecules becomes less and less. The vapor will condense into droplets, forming a liquid. The liquid, although having less space between the molecules than a gas, will still contain empty space between the molecules.

If the velocity of the molecules in the liquid is slowed by the removal of heat, which in turn will lessen the activity of the molecules, they will become more and more involved with each other and will form into a distinct relationship in the form of a solid.

The determination of whether a substance is a solid, liquid or gas is a function of the energy level of the molecules and the composition of the original matter. Most substances can exist in all three stages. For example, water can exist as a solid (ice), a liquid (water) and a vapor (steam). Some substances, such as iodine, can exist as only a solid (iodine crystals) or a vapor, but not as a liquid. When heated, it goes from the solid stage to the vapor stage without passing through the liquid stage. Other substances such as wood or other *organic* materials can exist in only one stage and are broken down chemically when any attempt is made to reduce them to another stage. Wood is broken down from a solid substance composed of lignin and cellulose to its elemental components.

The wood is pyrolized (burned)–broken down into volatile flammable gases–leaving a pile of ashes as the only evidence of the original wood. Gas emitted from the wood burns, not the wood.

Liquids are burned in a similar manner. The liquid is heated to the point where it becomes a vapor. The vapor is ignited as it combines with oxygen and is burned off. Gases are burned by combining the flammable gas with the appropriate amount of oxygen to form a flammable mixture. All liquids must be vaporized to burn.

Solids

Solids are of interest to the fire investigator because the buildings of today are constructed primarily of solid material–combustible material such as wood, cellulose, plastic, fabric, roofing material and insulation material. Also of interest are the noncombustible materials such as concrete, steel, iron, cast iron, bronze, brass, lead, zinc, tin and pot metal.

These are primarily of interest because the fire investigator can use their distinctive melting points as an indication of the temperatures attained.

Density

The density of a combustible or noncombustible material is of importance as it reflects other characteristics, such as the expansion rate, conductivity and the ability to resist the forces of a fire.

Coefficient of Expansion

The coefficient of expansion concerns the rate at which a solid is affected by heat. For example, at an equal rise in temperature, aluminum will expand more than steel. This greatly increases stresses imparted upon an aluminum structure subjected to heat. It also affects the action of electrical conductors when they are heated to the point that they start to put strain on their components as they expand and change shape.

Thermal Conductivity

Thermal conductivity of solids is a measure of their rate of reaction to change in temperature with respect to the energy levels within the material. For example, when a good thermal conductor is heated on one end, it will quickly transfer the heat energy (activity of the molecules) from the heated end to the cold end. A poor thermal conductor will prevent the transfer of energy from one end to the other. Thus, a piece of steel would be a better thermal conductor than a piece of wood. Should a fire heat one side of a sheet of steel, the steel would conduct the heat and ignite combustible material in contact with the other side much more rapidly than a sheet of plywood. The plywood may burn through in a short period of time, but the heat would not be conducted through the plywood until it had been destroyed. Heat can be conducted through a steel plate and ignite nearby combustible material without destruction of the steel plate.

Ignition Temperature

The ignition temperature of a substance is the minimum temperature needed to cause self-sustaining combustion of the substance. The ignition temperature is a variable, depending upon the rate and temperature of air flow and the air-gas ratio of the combustible. Therefore, when ignition temperatures of flammable gases and liquids are given in tables, it should be remembered that these tables are based upon tests conducted at a standard temperature and pressure within a given rate of air flow and heating.

The size and shape of the solid or liquid vapor droplets also influence the ignition temperature, and therefore, these variables should be considered when making laboratory comparisons with actual field incidents.

Although the ignition temperature is commonly related and applied to gases or liquids, it should be remembered that very few solids "burn." It is the vapor which has been pyrolized from the wood which burns, and the ignition temperature of the gas extracted from the wood should not be confused with the kindling temperature of the wood. The latter is the temperature at which the wood is heated so that gases are given off and ignited.

Kindling Temperature

Kindling temperature is another term for ignition temperature but is normally applied to the ignition temperature of wood. It is the temperature to which the wood must be raised to be ignited and sustain combustion. An example of kindling temperature familiar to everyone comes from the simple act of striking a match. Friction of the match head against the striking plate creates heat which, in turn, causes a chemical reaction in the match head which raises the temperature of the match above the kindling temperature as the match ignites.

If you strike a match, you will notice that it will flare for a brief period of time. This is a chemical reaction occurring in the coated match head, required to raise the kindling temperature of the wood or paper of the match stick so that the vapors evolved will sustain combustion.

The ignition and kindling temperature of a substance is useful to the fire investigator in determining temperature levels reached in a room. For example, if it is known that a substance ignites at 600°F, and the material shows signs of scorching, then the temperature approached 600°F. If there is an unburned portion of this same material, then the line of demarcation is near the range of the kindling temperature of the substance.

Each combustible solid can be rated by the amount of energy contained within the substance. The heat content of common solid fuels is shown in Table II.

From the table, the fire investigator can estimate the quantity of heat available (measured in BTUs) at a fire scene. This makes it possible to determine whether the damage sustained by the building corresponds with the amount of fuel available, or whether it appears that extra fuel was provided in the form of an accelerant or additional fuel.

Burning Rates

Underwriters Laboratories, Inc., and other agencies have established the burning rates of combustible materials normally found

TABLE II

PHYSICAL CHARACTERISTICS OF COMMON FUELS

Material	Density Lbs./ft³	Heat Content BTU/ft³	Flash Point Degrees F	Ignition Temperature Degrees F
Wood	40 (S)	320,000 (S)	—	500
Coal	50 (S)	650,000 (S)	—	600
Natural gas	8.82 (G)	462,000 (L)	—	900-1170
Propane	22.79 (G)	685,700 (L)	—	871
Butane	30.58 (G)	775,900 (L)	—	761
Gasoline	45.86 (L)	965,350 (L)	−45	—
Kerosene	50.50 (L)	1,010,000 (L)	110	—
Diesel fuel	52.97 (L)	1,035,560 (L)	110	—
#1 fuel oil	50.80 (L)	944,170 (L)	110	—
#2 fuel oil	53.35 (L)	980,786 (L)	110	—

(G) gas.
(L) liquid.
(S) solid.

in buildings in the United States. These are useful in evaluating the total fire scene for the amount of damage incurred in the building and the rate of fire spread after initial ignition. The burning rates are identified as the *Flame Spread Index.* The rule of thumb that wood burns at a rate of ¾ inch per hour is so variable that it should not be used in court. The burning rate should be determined by laboratory tests.

Melting Point

The melting point of a substance is the temperature at which it changes from a solid to a liquid. As more energy is applied in

TABLE III

MELTING POINTS OF MATERIALS FOUND IN A FIRE*

Alphabetical Listing		Listing in Order of Decreasing Temperatures	
Material	*Temp.* ° *F*	*Material*	*Temp.* ° *F*
Aluminum	1150	Tungsten (light bulb filament)	6100
Aluminum alloy (zinc)	728	Steel	2760
Brass, naval	1625	Iron, wrought	2750
Brass, red	1810	Steel (stainless)	2550
Brass, yellow	1710	Iron, cast	2150
Bronze	1910	Copper	1980
Copper	1980	Bronze	1910
Fuses, electrical	700	Brass, red	1810
Glass	1200	Brass, yellow	1710
Iron, cast	2150	Muntz metal	
Iron, wrought	2750	(Cu 60%, Zn 40%)	1660
Muntz metal		Silver (sterling)	1650
(Cu 60%, Zn 40%)	1660	Brass, naval	1625
"Pot Metal" (lead-based alloy)	550	Silver (coin)	1600
Silver (coin)	1600	Glass	1200
Silver solder		Aluminum	1150
(Pb 97.5%, Ag 2.5%)	580	Zinc	780
Silver (sterling)	1650	Aluminum alloy (zinc)	728
Soft solder (Sn 50%, Pb 50%)	421	Fuses, electrical	700
Soft solder (Sn 60%, Pb 40%)	374	Silver solder	
Steel	2760	(Pb 97.5%, Ag 2.5%)	580
Steel (stainless)	2550	"Pot Metal" (lead-based alloy)	550
Tin	450	Tin	450
Tungsten (light bulb filament)	6100	Soft solder (Sn 50%, Pb 50%)	421
Zinc	780	Soft solder (Sn 60%, Pb 40%)	374

* From Robert Perry and C. H. Chilton, *Chemical Engineers' Handbook,* 5th ed., 1973. Courtesy of McGraw-Hill Book Company, New York.

the form of heat, the activity of the molecules is such that the distance between them is increased until the solid can no longer sustain its form and becomes a liquid. A liquid will assume the shape of the container in which it is held, but unlike a gas, it has a relatively constant volume. The temperature at which this change occurs is called the *melting point*.

Some substances of particular interest to the fire investigators have no definite melting point. One of these is glass. Glass will soften at approximately 1000°F and flow at 1300°F. The melting points of common materials found in fires is given in Table III.

The melting point is a useful datum for determination of the temperatures experienced in a fire. For example, by noticing that aluminum has melted but brass fittings have not, the temperature range of a fire at a given point (where the two substances were found) can be bracketed. This is useful in determining if an accelerant was used in the fire. By knowing the fire load of the building, i.e. the material available for the creation of heat, a reasonable approximation of the highest temperatures attained can be made and compared with the temperatures to be expected had an accelerant been used.

The Effect of Heat on Construction Materials

Construction materials react differently according to their composition and use in the construction of a building. Wooden paneling, being exposed to the interior of a room, will burn rapidly, but the fire will not affect the wooden studding to which the paneling is attached until the paneling itself has been virtually destroyed.

Steel structures, unless protected by a thick layer of fireproofing, are sensitive to extreme temperatures and will lose their structural strength long before the melting point of the steel is reached. Concrete, stone and other masonry products have been known to explode due to trapped moisture within the material. The heat of the fire will convert the moisture to steam, exerting tremendous pressures within the material and causing spalling or explosions.

Wood

Wood must be heated thoroughly to or near the kindling temperature before pyrolysis will take place. Pyrolysis will then progress rapidly or slowly, depending upon the availability of sufficient air and heat. The strength of the wood is unaffected by extreme temperature. As the fire destroys the wood from the outside inward, the structural strength of the wood section is decreased until finally the structure collapses. Since structural timber is designed with a safety factor of about four to one, the cross section of the components of a truss can withstand considerable destruction before collapse occurs. The ability of wood to withstand high temperatures without structural failure is one reason that a wood truss is considered more fire resistant (not fireproof) than a steel truss.

The high temperatures generated in a fire accumulate toward the roof and expose the trusses of the building to higher temperatures than any other portion of the building. The heat source may be the material on the floor of the building, such as in a warehouse. The heat rises into the space occupied by the trusses. A wood truss, if treated with fireproofing material, can withstand considerable heat before destruction by burning of the truss. A steel truss, in the same situation, would fail through reduction in the strength of the steel components of the truss (see Fig. 6).

Metal

The melting point of carbon steel, such as used in building construction, is about 2700°F. This is far above normal temperatures encountered in a fire. If accelerants or other chemicals are present, temperatures can reach higher than indicated on the *Standard Fire Curve* (see Fig. 10). If the fire is of long enough duration, temperatures can reach a range of 2000°F. It is seldom that they reach the melting point of steel, except in rare isolated instances, such as near a jet of burning gas or other substances.

The temperatures have a profound effect on the steel construction members. The steel commences to lose its strength in proportion to the temperature, as indicated in Figure 6.

The strength of the steel, as indicated in Figure 6, will increase slightly from normal temperatures to a maximum of about 500°F. At 700°F, the loss in strength is inversely proportional to the increase in temperature. As the temperature approaches 1600°F,

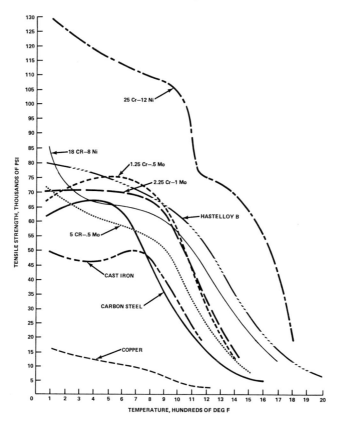

Figure 6. Tensile strengths of some common piping materials. From Sabin Crocker and R. C. King, *Piping Handbook,* 5th ed., 1967. Courtesy of McGraw-Hill Book Company, New York.

the strength is reduced to less than 10 percent of its original strength (from 65,000 to 5,000 psi).

In addition to structural failure, pipes containing liquids or gases will also fail, due to the increase in internal pressure resulting from heating of the contents and decrease in strength of the pipe.

Most piping systems are equipped with pressure relief valves. These may open but cannot relieve the pressure faster than the heat of the fire can build pressure within the pipes. This, coupled with the decrease in the strength of the pipe, leads to blowouts of the pipelines from internal pressure.

The pipe material may be copper, aluminum, brass or some alloy. Although the initial strength of the pipe may vary, the material will follow the same general curves as indicated for carbon steel in Figure 6.

Reinforced Concrete

Reinforced concrete is a composite structural material which utilizes the compressive strength of the concrete to its fullest by supplying reinforcing steel to provide the necessary tensile strength. Tensile strength is the ability of a material to resist a "pull." Concrete has virtually no tensile strength, so that the reinforcing steel is a necessity in order to utilize the otherwise highly efficient concrete.

To protect the reinforcing steel from atmospheric corrosion and heat, a covering of 1 to 3 inches of concrete is used. In a fire, the thermal conductivity of the concrete may permit the steel to be heated to the point where it will lose its tensile strength. At approximately 1000°F, the reinforcing rods have lost approximately 50 percent of their strength and are in danger of producing failure of the concrete slab. Factors affecting the thermal transmission coefficient of the concrete cover over the steel are the type of aggregate, the density of the concrete and the moisture content.

If the moisture content of the concrete is high, the heat may cause expansion of the trapped moisture, converting it to steam and causing a steam explosion. This causes spalling of the concrete cover on the reinforcing steel. The result is that the reinforcing steel will lose its protective cover, heat more rapidly and lead to early failure of the structure.

Masonry

Masonry, if reinforced, will react similarly to reinforced concrete. Masonry is more susceptible to spalling due to more trapped moisture. Spalling can also occur in masonry if the buildup of heat is so rapid that it causes unequal expansion of the surface of the masonry. Should this occur, the disproportionate stresses on the surface of the masonry will cause it to spall due to temperature differentials. All types of masonry will commence to fail at temperatures exceeding 1400°F.

Liquids

The physical properties of liquids differ significantly from those of solids. The main difference is the ease of kindling flammable liquids. It is necessary to chemically break down the components of solids by the application of heat—drive off a volatile flammable gas, mix it with the proper amount of air and provide ignition. This normally requires considerably more energy than is required to ignite flammable liquids. With liquids such as kerosene, gasoline or oil, the material is raised to the flash point and then the fire point. At these points, the vapor from the liquid, mixed with the proper amount of air, will ignite if provided a source of ignition. After the fire point is reached, the fire will be self-sustaining.

Boiling Point

The boiling point of a liquid is the temperature at which the liquid becomes a vapor. It is important to realize that the boiling point of a liquid can be reached either at the surface or bottom of the liquid. This is a function of the point of application of heat.

This can be readily observed in an open pot containing kerosene. As the boiling point is reached, tiny bubbles form on the bottom of the pot, which is closest to the source of heat. As these bubbles form, they break away and, being lighter than the liquid displaced by the vapor formed at the bottom of the kettle, rise to the surface. As more and more energy causes more and more bubbles, the rapidity of the bubble formation and the rapid rise of the bubbles causes the common phenomenon of *boiling*.

However, if the source of heat is applied at the top of the liquid, bubbles will form on the top surface of the liquid. This will permit the gas to immediately release into the atmosphere where it can, and usually does, form a flammable mixture. This mixture can be ignited from the same source of heat which started the boiling on the surface of the liquid. Immediately below the surface of the liquid, the temperature will cool rapidly, due to the low thermal conductivity of the liquid. Boiling will not occur until the entire body of liquid is heated to the *boiling point*. This is difficult to do from the top.

Until the liquid has completely vaporized, the surface upon

which the liquid is resting will not be heated, due to the protective layer of liquid above the surface.

For example, if heavy lubricating oil is poured on a floor and the room burns, it is quite possible that the entire room (except the floor) can be destroyed before the oil is completely burned to the underlying surface. If the oil is not completely consumed in the fire, the thin layer of oil will protect the underlying surface from burning, due to its low thermal conductivity qualities.

Flash Point

A solid or a liquid, if heated sufficiently, will give off a vapor capable of ignition. The temperature at which the application of heat will cause this vapor to ignite is termed the *flash point*. The flash point is recorded as the temperature at which ignition occurs to the vapor given off by the solid or liquid. It is the minimum temperature which will ignite the vapor. It is *not* the temperature at which the vapor will sustain burning.

The flash point gives the fire investigator a *minimum* temperature which must have existed at the time the vapor ignited, if this was to be the cause of the fire. If it can be shown that temperatures exceeding the flash point were available, then the material may have been the first material to ignite from the source of ignition. Thus, the flash point may assist the fire investigator in determining the material first ignited and the source of ignition.

Two types of tests are commonly used to determine the flash points of liquids. The first test is the ASTM D56 and is commonly called the Tagliabeu (TAG) closed-cup test. It is used for liquids with flash points at or below 200°F. For liquids with flash points of 200°F or higher, the Pensky-Martens closed-cup test (ASTM D93) is used.

The other type of flash point test is the Tagliabeu (TAG) open-cup test, ASTM D1310, or Cleveland open-cup test (ASTM D92). As the name implies, the closed-cup test is performed by warming the liquid in a closed cup and passing an open flame just above the liquid. When the vapor ignites, a flash is observed and the temperature duly recorded. This establishes the flash point.

In the open-cup test, the same apparatus is used except that the cup is exposed to the atmosphere. As a result, the vapors rise much more rapidly by convection and a higher rate of vapor production is required to form a combustible mixture at the surface of the liquid. The open-cup temperatures are usually higher than closed-cup temperatures. Application of flash points to the everyday fire should be handled with discretion. It should be remembered that the flash point, as established in Table IV, is a laboratory measurement under controlled atmospheric pressure and temperature. In addition, the normal oxygen content of the atmosphere is provided.

In an actual fire, the pressures may be higher (due to confinement of the heated atmosphere) and the oxygen content may have been depleted (by a smoldering source of ignition) long before the flash point of the liquid involved has been reached. Lack of oxygen would certainly interfere with the ignition of the vapors at the ASTM flash point. Should these conditions arise, the liquid should be tested by simulating the conditions which existed at the fire at the approximate time of ignition of the liquid.

Fire Point

The *fire point* is defined as the lowest temperature at which vapors are burning at the same rate that they are generated. At the fire point, the temperature is high enough to support continuous combustion. It will be from 10° to 50°F above the flash point indicated in Table IV.

Although the fire point is the point usually associated with continuing combustion, the open- and closed-cup tests are the tests most commonly available. The flash point for most substances has been established and is readily available in the literature. The fire point for many substances is also available but is not as widely published as the flash point. The flash point is not as significant to the fire investigator as the fire point. When the outcome of a case hinges upon the establishment of the ignition of a substance, the best procedure is to run a fire point test under as closely reconstructed conditions as possible.

TABLE IV
PROPERTIES OF FLAMMABLE LIQUIDS, GASES AND SOLIDS

Name	Flash Point Degrees F		Explosive Limits In Air Percent by Volume		Auto-Ignition Temperature Degrees F	Specific Gravity (water=1)	Vapor Density (air=1)	Melting Point Degrees F	Boiling Point Degrees F
	Closed Cup	Open Cup	Lower	Upper					
Acetone	0	15	2.1	13.0	1000	0.788	2.00	-137	134
Acetylene (air mixture)	Gas	Gas	2.5	81	571	0.621	0.91	—	-119
Asphalt (typical)	400+	535+	—	—	905	0.95-1.1	—	180-220	> 700
Automotive gas (premium)	-50±	—	1.3-1.4	6.0-7.6	700	0.71-0.76	3.0-4.0	<-76	91
Automotive gas (regular)	-50±	—	1.3-1.4	6.0-7.6	700	0.70-0.75	3.0-4.0	<-76	91
Benzene	12	—	1.4	8.0	1044	0.885	2.77	42	176
Brake fluid (heavy duty)	—	220	—	—	—	—	—	—	—
Butane, n-	-76	Gas	1.9	8.5	761	0.584	2.06	-217	31
Butane, iso-	-117	Gas	1.8	8.4	864	0.563	2.06	-255	11
Camphor	150	200	0.6	3.5	871	0.999	5.24	345	399
Camphor oil (light)	117	125	—	—	—	—	—	—	347
Carnauba wax	540	595	—	—	840	0.96	—	14	595
Charcoal	—	—	1.3	8.0	—	3.51	—	> 6300	7600
Coal tar oil	60-77	—	—	—	—	< 1.0	—	—	—
Coal tar pitch	405	490	—	—	—	< 1.0	—	—	—
Corn oil	490	490	—	—	740	0.92	—	14	—
Cottonseed oil (refined)	486	550	—	—	650	0.925	—	23-32	—
Creosote oil	165	185	—	—	637	> 1.0	—	—	392
Denatured alcohol, 95%	60	—	—	—	750	0.82	1.60	-60	175
Dry cleaning (naphtha)	100-110	—	0.8	5.0	440-500	0.8	—	<-50	300
Dry cleaning solvent, naphtha, 140°F	138	—	0.8 @ 302°F	—	451	—	—	—	<358
Ethyl alcohol	55	71	3.5	19	737	0.791	1.59	-173	173
Ethylene	-185	—	2.7	34	842	0.566	0.975	98	469
Fish oil	420	—	—	—	—	—	—	—	—
Formaldehyde gas	Gas	Gas	7	73	806	—	1.07	-134	-6
Fuel oil no. 1	114-185	—	0.6	5.6	445-560	0.78-0.85	—	—	340

Fuel oil no. 1D	>100	—	1.3	6.0	350-625	<1.0	—	—	<590
Fuel oil no. 2	126-230	—	—	—	500-705	0.80-0.90	—	—	340
Fuel oil no. 2-D	>100	—	1.3	6.0	490-545	0.81	—	—	380
Fuel oil no. 4	154-240	—	1	5	505	0.84-0.98	—	—	425
Fuel oil no. 5	130-310	—	1	5	—	0.92-1.06	—	—	—
Fuel oil no. 6	150-430	—	1	5	765	0.92-1.07	—	—	—
Gas, natural, 103 BTU	Gas	Gas	3.8-6.5	13-17	>1000	—	0.61	—	108
Gasoline, aviation, commercial	-50±	—	1	6.0-7.6	800-880	0.70-0.71	3.0-4.0	<-76	144
Jet fuel, JP-4 referee	<34	—	—	—	—	0.785	—	<-76	370
Jet fuel, JP-5	105	—	0.6	4.6	400	0.82-0.835	—	<-40	250
Jet fuel, JP-6	127	—	—	—	500	—	—	—	350
Kerosene	110-130	—	0.6	5.6	440-560	0.81	4.5	—	—
Lacquer	0-80	—	—	—	—	<1.0	—	28	—
Lard oil (commercial)	395	—	—	—	833	—	—	—	—
Linseed oil (boiled)	403	—	—	—	—	—	—	—	600+
Linseed oil (raw)	435	535	—	—	650	0.93	—	-2	680
Lubricating oil (mineral)	—	275-500	—	—	500-700	0.83-0.90	—	—	—
Methane	Gas	Gas	5.3	13.9	999	—	0.554	-296	-259
Mineral spirits	100	110	0.77 @ 212°F	6.0	475	0.80	3.9	21	300
Olive oil	>437	—	—	—	650	0.910	—	—	—
Paint liquid	0-80	—	—	—	—	—	—	—	—
Petroleum, crude	20-90	—	—	—	—	0.78-0.97	—	—	—
Propane	<-156	Gas	2.2	9.6	871	0.508	1.56	-306	-44
Rosin, gum	370	430	—	—	—	1.08	—	212-300	—
Rubber cement	<50	—	—	—	—	—	—	10-25	110
Soybean oil	250-600	—	—	—	833	0.9	—	—	—
Sulfur	405	440	—	—	450	2.046	—	234	832
Tallow	509	—	—	—	—	0.895	—	88-100	—
Transformer oil	—	295	—	—	—	0.9	—	—	—
Tung oil	552	—	—	—	855	0.936-0.943	—	88	—
Turpentine, spirits of	95	115	0.8	—	488	0.854-0.868	4.84	—	309
Varnish	10-80	—	—	—	—	—	—	—	—
Varnish shellac	—	610	—	—	—	—	—	—	—
Vegetable oil, hydrogenated	40-70	—	—	—	—	<1.0	—	—	—
Whiskey	82	—	—	—	—	—	—	—	—
Wines, high	60-80	—	—	—	—	—	—	—	—

Auto-ignition Temperature

Auto-ignition temperature is that temperature to which the substance must be raised for vapors to ignite spontaneously without the presence of an independent source of heat. The auto-ignition temperature is a function of the degree of molecular activity of the vapor and is influenced by the rate of air flow, rate of heating, size and shape of the material, as well as the source of oxygen available for the chemical process to begin and become self-sustaining.

Again, caution must be exercised in the use of tabular values. Many variables can and do influence auto-ignition temperatures. Therefore, tabular values should be used as an indication only. Should the question of spontaneous ignition arise, the material should be tested, duplicating the circumstances of the fire to prove whether such an occurrence was possible.

Heat Content

The heat content of a substance is the number of BTUs per unit volume. The heat content is useful to the fire investigator in forming a mental image of the amount of heat liberated by the elements found in a fire. This gives a base with which to compare the burning and destruction afforded by a number of pieces of wood furniture, for example, compared with a gallon of kerosene or fuel oil.

Table II illustrates the variation in BTUs per unit weight and volume for various fuels. A quick examination of the fuel content, for example, of kerosene compared with gasoline, indicates the danger in using gasoline as an accelerant. The low flash point indicates a low fire point. This could indicate danger of an explosion from the use of gasoline. It also indicates that kerosene has a higher BTU content per gallon. This means that the arsonist would have to provide less kerosene to obtain the same results with the equivalent weight of gasoline.

Gases

As the molecular activity of a liquid increases, the attraction between molecules becomes less and less until each molecule becomes independent of the body of liquid, reaches escape velocity from the surface of the liquid and becomes a gas particle.

A gas is a collection of molecules with much more activity than a liquid. A gas assumes the shape of its container. The gas exerts pressure against the walls of its container as molecules are constantly in motion and striking the walls of the container. This exerts a force which we recognize as pressure within the container. This pressure is referred to as the vapor pressure of the gas.

Vapor Density

Since we live in a sea of air, a vapor is soon mixed with air due to turbulence of air currents and diffusion, or the natural activity of the vapor as it penetrates the air. The vapor-air density is the weight of the mixture resulting from the vaporization of the liquid compared with the weight of an equal volume of air. It is important to the fire investigator since it will govern the manner in which a gas behaves as it mixes with air and after it is mixed with air. The mixture will behave according to the temperature and density of the gas, temperature and density of the air, temperature and density of the final mixture and thermal and convection currents in the immediate area.

Flash Point

The flash point, fire point and auto-ignition temperature of liquids are based upon gas pressures. Therefore, the text referring to liquids also refers to gases.

Heat Content

The heat content of a gas is much smaller than that of a liquid. This is why gases are commercially compressed to form a liquid. By compressing the gas to a liquid, such as is done with liquid oxygen, propane, butane, methane and other gases, the volume of the gas is reduced approximately 30 to 1, so that the heat content of the liquid is increased proportionately. The volume of the material is decreased so that energy can be stored in much less space.

Explosive Limits

Most of us are familiar with the phenomenon of being unable to start a car because the carburetor was flooded, resulting in too rich a mixture (of gasoline and air), or the carburetor being out of adjustment, resulting in too lean a mixture.

This phenomenon indicates that the carburetor was supplying fuel to the car either above or below the flammable or explosive limits of the gas. Practically all flammable gases or liquids which can be vaporized have limits beyond which the mixture will not burn. These are called the *explosive* or *flammable limits*. The range between the upper and lower limit is called the *explosive* or *flammable range* of the substance.

The fire investigator will use the explosive limits by first determining the volume of air in a room where an explosion occurred. He will then determine the upper and lower limits of the gas involved and compute the upper and lower volumes of cubic feet of gas required to form an explosive mixture.

By calculating the flow rate, a time element can be determined to estimate either the time required for the gas to accumulate or the flow rate of gas required for a given period of time. This will be covered in greater detail in Chapters 7 and 8.

Diffusion Rate

A gas is typified by the rapid motion of its molecules. If two gases are placed in proximity to each other, the rapid movement of the molecules of the gas will soon cause them to become a homogeneous mixture. The process by which this occurs is called *diffusion*. The rate at which it is occurring is called the *diffusion rate*. Tests have been performed with various gases to give a basis for comparison and a means of calculating the time required for this to occur under different conditions.

One excellent source of diffusion coefficients is a treatise by Marero and Mason, entitled *Gaseous Diffusion Coefficients,* prepared for the National Bureau of Standards.

For fires or explosions caused by leaking gas, the diffusion rates are so slow that the main influence on the distribution and mixing of the gas will be controlled by thermal, convective and draft currents rather than by molecular diffusion. Diffusion rates would normally be used for closed containers where the mixing of two gases would result in an explosion.

Specific Gravity of Gases

The specific gravity of gases is defined as the ratio of the weight of a standard volume of pure vapor divided by the same volume of dry air at a standard temperature and pressure. It can be calculated as the ratio of the molecular weight of the gas to that of air. Therefore, since air divided by air would be a ratio of 1, if the vapor were heavier than air, the vapor density of the gas would be greater than 1. If the vapor were lighter than air, the ratio would be less than 1.

This is a common phenomenon that everyone has experienced. For example, if you blow up a toy balloon, the density of the gas within the balloon is the same density as the air outside the balloon. Once the two gases reach the same temperature, the balloon will fall to the ground due to its own weight. However, if the same balloon were filled with helium, which has a vapor density of 0.09, the balloon would rise, being lighter than the air it displaced.

By the same token, if you filled one balloon with propane, which has a specific gravity of 1.53, it would fall to the ground. If it were filled with natural gas (specific gravity of 0.53) it would rise.

This information is useful to the fire investigator in determining how and where an explosion occurred and how and where gases would migrate once a leak occurred. However, since 90 percent of a commercial flammable mixture is air, the specific gravity of the gas has little effect on the movement of the final flammable mixture. The movement is governed by convection and thermal currents, or drafts.

PRINCIPLES OF FIRE

Fire is commonly defined as the rapid oxidation of a substance. This definition infers, correctly, that oxidation can also take place at a slower rate. A typical example of slow oxidation is the rusty nail, which is the oxidation of iron with air. Other signs of slow oxidation include the change in the color of copper as it weathers.

However, the principle in which the fire investigator is most interested is the slowest rate of oxidation which can take place and

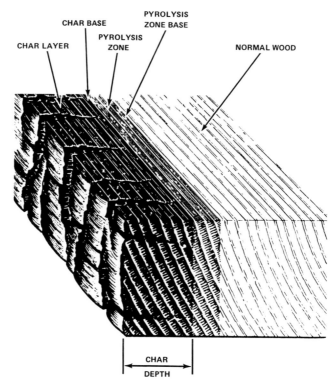

Figure 7. Degradation zones in a wood section. Courtesy of U. S. Forest Products Laboratory, Madison, Wisconsin.

cause a fire. The slowest rate is the process of *pyrolysis.* Pyrolysis is defined as the destruction of wood through the application of heat in the presence or absence of oxygen. If no oxygen is present, charcoal is formed; with oxygen present, pyrophoric carbon is formed.

Pyrophoric Carbon

Pyrophoric carbon is formed by the application of a low heat source to a combustible material and has been known to take as long as ten to fifteen years before the ignition temperature of the material is lowered to the point where spontaneous combustion occurs.

The process is as follows. The application of low heat drives off the volatile substances from the wood at such a slow rate that no flammable mixture is formed. (The gas mixture is below the lower flammable limit.) Therefore, even though ignition may be available, no fire results. The wood pyrolizes to the point where it is completely transformed into pyrophoric carbon. In this process, the structure of the wood changes; the kindling temperature is lowered until the wood ignites through auto-ignition. Generally, the auto-ignition temperature is at the temperature of the heat applied.

The application of heat at slightly higher temperatures will speed up the process so that a shorter period of time will produce the same results. Characteristics of pyrophoric carbon formation are the application of low heat (temperatures of 200° to 300°F) over extended periods of time.

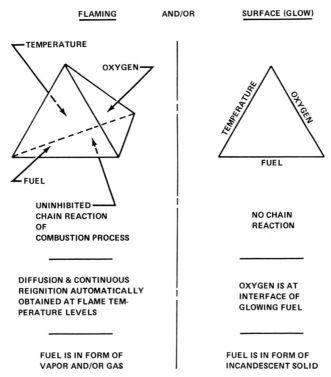

Figure 8. Basic fire system mode requirements. From Gordon P. McKinnon and Keith Tower (Eds.), *Fire Protection Handbook,* 14th ed., 1976. Courtesy of National Fire Protection Association, Boston, Massachusetts.

Slow Fires

A second phenomenon resulting in slow burning, or prolonged chemical reactions without burning resulting in a fire, is called *spontaneous combustion*. In spontaneous combustion, an exothermic reaction between the air and the substance occurs. The heat generated is not dissipated as fast as it is generated, resulting in a rise in the temperature of the substance, which in turn causes an increase in the rate of reaction of the substances. This results in a chain reaction that raises the temperature even faster. In a matter of days, hours or sometimes minutes, the substance will burst into flames through auto-ignition.

Glowing Combustion

The next fastest rate of reaction is *glowing combustion*. This is characterized by the rapid evolution of volatile gases from wood or other substances so that no flame is apparent. Such combustion is apparent in the coals seen in a fireplace after the logs have burned down to glowing embers, or a charcoal fire where a forge is used for heating metal objects. The process is characterized graphically in Figures 8 and 9. Figure 8 is the familiar fire triangle. It is applicable to glowing fires but not to flame fires. Figure 9 shows the feedback required to sustain combustion.

Flame Fires

It might be noted at this time that very few solids or liquids burn. When heated to the point of vaporization, they give off a gas which forms a flammable mixture; this is the material which burns. Evolution of these gases continues until the wood has completely decomposed, leaving only an ash, or the liquid completely vaporizes, leaving no residue. These fires are called *flame fires*. The flame fire is shown graphically in Figures 8 and 9.

This can be easily confirmed by observing the surface of a kerosene fire in a jar. Fill the glass jar nearly to the top with kerosene, and light the surface. Then place your eye level at the top of the kerosene, and observe the interaction between the flame and the surface of the kerosene. You will note a very minute space between the actual flames and the surface of the kerosene. This

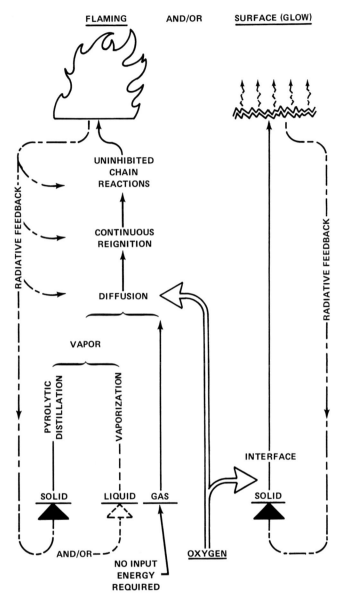

Figure 9. Basic fire system modes. From Gordon P. McKinnon and Keith Tower (Eds.), *Fire Protection Handbook,* 14th ed., 1976. Courtesy of National Fire Protection Association, Boston, Massachusetts.

space is the area where the kerosene is changing from a liquid to a vapor; the vapor is mixing with air and burning as the mixture reaches the flammable limit.

It will also be noted that the flame rises a considerable distance above the kerosene. This is because, as noted before, the kerosene has to be within the flammable limits in order to ignite. Therefore, the kerosene vapor has to mix with air to form a flammable mixture. This mixture rises as it burns, and the time for the reaction to take place is the distance the flame travels from the time it starts to burn and the time the reaction has been complete for each molecule of oil.

Flaming Combustion

The final stage of fire is flaming combustion (see Fig. 9), the most familiar type, in which flames are sent leaping into the air. The visible flames are evidence of the chemical reaction wherein oxygen is combining with the reactant, which is the distilled vapors from the kerosene, wood, gasoline or plastic combustible. The flame emits heat and light, evidence of the exothermic character of the chemical reaction.

The rate of generation of the flammable vapors and the reaction dictate how high the flames leap into the air. It is a chemical phenomenon that the rate of burning doubles with every 10 Centigrade degree rise in temperature of the reactants. Therefore, as can be imagined, a chain reaction sets in which would be uncontrollable were it not for the physical limitations of the fire.

As the fire starts to rage out of control, there is usually sufficient fuel and the avenues of escape are few, such as in a closed room. However, as the fire spreads it soon envelops the entire structure so that the additional heat generated finds no more material to feed upon and the heat is lost by radiation, conduction and convection as all the fuel is consumed.

Radiation

From a physical standpoint, heat is energy; this energy is conducted by three methods. Radiant heat is the transmission of heat energy through space in the form of waves. The most common phenomenon known to us is the heat which arrives daily from the sun. This heat travels millions of miles through empty space in the form of radiant energy.

Convection

The second most common method of heat loss is convection. The energy is lost through air currents as the air is heated by proximity to the fire to the point where the volume of the air, due to expansion of the heated gases, becomes lighter than the surrounding air. It moves upward, carrying with it the heat which resulted in the expansion of the gas, causing it to become lighter.

Conduction

The third method is conduction. Conduction is the transmission of heat through a solid by molecular activity. Each material has a characteristic conductivity which governs how rapidly heat travels from one end of the substance to the other. The higher the conductivity, the faster the heat travels. A copper bar will heat up much more rapidly than a steel bar, and a steel bar heats up much more rapidly than a piece of wood.

Heat Balance

As a fire rages out of control, it soon establishes a heat balance where the heat generated is balanced by the amount of heat lost through radiation, convection and conduction. This is illustrated in Figures 8 and 9. Once this heat balance is maintained, the fire will continue to burn at the same rate until either a change in the heat loss or a change in the amount of fuel occurs. Usually all of the available fuel is consumed, and the fire will burn itself out or be extinguished by external changes that reduce the temperature of combustibles or eliminate oxygen.

Flame-Spread Index

It has been found that various construction materials burn at different rates. In order to have a definitive and comparative base, a Flame-Spread Index has been established for construction materials. The Flame-Spread Index, as established by Underwriters Laboratories, Inc., and others, gives a comparative rate to the burning of construction materials, based upon red oak flooring as 100 and asbestos fibers as 0.

Fire and Arson Investigation

Using this Flame-Spread Index (available from Underwriters Laboratories, Inc.), a fire investigator can determine the comparative rate of how fast a fire should or should not have spread under normal circumstances by comparing the burning rates of known fires and the Standard ASTM (American Society of Testing and Materials) Time/Temperature Fire Exposure chart shown in Figure 10.

Figure 10 shows the temperature acquired as a function of time, which has been found to be the average temperature eight feet off the floor. It is used as a basis in comparing the burning rates of structures.

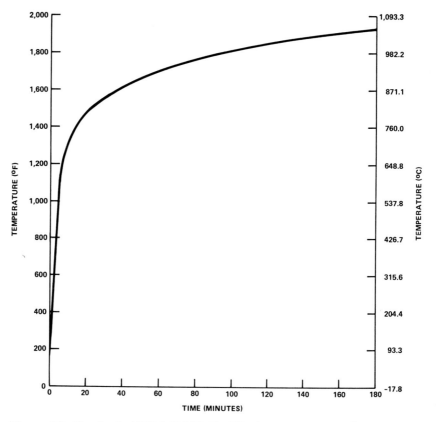

Figure 10. Standard ASTM E-119-61 (2) time-temperature fire exposure. From *U. S. Forest Service Research Paper FPL 60,* Forest Products Laboratory, Madison, Wisconsin. Reprinted by permission of the American Society for Testing and Materials.

Flashover

Another phenomenon associated with the chemical and physical aspects of fire is *flashover*. Flashover is the rapid involvement of the combustible material in a room, due to the gradual rising of temperatures to the ignition points simultaneously by radiation, convection and conduction of energy from an adjacent room. The usual manner is by radiation and convection.

It can occur in rooms adjacent to burning rooms as the generated gases burn, forcing the heated air into the adjacent room to

Figure 11. This room was ignited by flashover. The line of demarcation between the char and unburned paneling is level and uniform throughout the room. The small amount of smoke staining indicates a fast fire.

seek areas of expansion. This results in a rise in temperature of the combustibles in the adjacent room until the ignition points are reached.

Since there is little difference between the temperatures along the ceiling from one end of the room to the other, the entire ceiling will appear to burst into flames simultaneously, although it usually happens that the area closest to the source of heat will ignite first and very rapidly involve the entire ceiling, working to the extreme end of the room in a matter of seconds.

Compare Figure 11 (flashover) with Figures 19 and 20. Note that the slope of the edge of the burn pattern points toward the origin of the fire. In Figure 11, the edge of the burn pattern is level, indicating flashover.

BEHAVIOR OF FIRE

While the primary concern of the fire investigator is to determine the origin and cause of a fire, an understanding of the way fire behaves once it has started will assist in understanding and interpreting burn patterns. Accidental fires can start slowly, as in the case of a smoldering cigarette fire, or instantaneously, as in the case of an explosion with a resultant fire.

In the first case, the smoldering fire will emit copious volumes of heavy smoke. The smoke, being a heated gas, will rise to the ceiling and spread outward and downward, staining the walls and windows as it builds up. If the fire is extinguished before it destroys this smoke staining, the staining is confirming evidence of the smoldering characteristic of the fire.

In the second case, the explosion and rapid involvement of the building by heat and flames raises the temperature of the room so rapidly that little smoke staining is apparent in the room of fire origin. However, the smoke from the room of origin is carried into adjacent rooms, where smoke staining takes place until flashover occurs.

In the combustion process, great volumes of water vapor are the byproduct. The warm water vapor condenses on the cold surfaces of windows in adjacent rooms.

Between these two extremes are all of the various rates of fire beginnings. Whether the fire was started by incendiary methods or not, once the fire commences it will be controlled by two variables: fuel and ventilation. Although other chemical reactions are possible, the normal reactants concerned are air and oxygen. The amount of air feeding a fire is controlled by the ventilation characteristics of the building in which the fire is contained.

Fuel-controlled Fires

When most fires begin, an abundance of both fuel and air is available. However, as the fire increases in size, it soon involves the entire contents of a room and becomes dependent upon ventilation for survival. Due to the confinement of a room, adequate heat feedback is assured so that all of the fuel available becomes involved in a short time. At the time of maximum fuel involvement, the fire (unless adequate ventilation is present such as open doors, windows and roof openings for expanding gases to escape and be replaced by fresh air) will be ventilation controlled. However, once the fuel has been consumed, the fire will diminish due to lack of fuel. Most rooms contain adequate fuel so that the fire becomes ventilation controlled.

Building Contents

The greatest variable of the fuel available to a fire is the building contents. In normal residences, the contents vary little from one home to another. However, in industrial or commercial establishments, the building contents range from highly combustible wooden furniture stacked in tiers or cartons, so that the fuel load is very great, to warehouses stocking noncombustible materials such as metal parts stored in metal bins, where the only flammable material is the paint covering the bins, floor, walls and ceiling.

The contents of each room prior to the fire will have a profound influence on the size, temperature and heat flux of the fire.

Building Construction

If the building is constructed of wood with wooden walls, open studding, open rafters and ceiling, the building will burn more rapidly than an all-concrete fireproof building. The construction materials play a vital role in the rate at which the fire burns.

One-quarter inch plywood walls have a burn rate of approximately fifteen minutes per sheet. The plywood, once burned through, will provide enough heat to ignite the studding. The heat generated by the plywood will be more than enough to ignite the ceiling and eventually the roofing material.

The burning rate of building materials will be lowered significantly by the application of fire-retardant materials. Modern paints may have a fire-retarding component which may materially slow the spread of the fire. The presence of fire-retardant material may completely destroy a fire investigator's timetable of the fire spread.

Ventilation-controlled Fires

Most building fires are ventilation controlled. Unless it is an incendiary fire and the arsonist has provided adequate ventilation, most buildings are closed to prevent heat loss and theft. Windows and doors in business establishments are usually closed, particularly during the evening hours. It is not uncommon to find a fire of accidental origin which smothered itself due to lack of oxygen.

Modern buildings with forced-air circulating systems defeat the self-extinguishment of fires by circulating throughout a building the air which will feed a smoldering fire and provide fresh air. Not only will the building itself provide sufficient ventilation to support a fire, but firemen in their efforts to extinguish a fire must, by necessity, open the building to eliminate smoke from a smoldering fire. In so doing, quite often they will provide enough additional ventilation to bring the fire to life, even causing small *blow-backs* (ignition of overrich gas mixtures by sudden dilution caused by ventilation) or minor explosions in the process.

Before Extinguishment

Before the firemen arrive, a smoldering fire may have reached a point where it is ventilation controlled and producing voluminous smoke and carbon dioxide. This will cause heavy staining, as previously mentioned. The heat generated by the fire will raise the temperature of the room and within the building to the range of 1000° to 1200°F. At this point, the carbon dioxide reacts with smoke to produce carbon monoxide.

When the firemen open up the roof and windows for smoke removal and lower the building temperature for safe entry, the additional air can cause ignition or explosion of the carbon monoxide. Prior to opening the building, the fire will have moved at a diminishing rate as less and less oxygen is available to sustain the fire growth. Once temperatures of 1500°F are reached, conversion of carbon dioxide to carbon monoxide occurs which can result in blow-back.

During Extinguishment

When the fire service has opened the building and started to apply water, they will attack the hottest spots first. This leaves the cooler spots time to grow and increase in size until the hotter spots have been cooled, at which time firemen will go to the new hot spots until the entire building is cooled down and under control. At that time, the entire fire will be extinguished.

If the building is large, it is imperative that the fire investigator be aware of the procedures used by the fire service in extinguishing it, as well as their methods of ventilation. This knowledge will enable the fire investigator to evaluate the burning rates and times of the various areas of the building. It will assist in determining the area of origin.

Weather Effects

The path and behavior of the fire is primarily a function of the fuel and ventilation of the fire. To some extent, these factors are affected by the outside weather. The most effective of these is wind. Rain has a mild effect; temperature has little effect except on the efforts of the fire service.

WIND. A strong wind will have a profound effect on the ventilation of the building. Before the fire service arrives on the scene, any open windows on the leeward side will act as a draft or exit for heated gases escaping. Any open windows or doors on the windward side of the building will provide an entrance for wind and air and will provide the fire with adequate oxygen. The wind will also affect any roof openings (vents, stacks or openings cut by firemen for ventilation).

A strong wind blowing across a flat roof will cause a suction on the openings at the top of the roof and increase the effectiveness of a vent stack or ventilation opening. On a pitched roof it will have the same effect on the leeward side but may have an adverse effect on the windward side if the roof pitch exceeds 30°. By creating a downdraft, the wind may reverse the flow of ventilation up a stack, stack vent or ventilation opening cut in the roof.

RAIN. Rain will have very little effect upon building fires, since the normal rate of rainfall is insufficient to have any extinguishing effect on the fire. Unless a cloudburst occurs, the effect of rain can be ignored, particularly if the fire is contained within the building and the roof is still functioning, preventing the rain from reaching the fire.

TEMPERATURE. The ambient temperature has little effect on fire, since fires are generally confined to the rooms of a building and are ventilation controlled, provided there is adequate fuel available. The most profound effect of temperature is in subzero weather when firemen experience freezing lines and water hoses so that efforts to extinguish the fire are hampered. The extra time needed to extinguish the fire is due to the problems encountered in getting the necessary water onto the fire, not because of the severity of the fire.

Comparison of extinguishment times should be made on the basis of similar climatic conditions in comparing one fire to another. Consideration should be given to the atmospheric conditions under which the fire service was operating when making this comparison.

HAZARDOUS MATERIAL

One source of information on the chemical characteristics of substances is found in a publication by the National Fire Protection Association, *Fire Protection Guide on Hazardous Materials,* Sixth Edition. Included are the flash points for more than 8,800 trade name products. Also included are the fire hazard properties of more than 1,300 flammable substances. These are listed alphabetically according to their chemical identification.

In NFPA Standard 49, *Hazardous Chemicals Data,* which is also included as part of the aforementioned publication, the data for 416 chemical compounds is given on relative hazards in relation to fire, explosion and toxicity. The chemicals are arranged alphabetically by names and synonyms. Also included are 3,550 mixtures of two or more chemicals considered to be potentially dangerous as a possible cause of fires, explosions or detonations at standard or higher temperatures.

Additional information may be found in Manual 704-M, which is included in NFPA 49. This is the recommended system for identification of fire hazard of materials. The system is a simplified method for the immediate identification of the hazard involved in the storage, transportation and fire fighting of chemical compounds. It is useful to the fire investigator in identifying possible causes of fires through chemicals which may have reacted to an accidental fire, giving the impression of an incendiary fire, or vice versa. Table V provides information, condensed from the above sources, regarding hazard identification. The three main hazards identified are those relating to *health, susceptibility to fire* and *reactivity* or susceptibility to release of energy of the stored materials.

Health Hazard

The *health* hazard is identified in the NFPA material by a color code of blue and is rated from *4* decreasing to *0* in respect to the type of possible injury. For example, a *4* would indicate materials which, on very short exposure, could cause death or major injury even though prompt medical treatment was given. A *3* would indicate materials which, on short exposure, could cause serious temporary or residual injury even though prompt medical treatment was given. A *2* would indicate materials which, on intense or continued exposure, could cause temporary incapacitation or possible residual injury unless prompt medical treatment is given.

A *1* would indicate materials which, on exposure, would cause irritation but only minor residual injury even if no treatment is given.

A *0* would indicate materials which, on exposure under fire conditions, would offer no hazard beyond that of ordinary combustible materials.

The health hazard is primarily of interest only after the fact of

TABLE V

HAZARD IDENTIFICATION*

Substance	Health	Flammability	Reactivity	Flash Point Degrees F	Water Soluble	Fire Fighting Phases†
Acetone	1	3	0	0	Yes	2
Asphalt (typical)	0	1	0	400+	No	7
Butane	1	4	0	Gas	No	1
Corn oil	0	1	0	490	No	7
Diesel fuel oil #1	0	2	0	100 minimum or legal	No	5
Diesel fuel oil #2	0	2	0	125 minimum or legal	No	5
Diesel fuel oil #4	0	2	0	130 minimum or legal	No	5
Ethyl alcohol	0	3	0	55	Yes	2
Ethylene glycol	1	1	0	232	Yes	6
Fuel oil #1	0	2	0	100 minimum or legal	No	5
Fuel oil #2	0	2	0	100 minimum or legal	No	5
Fuel oil #4	0	2	0	130 minimum or legal	No	5
Fuel oil #5	0	2	0	130 minimum or legal	No	5
Fuel oil #6	0	2	0	150 minimum or legal	No	5
Gasoline	1	3	0	-45	No	4
Glycerine	1	1	0	320	Yes	6
Hydrogen	0	4	0	Gas	Slightly	1
Boiled linseed oil	0	1	0	403	No	7
Lubricating oil	0	1	0	300-450	No	7
Propane	1	4	0	Gas	No	1
Turpentine	1	3	0	95	No	4

* Adapted from NFPA Publication No. 49, *Hazardous Chemicals Data*, 1973, Table I (Appendix). Courtesy of National Fire Protection Association, Boston, Massachusetts.

† Descriptions of the various fire fighting phases are available from NFPA Publication No. 49, *Hazardous Chemicals Data*, 1973. p. 49-309. This material is also contained in NFPA, *Fire Protection Guide on Hazardous Materials*, 6th ed., 1977.

injury to persons involved in extinguishing the fire and as a result of an incendiary fire. Of more interest to the fire investigator is the *flammability hazard*.

Flammability Hazard

The identification of *flammability* is by the color code red, indicating the *susceptibility of materials to burning*. Again, these are graduated from *4* decreasing to *0* in susceptibility.

A *4* would indicate materials which will rapidly or completely vaporize at atmospheric pressure and normal ambient temperature, or which are readily dispersed in air and will burn readily.

A *3* would indicate liquids and solids which can be ignited under almost all ambient temperature conditions. A *2* would indicate materials that must be moderately heated or exposed to relatively high ambient temperatures before ignition can occur.

A *1* would indicate materials which must be preheated before ignition can occur. A *0* would indicate materials which will not burn.

These indications are of the most significance to the fire investigator and give clues to the possible cause of a fire. Any time these symbols indicate the presence of highly flammable substances, the substance should be properly identified and the storage and circumstances involved in their use should be thoroughly investigated.

Reactivity Hazard

The identification of *reactivity* of a material is by the color code yellow. It indicates the *susceptibility of a material to the release of energy*. Again, these are graduated and identified in decreasing susceptibility from *4* to *0*.

A *4* would indicate materials which in themselves are readily capable of detonation or of explosive decomposition or reaction at normal temperatures and pressures.

A *3* would indicate materials which in themselves are capable of detonation or explosive reaction but require a strong initiating source, or which must be heated under confinement before initiation, or which react explosively with water.

A *2* would indicate materials which in themselves are normally unstable and readily undergo violent chemical change but do not detonate. This also includes materials which may react violently with water or may form potentially explosive mixtures with water.

A *1* would indicate materials which in themselves are normally stable but can become unstable at elevated temperatures and pressures, or may react with water with some release of energy but not violently.

A *0* would indicate materials which in themselves are normally stable even under fire exposure conditions and are not reactive with water.

These identifications are of prime interest to the fire investigator for possible causes of primary or secondary (or both) explosions or fires which may or may not be of incendiary origin. The higher the number, the more possible the involvement in the fire and the more thoroughly the substances should be investigated.

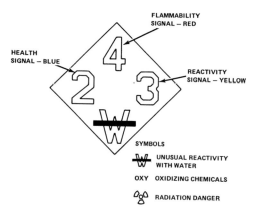

Figure 12. Identification of materials by hazard signal arrangement. From NFPA, *Fire Protection Guide on Hazardous Materials,* 6th ed., 1977. Courtesy of National Fire Protection Association, Boston, Massachusetts.

CHAPTER 4

Origin and Cause of Fires

A BRIEF GLANCE at the statistics indicates the need for determining the origin and cause of fires. If we understand the nature and cause of fires, we are better able to control and eliminate their causes. The fire investigator makes an important contribution to society each time he makes an investigation and a positive determination as to the cause of a fire. The degree of his contribution is measured by the accuracy of his determination of the true cause of the fire.

Fire was with us before man crawled out of the primeval ooze onto dry land. The first fires were caused by lightning; it still exists as one of the natural causes of fire.

NATURAL CAUSES

Natural causes are easily identified, as they are usually accompanied by unusual or near-catastrophic natural occurrences. Although they are not listed in the NFPA tabulation, they should be considered as a possible cause, particularly in arson cases. Unless they are eliminated, the defense may claim (and prove) that the true cause of the fire was a natural cause.

Lightning

Heading the list of natural causes of fires is lightning. Lightning accounts for 2 percent of all reported fires. Fortunately, it is one of the most easily detectable fire causes.

The first requirement is that meteorological activity be present. This can be easily verified by contacting the local airport weather service or local meteorologists at television stations, or by contacting the National Oceanic and Atmospheric Administration, Environmental Data Service, National Climatic Center, Federal Building, Asheville, North Carolina 28801. Upon request, a certified copy of the official weather records for conditions existing at the time of the fire will be sent at no charge.

When lightning strikes, sufficient electrical energy can be involved so that temperatures are raised and a fire may ensue. Lightning is always accompanied by thunder as the vacuum created by the electrical discharge of the lightning stroke is filled by the surrounding air. Therefore, many witnesses should be available to confirm the presence or absence of a lightning stroke.

Lightning will usually strike a building at its highest point and travel through the building, following the path of least resistance, which at times may be the wooden structure of a house, causing moisture in the wood to explode. At times, the electrical discharge will enter the electrical system, causing wires to overheat, melt and start a fire in the process. The odor of ozone is commonly detected at the scene of a lightning fire.

Not all lightning strokes start fires. The reason is that the duration of the lightning stroke may not be long enough to raise the temperature of the struck object to the kindling temperature. Laboratory tests show that a current discharge of long duration will usually start a fire. The time required for the start of a fire has been determined to be in excess of 40 milliseconds. It was also found that approximately 50 percent of all lightning strokes are of durations long enough to start a fire.

Lightning is caused by convective currents occurring in thunderclouds containing cold, dense air aloft and warm, moist air at the bottom. This results in strong updrafts, and the cold air descends as the warm air rises. This produces a positive charge in the upper portion of the thundercloud, while the lower portion develops a negative charge; the result is an intracloud discharge. When the clouds are close enough to the earth, a positive charge on the ground is close to a negative charge on the cloud, and a discharge from the earth to the cloud will occur.

While the majority of lightning strokes occur in summer thunderstorms, they are by no means limited to this season. It is not uncommon to have lightning strokes during snowstorms, sandstorms, or even on a cloudless day, as uncondensed moisture in the air can produce the static electricity necessary to form a discharge path.

In a typical storm, 50 percent of the lightning strikes exceed

10,000 amps with a maximum approaching in excess of 20,000 amps. With this type of energy being released, it is understandable how a substantial amount of damage can occur, including a fire.

Lightning Arresters

Lightning rods and lightning arresters are electrical devices installed as protection against lightning strikes. Article 280 of the *National Electrical Code* provides that lightning arresters shall be installed in industrial locations as protection against lightning. The lightning arrester system consists of a low resistance course from the top of the structure to the ground. However, the presence of a lightning rod or arrester system does not guarantee that a fire will not occur. Table VI indicates losses by lightning for the years 1970 to 1974, inclusive.

As shown in Table VI, 40 percent of lightning losses occurred on buildings equipped with lightning protection, wiring arresters and capacitors. The table also shows that 52 percent of the dollar value of the loss occurred on buildings equipped with lightning rods. Therefore, the presence of a lightning arresting system does not guarantee that a fire will not be caused by lightning.

Wind

Among other natural causes of fires is the action of wind. The

TABLE VI

LOSSES BY LIGHTNING FOR 1974 AND ACCUMULATED
FIVE-YEAR EXPERIENCE*

Type of Lightning Strike	1974 No. of Losses	1974 Net Loss Paid	1970-1974 Inclusive Net Loss Paid
Lightning entering on wiring, arresters and capacitors installed	88	$189,723	$337,338
Lightning entering on wiring without arresters and capacitors installed	137	91,345	327,266
Lightning, direct hit, building not rodded or grounded	27	17,813	195,331
Lightning, direct hit, building rodded or grounded	18	11,176	234,979

* Courtesy of Mill and Elevator Rating Bureau, Chicago, Illinois.

wind can blow a building apart, shorting out an electrical wire, which in turn may start a fire. Wind can also sway electrical lines, causing them to oscillate until they make contact, shorting them out and causing a fire.

Since power companies are aware of the danger of wind-induced swaying wires, it would take an unusually strong wind to cause wires to sway and make contact. Therefore, the weather conditions existing at the time of the fire should be checked for strong or unusual winds.

A strong wind can cause a downdraft and a reversal of the flame in furnaces and water heaters. The flames can be forced to reverse their normal path and exit through the combustion air intake path, igniting any nearby combustibles. Unattended fires can also be blown out of control by stray wind currents.

Rain

Rain has been known to leak into a building, short out electrical wires and cause fires. Wind-blown rain can achieve the same result. Rain can enter and penetrate containers which contain water-reactive chemicals, resulting in exothermic reactions and causing fires. Confirmation of the weather conditions can be obtained from eyewitnesses, including the fire department involved, local weather stations and the Climatic Center noted previously.

Flood

Floods are a common cause of fires. They can cause sewers to back up until they short out electrical circuits. As the flood water recedes, the evidence of the flood is destroyed by water from the firemen's hose and no trace is found of the sewer flooding and stoppage.

The clues to look for in this case would include a check with the neighbors to see whether their sewers had backed up. Check any electrical appliances or circuits in the lower part of the basement which could have shorted out and caused the fire. These fires would be classified as electrical fires, but the cause of the fire could be the backing up of the sewer or other types of floods.

Natural floods from rainfall or rivers are common occurrences. No difficulty would be encountered in associating a natural flood with a fire. However, in sewer backups or stoppages, this source is sometimes overlooked.

Included with floods should be consideration of broken water lines, cracked toilet bowls or similar sources of water in a residence or building which can extinguish pilot lights, furnace fires or the function of other appliances. The gas, due to flooding of the controls, may continue to flow until a source of ignition is found, resulting in an explosion and fire.

Building Settlements

Buildings can settle to such an extent that pipes conveying gas will crack or break, causing leaks which result in an explosion and fire. Most buildings are constructed with an allowance for settlement regarding pipes. However, once this settlement has occurred, the load of the building is transmitted directly to the pipe and can cause fractures. Careful examination of this type of rupture will usually reveal the cause of the broken pipe (see Fig. 40), since pipes in earth, or where they enter a building, do not burn and are not affected by the fire.

Settlement can be caused by the passage of time or can be induced by heavy construction equipment or other vibrations near a foundation which will aggravate the settlement of the building. This can cause settlement within a short time which normally would not occur over a twenty-year period.

Animal-caused Fires

Wild and domestic animals have been known to chew on the insulation of electrical wires, particularly in older homes which contain knob-and-tube wiring (wiring with only one layer of insulation on each wire). After chewing off insulation from one wire, they will sit on the bare wire while chewing on insulation of an adjacent wire. Once contact is made with the bare copper wire, the animal's body acts as a conductor to short out the circuit. The fur will sometimes ignite and spread the fire to nearby combustible materials.

ACCIDENTAL CAUSES

Accidental causes of fires are those which are not natural fires or deliberate acts of man. An incendiary fire is one which was deliberately set by man. The process of elimination is used to narrow the probable cause of the fire in the shortest time possible to one of the three types of fires.

The process of elimination is a logical and philosophical method of arriving at a conclusion. In this process, the most obvious candidates for elimination are considered first. A statistical or probabilistic approach is used. Using this approach, the natural causes are eliminated first. These can be rapidly considered. The weather on any particular day is a matter of record and can be quickly, easily and accurately verified by witnesses and official records.

The next question to be answered is whether the fire is of incendiary or accidental origin. Again, taking a statistical approach, since less than 25 percent of fires are incendiary, the process of elimination dictates that this possibility should be eliminated next.

In order to prove or disprove that a fire is incendiary, *all* accidental causes must be eliminated, or the probability of an incendiary origin must be eliminated. It is quite often in this process of eliminating arson that the true cause of the fire is discovered and enough circumstantial evidence is found to eliminate the question of arson.

Before an accidental cause can be eliminated, it has to be a possible cause. A possible cause is usually located in the immediate area of the origin of the fire. Once the origin is determined beyond a reasonable doubt, a list of the possible causes can be made. The most improbable possible causes can be rapidly eliminated by some of the tests outlined in the chapters appropriate to the particular type of fire under consideration. It should be emphasized that the origin of the fire must be positively identified. If the origin is incorrect, the cause cannot be correct.

Electrical Causes

The most common cause of electrical fires is the overloading of a circuit. Overloading may occur if an undersized conductor is used for the load carried. For example, a 10 amp current can overheat a conductor if the conductor is too small to carry 10 amps. The circuit breaker may be designed for 15 amps, and therefore, the circuit breaker cannot be relied upon for protection.

Other causes of electrical fires include massive shorts. Common shorts involving 15 or 20 amp circuits will usually not generate enough heat or be of long enough duration to initiate a fire. However, shorts in the vicinity of fuse boxes or distribution systems, where much higher amperages are found, are of adequate amperage to cause fires by shorting. Other causes of electrical fires include overheating of motors or space heaters utilizing electrical current as the energy source.

Gas Fires

Utility gases such as propane, butane, methane and natural gas usually cause fires by escaping through a leaking system. The leak may be caused by building settlement, construction mishaps, appliance malfunctions or numerous other causes. As a rule, the gas escapes in sufficient volume to form an explosive mixture which finds a source of ignition, ignites and causes a minor or major explosion, resulting in a fire. As a result of the explosion, the gas line may be ruptured, with the gas line supplying additional fuel to the fire.

The majority of gas fires are caused by failure of the gas controls. Despite the efforts of manufacturers to create fail-safe devices, they still malfunction and permit gas to escape, finding a source of ignition which results in an explosion.

Appliance Malfunctions

Appliance malfunctions include the failure of gas control devices as well as electrical and mechanical overheating of various appliances. The usual cause of a fire is the mechanical failure or lack of lubrication of a mechanical part which permits overheating of the appliance and the resultant heating of combustible materials in the vicinity of the appliance.

Sponaneous Combustion

Spontaneous combustion is nearly always determined by reconstruction of what was present at the origin of the fire. If the right combination of chemicals, volatile substances, combustible materials and ventilation are found, and all other possible causes have been successfully eliminated, then spontaneous combustion is the only logical answer.

Explosions

Explosions can be divided into three types: *gas* explosions, *dust* explosions and explosions of *explosives*. Chapter 8 describes clues to look for in an explosion with details and examples. Explosives, by nature and definition, require a detonator in order to propagate. Therefore, determining the cause of an explosive's detonation requires determining the source and origin of the detonator and the circumstances involved in producing this detonation which touched off the explosion.

Miscellaneous Causes

This category includes fires related to smokers and tobacco users. These are usually caused by misplacement of hot smoking materials in the presence of combustible materials such as upholstered furniture, paper, waste material or other similar materials.

It also includes fires started from open flames and sparks such as welder's torches and arcs, sparks from machinery, fireplace fires, open fires and chimneys. This classification also includes children playing with matches or other sources of fire.

INCENDIARY FIRES

An incendiary fire is one set for the willful destruction of property by fire. Arson is the crime associated with an incendiary fire. Quite often before a fire investigation has been completed and even before the origin of the fire is found, indications of an arson fire are detected. Should this be the case, the fire investigator should immediately concentrate his efforts on eliminating or confirming arson as the cause. If arson cannot be eliminated, it should be reserved as a probable cause and the investigator should proceed with the elimination of accidental and natural causes.

The incendiary fire is usually set by a person falling into one of three classifications: the pathological firesetter, or pyromaniac; the emotional firesetter; the arsonist.

The Pathological Firesetter

The pathological firesetter is characterized by the emotional satisfaction he receives from setting fires. He may have no other motive than to satisfy his own pyschological need to see a fire, the excitement of firemen battling the flames and the destruction incurred by his act, all of which gives him a feeling of power.

The Emotional Firesetter

The emotional firesetter is a person who destroys property in an act of revenge, spite, hatefulness or to strike out at persons whom he feels have wronged him. This would include the housewife who burns the house after her husband left her for another woman. It also includes the high school student who burns down the high school after being dropped by the basketball team for low grades. Another case could be a teenager who deliberately burned her father's car to conceal that she accidentally burned a hole in the front seat with a forbidden cigarette.

Fraud Fires

Fraud fires are started by profit-seeking arsonists who wish to benefit from the proceeds of a fire by the insurance funds. The amateur arson attempt is usually characterized by ill-founded attempts to totally destroy the building using inept methods which are easily detected and proven.

The professional arsonist, on the other hand, will usually burn a building to the point of total destruction with no obvious evidence of arson. He will have a well-established alibi to explain his whereabouts at the time of the fire.

If total destruction is not the case, the fire will appear to be of accidental origin with selective damage to the insured's property or goods. These fires are a challenge to the fire investigator.

DETERMINATION OF THE ORIGIN OF FIRES

The *origin* is the physical area where the fire started. It is the physical location within the building where combustible or flammable materials were exposed to heat and a form of ignition.

For example, if a cigarette was dropped in an upholstered sofa, the combustible material would smolder until sufficient heat was generated to produce a flaming fire. The fire would then spread from the sofa to the wall, then to the floor of the room, until the entire room was involved. The general area of origin would be the room; the specific area would be the two or three inches occupied by the cigarette on the sofa.

Not all fires have a single origin, nor is the origin always that well defined. For example, a gas leak can develop in a natural gas line one-half mile from the nearest residence. The leaking gas will follow the trench of the pipeline to the branch line leading to the residence, follow the branch line into the residence, enter the home and gravitate into rooms until a flammable mixture is accumulated, probably in an upper story.

Assuming that the gas entered the house at the basement level, gravitated up the open stairway and accumulated in the kitchen, an explosion and fire would ensue, with the major damage incurred in the kitchen located at the head of the stairs from the basement. Since the explosion involved the entire room, an oversupply of gas would result in ignition of combustible substances throughout the room. Once the room had ignited, it would quickly spread to other rooms, but the *origin* would be the entire kitchen

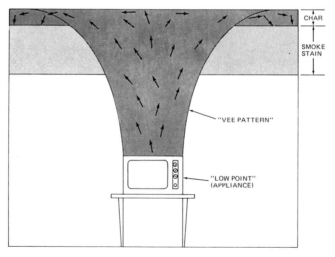

Figure 13. Accidental fires will characteristically have a well-established low point where the fire started. Rising from this low point in a "V" will be the burned material ignited from the heat source. Then, as the fire reaches the ceiling, it spreads horizontally and starts downward in a uniform manner. Preceding the char will be smoke staining, dependent upon the intensity of the fire.

area rather than any specific cabinet or other physical location in the room. The *cause* would be the gas leak, one-half mile from the kitchen.

Arsonists sometimes create *multiple origins* by involving an entire floor of a building in the start of the fire. This is normally accomplished by using accelerants and *trailers.* A trailer is a continuous path of highly ignitable material leading from one area to another to ensure rapid spread of the fire. It may consist of newspapers, wood shavings, excelsior, sawdust and/or flammable liquids such as kerosene, motor oil, fuel oil, diesel oil or gasoline. When an arsonist ignites a fire, the trailers rapidly spread the fire so that burn patterns in each room are identical. These are classified as *multiple origins.*

The determination of the area of origin of a fire is made by a careful examination of the entire building, working from the outside inward and examining the burn pattern in respect to its horizontal and vertical spread in relation to the building.

Burn Patterns

The *burn pattern* is the record left by the fire of its progress, burning rate and duration, and its effect upon the structural components of the building and its contents. Fire will characteristically burn vertically, spreading slightly horizontally, until it meets resistance, such as a ceiling. At that point, it will spread horizontally and start to burn down the walls near the point of origin.

The "V" leading from the point of origin, as in the case of a television set which started a fire near a wall, is shown in Figure 13.

The "V" rising from the origin of the fire will spread as it rises. The lowest point of the "V" is called the *low point* of the fire. The *cause* of the fire can usually be found at that point.

Not all low points are the origin. As mentioned earlier, arsonists will create many origins and each may have one or more low points. In addition to deliberately-set fires creating low points, accidental low points are not uncommon.

A low point can be caused by dripping, flaming material from the ceiling, such as melted plastic light fixtures which may ignite and drip a pool of the molten material on the floor. This can cause a burned spot in the floor directly beneath the source.

These sources are readily identified unless the source of the molten material and its supporting structural members have been destroyed. Electrical wiring, where an entire circuit becomes overheated due to an overload, can create low points along its path, dependent upon its proximity to combustibles. When a number of low points are discovered and multiple origins are apparent, the most common methods for locating the true origin of an accidental fire are by eyewitnesses who can testify what burned first, or by measuring the depth of char near each low point.

Depth of Char

The *depth of char* is the depth of the pyrolyzed or charred wood surface. It is easily measured by penetrating the soft charred surface with a penknife or a depth gauge, such as used to determine the depth of tread on tires. The depth of penetration is measured at the same elevation throughout the room, commonly eye or waist level. Near a low point, the depth of char will take a sudden increase and be readily apparent by comparison of the char pattern. If the low point terminates at the top of a table or piece of furniture, the underside of the furniture should be carefully examined to determine whether the fire burned upward from the floor or downward from the top of the table.

If the fire burned upward from the floor, the underside of the furniture would have the same effect on the heat as the ceiling. The bottom of the table would deflect the heat so that a char pattern would appear on the underside of the top of the table. This would indicate that the fire had started below the table, burning upward. This principle applies to any piece of furniture or combustible equipment found in a room.

The same principle applies to joists. If the fire burned downward, the point of penetration on the flooring and joists will be characterized by an oval burn pattern located directly at the point where the penetration of the flooring was made.

The low point of the fire can usually be found by tracing the depth of char back to the point of greatest destruction or longest burning. By starting from the rooms which are smoke stained with no apparent burning and moving toward the rooms of greatest destruction by the fire, the process of elimination will soon bring you to the only possible room where the fire could have originated. Within that room, checking the depth of char will locate the precise point (if there is one) where the fire originated. The depth of char will determine the exact origin.

The various depths of char should be recorded until the origin of the fire has been resolved. The origin of the fire, determined by the depth of char, must be in agreement with eyewitness reports and all other evidence. Unless the origin of the fire is located, any efforts to determine the cause will be futile.

Since the depth of char plays an important role in the determination of the true origin of the fire, the variables which may affect the depth of char should be fully understood to make allowances for their influence.

Effect of Ventilation

When a fire is first initiated, whether it is accidental or incendiary, adequate fuel is available. The fire is, therefore, controlled by ventilation. If adequate ventilation is supplied, the fire will rapidly involve all combustibles within the structure. For example, full-scale tests run by the National Bureau of Standards show that an entire two-story wooden frame residence could be completely involved in a fire within twenty-six minutes after the start of the fire.

A professional arsonist will use as little accelerant as possible to guarantee the start of a fire (5 gallons or less, or 5,000 BTUs). The average residence of 2,000 square feet will contain 14,000,000 BTUs in furnishings alone; the walls, ceiling and floors will more than triple this amount. The accelerant would represent less than 0.001 percent of the total fuel available.

The arsonist uses an accelerant to guarantee rapid involvement of the entire structure before the fire is detected to ensure total destruction of his handiwork. The accelerant plays a very small role in initiating the fire. Ventilation plays a much more important role. Regardless of whether the fire is of accidental or incendiary origin, it requires adequate ventilation. Accidental or incendiary fires can be more than adequately ventilated if the furnace fan is operating, the air conditioning fan is operating or if the fan is operating merely to circulate fresh air. It is *not* necessary to open doors or windows.

The condition of all mechanical devices used to move air should be checked, such as window air conditioners, humidifiers and furnaces, whether they are operating or not. The fan can be operating on a furnace even when the furnace does not call for heat. Also examine portable fans, exhaust fans, exhaust attic fans and fans installed for removal of condensation, odors and other special-purpose fans.

Finding the Origin

A first step to take in determining the origin of the fire is to obtain all information available from eyewitnesses as quickly as possible. Interviews should be conducted with persons who discovered the fire, turned in the alarm, witnesses who saw the fire before the arrival of the firemen, other onlookers and building occupants, as well as the fire dispatcher and fire fighters.

Quite often an eyewitness account of the cause and origin of the fire aids the investigation. Physical evidence should verify the story related by the eyewitness. Whether or not interviews yield a cause, the next step is a physical examination of the building. The best procedure is to work from the outside inward.

Once the origin has been identified, however, that area should be protected from trespassers; overhaul activities by firemen should also be halted.

Exterior

A quick tour of the exterior perimeter of the building can reveal whether the fire started on the exterior. In addition to an examination of the exterior for a possible origin, many valuable clues can be obtained by this examination.

All openings into the building must be carefully examined to determine their status at the time of the fire. Evidence of forced entry should also be sought.

Doors. The status of each door opening during the time of the

Figure 14. Vandals entered the porch of this residence and started a fire at the left side of the porch between the end of the building and window. The low point was at the porch floor. Notice the progressive destruction of the porch and the spread to the upper story.

fire should be determined. Interviews with the fire service will reveal if forced entry was necessary by them and the manner in which forced entry was made. For example, the firemen may have kicked in the doors. If prybar marks are found on the door jambs, it can be assumed that an arsonist made forced entry, then barred the door, requiring the firemen to use forced entry to gain access. The arsonist would have then left by another exit.

The same logic applies to multiple and dissimilar prybar marks on door jambs where the fire service used a prybar to gain entrance. A set of prybar marks not attributed to the fire service could indicate a forced entry.

Skylights should also be examined, particularly a combination skylight and vent. These should be inspected to determine whether they were left open to provide ventilation for the fire or whether the vent was normally open by its owners or occupants in order to provide ventilation. If arson is suspected, the area should be dusted for fingerprints.

WINDOW GLASS. The condition and location of broken window glass at each window should be carefully noted. If an explosion had occurred, glass would be scattered over a large area outside of the window and would be unstained. However, if the explosion occurred after the fire started, the glass, scattered some distance from the building, could be smoke stained from the early stages of the fire. The presence or absence of smoke staining will give clues as to when the explosion occurred in relation to the time of the fire.

As a fire intensifies, it builds internal pressures capable of breaking large single-strength windowpanes. This forces the glass outward and away from the building short distances (ten to twelve feet). Glass from a pressure-broken window pane will normally be smoke stained with a buildup of smoke and soot, indicating the length of time necessary to build up pressure sufficient to break the windowpane. Shards of glass from the lower portion of the window may be clear if the smoke level had not descended to that area at the time the windowpane failed.

During fire extinguishment the fire service may break windows to gain access for their water streams. These windows are normally broken by smashing them from the exterior. This will cause glass to fall on the inside of the building. An arsonist may also gain forced entry into a building by breaking windows from the outside. This would also cause shards of glass to fall on the interior. However, as a rule the glass panes broken by the firemen will already have been smoke stained, particularly at the top of the window panes; the windowpanes broken by an arsonist would not.

By careful examination of the broken glass lying near the window, the absence of staining on the glass could indicate forced entry at that window. This must be verified by comparing the staining on the broken windowpane with windowpanes subjected to the same fire exposure. Suspicious glass should be examined for fingerprints.

Interior

A quick tour of the building interior will reveal rooms in various stages of destruction. The rooms which are merely smoke stained, with no apparent scorching of woodwork, can be quickly eliminated as possible origins. Rooms where the fire has ignited the ceiling and burned partially downward can also be eliminated, as can rooms ignited by flashover. A quick clue to the sequence can be found above the door jamb in each room.

DOORS. Immediately above the door jamb in each room where fire entered through that doorway, severe smoke staining, char or burning will be visible in relation to the remainder of the room. This is due to the heat buildup in the adjacent room which accumulates at the ceiling and increases in temperature proportional to the elevation in the room. The highest temperatures will be attained at the ceiling.

As this heated air accumulates, it will expand and seek an exit. When it finds a doorway, the heated air flows under the top of the door jamb and upward into the adjacent room. That concentrated flow of heated air ignites the wall and ceiling directly above the top of the door jamb, leaving its telltale mark. Heavy char or burning above the door jamb indicates that the fire entered the room through that doorway.

All openings should be examined to determine whether the doors were open or closed during the fire. This can be done by inspecting the hinges (where the door is destroyed) and visualizing the position of the door in relation to the hinges. If the two halves of the hinges are close together and parallel, then the door was closed during the fire. If the two halves of the hinges are at 90° angles to each other, the door was ajar at an angle of 90°.

Other clues indicative of the door position are the burn patterns on the door jamb, a difference in the char depth of both sides of the door jamb at the hinge and unburned paint where the door fits tightly against the door jamb on the hinged side of the door, as shown in Figure 15.

Fire doors should be checked to see that the fusible link melted and the door was not held open by a solid wire cable or chain to prevent the door from being closed. It could also be blocked by heavy objects to prevent its closing, although the fusible link separated. Fire doors are effective in controlling fire movement. If they fail to close, the fire investigator should determine why the fire door failed and the persons responsible.

Figure 15. The exterior of the residence shown in Figure 14. The depth of char on the outside of the door was twice that of the inside. This fact, plus the unburned edge of the hinged side of the door and its mating surface on the door jamb, indicate the door was closed during the fire and that the fire originated on the porch.

This same logic applies to any automatic, self-closing vents which should have closed to prevent spread of the fire. These should be carefully examined and their normal condition determined by interviews with occupants.

CONTENTS OF THE BUILDING. During the examination of each room the contents, in terms of the *fire load,* should be considered.

Fire load is the amount of combustible material per square foot contained in a room. If the fire service has removed furniture during the overhaul and cleanup process, the original position of each piece of furniture should be determined. In the room of origin, the exact location of each piece of furniture should also be determined. This will assist in locating the low point of the fire by examination of the underside of each piece of furniture.

The contents of each room should be recorded photographically and in your notes, with comments regarding the fire load of each room. The rate of burning and progress of the fire will be influenced by the contents of the rooms.

LOW POINT. The low point is the area where the fire started; therefore, it must be in the room of origin. Once all possible rooms of origin have been eliminated, the only remaining room must logically be the room where the fire originated. Since the low point and the room of origin are inseparable, the investigator should double-check the room of origin to ensure that the proper room has been located in which to seek the low point of the fire.

Figure 16. A close-up of the low point of the fire. The porch flooring has been completely consumed. The debris from the fire lay under six inches of snow and ice. The origin of the fire is at the top center of the photograph, where the fire burned through the exterior siding.

Figure 17. Once the fire burned through the flooring, it attacked the floor joists. At the origin, the char on the joist closest to the house was deeper than the char on joists further away. The underground tank contained no fuel and contributed nothing to the fire.

Conversely, once the low point of the fire has been determined, before attempting to discover the cause of the fire the investigator should check other possible rooms of origin for other low points. He should double-check and reverify that the true origin and low point of the fire have been discovered.

The fire investigator can rest assured that if he does not double-check the low point and origin that the expert retained by the opposing attorney most certainly will.

The low point of the fire will be identified as being the point of heaviest destruction from the fire, or the point of lowest burning, or possibly a logical source of energy which could have been the cause of the fire. These sources could include an electrical appliance, a gas appliance, or similar types of energy-using devices. Characteristically, directly above the low point of the fire, the fire will have risen vertically and spread horizontally, forming a "V," as shown in Figure 13.

As indicated in Figure 13, the "V" would be an ever-widening burn in the combustible surface of the wall directly above the low point. This is the result of heat from the origin of the fire rising vertically and spreading out as it feeds upon the combustible material of the wall. Once the flame reaches the ceiling, the flames start to burn down the wall. This will result in a much shallower "V," starting from the top of the "V" that originated at the low point.

A normal fire, consuming wood, plastic or electrical insulation, would burn with a "V" pattern of approximately 30° measured vertically. If an accelerant was used, or if highly combustible material was involved, the "V" would be narrower as the temperature of the fire increased, due to the additional heat content of the accelerant or flammable liquid. This would cause a faster rise of heat and flame, resulting in a "V" pattern of approximately 10°, depending upon the heat flux generated by the accelerant.

The wider "V" at the top of the room will encompass an angle of 160° more or less, depending upon the fire load of the material available for combustion.

Figure 18. The debris beneath the point of origin of the fire was carefully removed to the surface of the tank and the bare earth to ensure that all evidence of the fire cause would be examined. The snow and ice was melted with infrared heat.

The angle of the "V" will point downward so that the lower portion indicates the direction from which the fire came. When the upper "V," formed by the fire burning downward, reaches the floor, the two halves of the "V" will move away from each other as the burning progresses, consuming the entire wall. The lower portion of the sloping line will always be closer to the point where the fire originated.

MULTIPLE LOW POINTS. The discovery of a low point

Figure 19. This photograph illustrates the burn pattern associated with a fast-moving fire. The sloping line of the burn pattern indicates that the fire was burning toward the viewer.

Figure 20. This fire started at the ceiling directly above the extreme right of the photograph. The fire burned along the ceiling and downward, accounting for the flat line of demarcation of the totally destroyed lining. The fire traveled from right to left. Again, the sloping line of total destruction points toward the origin, although the origin is at the top of the room.

should not be considered the end of the search since more than one low point may be discovered. This is particularly true in arson fires. Therefore, the search should continue to identify other low points and determine the relationship of any other low points found to the origin of the fire. Depths of char, fire load, ventilation and other considerations should be weighed in assessing whether other low points are natural results of the fire or whether they were set.

Accidental causes of additional low points result from the transfer of fire from the ceiling to the floor by falling debris. Each low point should be examined carefully to determine whether something located directly above it could have contributed to its cause. Significant differences in char depths at two different low points would indicate an accidental low point.

Materials on open shelves may contain highly flammable substances which, when subjected to heat, burst and spill highly flammable contents which ignite and cause additional low points. These are not as obvious as they would appear on the surface and, at times, may be misconstrued as an arsonist's plant.

The modern office, home and industrial plant contains many solvents, cleaning agents, hair sprays and other highly flammable materials in containers varying from plastic bottles to pressurized aerosol containers. The normal storage area for these containers and their contents should be inspected before concluding that low spots caused by them are of incendiary nature.

Every effort should be made to determine whether multiple low spots are accidental or deliberate. If they have been set in an incendiary effort, these would be considered evidence of arson.

When faced with the problem of multiple origins, quite often the time factor will be the deciding element regarding the fire origin. Eyewitnesses may identify one room as being completely engulfed in flames before another room was full of smoke. Electric clocks located on different circuits in different rooms may stop several minutes apart, indicating which circuits were affected by the origin of the fire.

Clocks as Time Indicators. It is a common practice among investigators to examine an electric clock found at the fire scene, determine the time indicated by the clock and conclude that the fire started at that approximate time. Another indicator is a mechanical spring-wound clock which leads to the same conclusion.

As indicators of the general time of the fire, these are generally safe conclusions, but several factors can lead to an incorrect conclusion. Until all facts have been established, the only positive fact is that a clock was found that had stopped at some point in time. The most obvious questions are whether the clock was running and its accuracy. The owner of the clock may be able to state whether the clock was running and how accurate it was. A few tests may also establish the same results.

The electric clock is operated by a motor. If the motor was running at the time heat found its way to the clock, it is possible that the lead from the armature may have been thrown around the interior of the clock as the armature melted. This may have occurred before the electrical circuit was interrupted by the fire.

By examining the relationship of the clock to the origin of the fire, an approximate determination of how long it took the fire to reach the clock and the temperatures involved can be made. Thus a more accurate estimate of the starting time of the fire can be reached. In the event of a discrepancy between the start of the fire and the time indicated by the clock, new possibilities manifest themselves.

First of all, the clock may not have been running, or the power may have been interrupted between the time of the start of the fire and the time the fire reached the clock. This can be verified by questioning the local power company regarding outages preceding or during the time of the fire.

Another possibility is that the clock may be found in a room which was untouched by the fire. The clock could have stopped when the fire service cut the power lines. However, the clock may be connected into an electrical circuit which was at or near the origin of the fire and thus shorted out early in the fire, stopping the clock. Another possibility is that the fire impinged upon the clock itself, causing it to stop with no interruption of the electrical circuit.

It should be remembered that an electrical clock can be stopped by three different ways in a fire. The first and most obvious is interruption of electrical current. This can occur regardless of the location of the clock. The electrical current can be interrupted at the transformer where the power is diverted from the distribution system into the building containing the clock, or anywhere within the system.

The second way a clock can be stopped is by overheating of the electric motor, although current is still flowing through the motor. Finally, the clock can be stopped by falling or by falling debris which breaks the clock mechanism or interrupts flow of power to the clock. Debris can also disturb the hands of a clock so that they do not indicate the actual time that the clock stopped. All of these factors should be considered when relying upon a clock as an indicator of the start of the fire.

Mechanical clocks are subject to the same factors, with slight variations. The flow of power can be interrupted if the spring runs down on the clock coincidentally with the start of the fire. This would be difficult, if not impossible, to check except by questioning persons responsible for keeping the clock wound and operating.

Subjecting the clock to the heat of the fire can also melt the bearings, bringing the clock to a stop. This will only occur when the fire has heated the clock to temperatures high enough to raise the bearing materials to the melting point. Other forces which stop mechanical clocks include falling debris, which jams the gears or stops the clock hands from moving. All of these possibilities must be considered before giving the time indicated on the clock any credible probative value.

ROOMS OF ORIGIN. The location of the origin will give a clue regarding the cause of the fire. For example, most closet fires are caused by children playing with matches. Unless another source of energy is available (gun rags containing solvents which could cause spontaneous combustion, electrical panels or a gas line), the probabilities are very much in favor of finding that children were playing with matches.

Kitchen fires are usually caused by the failure of automatic controls on cooking appliances or negligence on the part of a housewife. A pot of food can boil over and extinguish a gas flame, permitting raw gas to flood the kitchen and resulting in an explosion when the gas finds ignition.

Probable causes of bedroom fires are usually limited to smokers, frayed lamp circuits or electric blankets. The frayed lamp circuit fires and fires involving electric blankets are rare. Therefore, most bedroom fires can be attributed to smoking.

Living room fires can be attributed to a combination of smoking and overloaded or worn electrical cords.

The use of the basement will govern the statistical record of how the fire originated. For example, if the basement has a party

TABLE VII

RESIDENCE FIRE CAUSES BY ROOMS

Probabilities of Origin and Cause

Cause	Basement		Work Shop	Main Floor					Upstairs	
	Furnace Room	Laundry Room		Living Room	Dining Room	Kitchen	Bath	Bed	Bed	Bath
Smoking	5	10	40	40	15	5	5	80	80	5
Electrical	40	30	30	10	10	25	5	10	10	5
Spontaneous ignition	25	5	50	5	5	10	5	5	80	5
Children and fires	5	5	5	5	5	5	5	80	80	5
Gas fires and explosions	80	80	80	25	5	50	5	5	5	5
Heating and cooking equipment	80	80	50	20	5	80	5	5	5	5
Trash burning	40	5	40	40	5	40	5	5	5	5

room, this would correspond to a living room and the same type of fire causes would apply and predominate. However, the presence of the furnace, water heater and other household appliances would also start to make inroads into statistical averages. Table VII gives the statistical distribution of fire origins by rooms.

DETERMINATION OF CAUSES OF FIRES

Once the origin and low point of the fire have been determined beyond reasonable doubt, the next step is to determine the cause of the fire. Most often in the area of the low point or the origin will be found the source of ignition, the material first ignited and how the two came together at that particular time and place.

Ignition

For a fire to occur, four requirements are necessary:
1. A combustible material must be available.
2. A constant source of oxygen, or air, must be present.
3. A source of ignition must be supplied.
4. Physiochemical feedback of the system must sustain the fire.

At the origin of the fire, the source of ignition will usually be found. However, three notable exceptions should be considered. These include the following:
1. Source of ignition for flammable vapors
2. Spontaneous combustion (see Chapter 11)
3. Electrical fires

These three types of fires can create multiple low spots. In the case of flammable vapors, the source of ignition need not be at the origin.

For example, flammable gases or vapors can travel considerable distances to find a source of ignition. Once the gas is ignited at a remote location, the trail of vapor can flash back to the leak where the vapors will be more concentrated so that the sustained burning of the excess vapors ignites combustible materials.

In the case of electrical fires, a short will usually be the source of ignition and be found at the origin of the fire. Where an overheated appliance or circuit is found, the uniform overheating of the circuit can cause multiple fires with multiple low spots and multiple origins. However, the overheating of the wire will be a single cause and will supply its own heat for ignition.

The same principle applies to spontaneous combustion fires. The origin may be very small or encompass an entire grain bin, for example. The origin will contain the source of ignition. The amount of material involved may be of large proportions so that the illusion of multiple origins will be found, although the cause and origin will be the same–spontaneous combustion–and the source of ignition will be the volatile liquid involved in the chemical reaction.

Static Electricity as a Source of Ignition

Static electricity becomes a source of ignition when a sufficient number of electrons have accumulated on a surface with enough energy to cause a significant spark. In the presence of flammable gases, only 0.25 millijoules of electricity are required to ignite the flammable liquid. Since the human body generates approximately 10 millijoules by walking on wool carpeting with leather soles in dry weather, it is apparent that very little static electricity is required to ignite a flammable gas and cause an explosion. When no obvious source of ignition is apparent involving flammable liquids, the presence of a human in the area should be investigated.

Other sources of ignition include overheated or normally heated appliances which are located too close to combustible materials. Examples would include a large light bulb contacting combustible material.

As indicated in the preceding paragraphs, the source of ignition can vary considerably. The source of ignition is vitally important to explain how the fire started and should never be taken for granted. It is as important as any other factor involved in the start of the fire and should be given as much consideration as other aspects of the fire. Without a satisfactory explanation of how the combustible material ignited, the fire investigator has not completed his task.

Process of Elimination

Examination of the area of the origin of the fire will yield a list of possible causes of the fire. These can be considered using the process of elimination until the most logical cause has been found or until all other possible causes have been eliminated, leaving the most probable cause.

The *process of elimination* is the most useful and efficient method of determining the origin and cause of the fire. The possibilities which seem most obscure are eliminated first until the most probable causes are the only ones remaining.

Then, by careful questioning of occupants, tenants, witnesses and fire personnel, their testimony, in addition to confirming physical evidence, may be enough to eliminate all possible causes but one. This would be the most probable cause of the fire.

Sequence of Eliminating Possibilities

The possible causes of a fire should be eliminated by considering the most remote possibilities first. The natural causes would be eliminated in the following order:

Natural Causes
1. Lightning
2. Animals
3. Floods

Typical accidental causes to eliminate would include the following:

Accidental Causes
1. Children playing with matches
2. Smoker's fire
3. Gas leak
4. Negligent act
5. Spontaneous combustion
6. Electrical fire
7. Other accidental fire cause

Typical incendiary causes to eliminate would include the following:

Incendiary Causes
1. Pyromania
2. Emotional fires

3. Fraud arson fires

4. Professional arson fires

In this chapter, the general principles to be used in determining the origin and cause of a fire have been discussed. In the following chapters, the techniques to use in determining specific fire causes and the characteristics unique to each type of fire will be examined.

Time of Investigation

It is impossible to make an investigation too soon. If possible, the investigator should be present as the firemen are making their overhaul. He can readily identify areas involved late in the fire and direct the firemen to those areas where the fire obviously did not start, thus preserving the origin of the fire.

This gives the investigator the added advantage of viewing the fire much like it was before the firemen arrived. It is true that the stream of the fire hoses may have rearranged the debris somewhat. However, the major furnishings will be in position, and this will save the investigator many hours of reconstruction time.

Should the investigator arrive after overhaul is completed, the investigation will be somewhat handicapped, but it should be re-membered that the major indicators of the origin of the fire are the burn patterns on the walls, ceilings and floors. These would usual-ly not be altered by the firemen.

The locations where the firemen have affected overhaul are usually obvious. They chop holes in the walls to determine if the fire gained access to the interior of the wall. They also break open doors and windows for ventilation. It is not uncommon to find holes ripped in the ceiling to determine if the fire is burning be-tween the ceiling and floor above. These openings made by the firemen will bear ax marks, and bare unburned wood will indicate marks made after the fire.

Should the investigator arrive long after the fire occurred, the burn pattern on walls, ceilings and floors will still be visible. Con-clusions based upon these factors will be as valid as if the investi-gation was made immediately after overhaul.

Debris found in a heap, even after overhaul, may still yield clues to the fire's origin, despite the fact that it may have been moved a considerable distance. Movement may or may not have an effect on the debris, depending upon the nature of the evidence. This applies to appliances where the components are fairly rugged and can stand considerable physical abuse without undue damage.

Furnace controls, tested after a violent explosion, have been found to function normally and without damage, despite complete destruction of the building. The only foolproof method is to test the controls. Should the controls malfunction, the investigator must determine whether the malfunction was caused by the explosion or was a cause of the fire. If the controls function normally after a fire or explosion, it does *not* mean that a one-time failure has not occurred.

Development of a Theory

After completing the investigation, the origin has been determined and, by the process of elimination, a cause of the fire has been found. As a result, a theory has been developed regarding how the fire originated. This has been done, based upon the evidence available at the scene of the fire and by information supplied by witnesses. It is at these moments, when a theory has been formulated, that one must be as ruthless as possible with theories and discard them as rapidly as they are proven wrong.

It has been said that "there is no more heinous crime than the murder of a fine theory by a hard fact." Verification of a theory will come from eyewitnesses intimately associated with the fire, including persons reporting the fire, eyewitnesses who first saw the fire, the dispatcher and the fire service.

Eyewitness reports of the first firemen on the scene would be the most important regarding the origin of the fire and its behavior. It is at this point that the determination is made as to whether the fire is of accidental or incendiary origin. This determination should be made as soon as possible. If arson cannot be eliminated as a possible or highly probable cause, the assumption is that it is an arson fire and the collection of evidence should proceed on that basis. If any doubt exists as to whether the fire is or is not an incendiary fire, evidence should always be collected on the assumption that it is an incendiary fire. The fire investigator should not presume it is an accidental fire. His mind should be open to *any* possible cause, but evidence should be collected as for an incendiary fire.

Confirmation by Witnesses

While it is pleasing to the ego to have a witness confirm a fine theory, the job is not complete until all known witnesses have been questioned and their stories cross-checked. To expedite the development of a theory, and as a graphic aid to the completion of questioning, the *Witness Matrix* (Fig. 21) can prove useful.

The Witness Matrix is a convenient bookkeeping method. The headings are selected based upon questions to be answered or facts to be verified. The first column is a vertical list of names, addresses and telephone numbers of witnesses. For convenience, these are all given numbers. Columns are added across the page to describe the substance of their statements in answers to standard questions which the investigator devises to fit the occasion and need.

In Figure 21, column B indicates the time that the witness first observed the fire. Column C reveals the location and density of smoke; column D indicates the location and intensity of flames; column E describes what was burning; column F describes where it was located. Column G is a statement to be verified, such as any statement which might support or destroy a theory which should be checked (such as "saw an unidentified male Caucasian wearing a blue jacket leaving the scene"). Column H is the checking column; it includes the category of "Statement confirmed by/Confirms statement of."

Under column H, the "P.E." indicates physical evidence. This would indicate, for example, that the witness stated that he saw heavy black smoke with bright red flames coming from the front room window. After examination of the front room, it was discovered that oil had been poured on the floor and had ignited, confirming the heavy black smoke at the scene. This was confirmed later by other witnesses and would be indicated by additional numbers in this box.

The usefulness of this matrix is in the ease of bookkeeping where numerous statements have been taken and are full of de-

Investigator _____   File No. _____   Date _____

	A	B	C	D	E	F	G	H
1	Witness name, address, phone	First observed fire; time and place	Smoke location and density	Flames location and intensity	What was burning	Where was it located	Statements to be verified	Statement confirmed by:/Confirms statement of: P.E.
2	Joe Blow Box 75 Skinned Elbow, Montana 610-456-7980	1:30 PM Driving by in car, noticed smoke and flames; called F.D.	Upper windows; heavy, black rolling smoke	Lower inside of building solid mass of flames	Unable to tell	Inside front rooms, both sides of building	All; especially 2E	3D
3	Edna Witles 324 E. 1st Bent Elbow, Montana 610-456-7981	1:35 PM Walking home from work; walked right by front of building	Upper windows; heavy, dark, billowing	Inside sales room aflame; yellow-red flames	Furnishings, walls and ceilings aflame	North, east and south walls of both sales rooms	Check with 2A	2D P.E.
4	Chief, F.D. Fred Fearless R.R. #1 Bent Knee, Montana 610-456-7982	Received fire call at 1:32 PM; responded; arrived 1:39 PM	Upper windows, roof; heavy, dark black	Lower floor aflame; upper window showing flashes	Furnishings, walls, ceilings; upstairs hallway and doors	Main and second stories	Check with other firemen	5A (S.S.) P.E.
5	Fireman Charles Courageous 326 N. 2nd Bent Thorax, Montana 610-456-7983	Same as 4B	Same as 4C	Same as 4D	Same as 4E	Same as 4F	Saw man running from scene; see written statement	Unconfirmed by 4A

F.D. - Fire Department   P.E. - Physical Evidence   S.S. - Signed Statement

Figure 21. Witness Matrix.

tails which need checking and cross-checking. The principle of the Witness Matrix can be used by changing the headings to suit the subject matter and is limited only by the investigator's imagination. The value of the matrix depends upon selecting key questions which can be answered with brevity and accuracy.

Should identical statements of two or more witnesses conflict with the physical evidence or with statements made by other witnesses, collusion and arson may be suspected.

Analysis of the matrix may reveal unconfirmed or conflicting statements which must be reevaluated and may require further investigation. All answers must be confirmed, considered as possible clues to arson or eliminated as "honest" mistakes or misunderstandings.

Confirmation by Tests

Frequently a search for witnesses will prove fruitless. When this occurs, reliance must be placed on physical evidence to develop and prove a theory regarding the origin and cause of the fire. To do this, physical, optical and chemical tests are often used to confirm or refute a theory. For this reason, before a fire investigator begins a case he should have a thorough working knowledge of the capabilities of laboratories available to him. In addition, he should know the proper procedures for taking samples which comply to the laboratory standards. Should there be any doubt in the investigator's mind of the proper procedure to use for extracting a sample, or how a test should be conducted, he should learn the proper procedure rather than to proceed and run the risk of destroying vital evidence which could mean the difference between a successful and unsuccessful court case.

Confirmation by Expert Opinion

Although the fire investigator is an expert in his field, quite often the technical aspects of how a fire started require help from other experts. The decision to use outside experts should be made only after the fire investigator has made reasonably certain that a theory is correct and that the expert will confirm the fire investigator's theory. The reason is twofold.

First, the expert is an independent professional who will render an independent opinion, regardless of the opinion of the fire investigator. In other words, the expert may disagree with the fire investigator and he may find himself being contradicted by his own expert. Should this occur, the fire investigator should make a ruthless examination of his theory to see if he has erred and reevaluate his case on the basis of the expert's opinion.

Secondly, the expert has been retained to render an opinion. At the time he is retained, he has no idea what his opinion will be. His opinion may help or contradict the fire investigator's interest. Should the expert's opinion favor the opposition, the expert could be called as *their* witness.

CHAPTER 5

Incendiary Fires

INCLUDED IN THE NFPA classification of *Incendiary, Suspicious Fires,* are fires known or thought to have been deliberately set, fires set for the purpose of defrauding insurance companies and fires set for emotional reasons by pyromaniacs or vandals.

The dramatic increase in incendiary fires over the last decade has led fire investigators and statisticians to the conclusion that approximately one-half of the fires of unknown cause are of incendiary origin. No other source of arson statistics is as complete or accurate as the NFPA figures. Figures based upon insurance studies agree with percentages indicated in Tables I and VIII.

Table VIII clearly indicates that incendiary fires account for at least 11 percent of all fires. If only 50 percent of the unknown causes are incendiary, as most experts believe, the percentage of incendiary fires would rise to 18 percent.

By comparing the dollar loss of arson fires and making the same assumption (that 50% of the dollar loss of the unknown fires is of incendiary origin), the significance of incendiary fire loss is underscored. Incendiary fires plus one-half of the unknown fires yields 36 percent of the total dollar loss of fires of incendiary origin, amounting to more than $1 billion per year (Table IX).

The assumption that the arsonist is "getting away with it" is based on the difference in percentages when comparing the number of fires with the dollar loss. The 36 percent dollar loss figure indicates the possibility that an even larger number of unknown fires and the larger dollar loss fires may be of incendiary origin.

When the difficulty of obtaining an arson conviction is considered, it is not surprising that statistics clearly show that the arsonist is "getting away with it." According to a recent study by the Law Enforcement Assistance Administration (LEAA), U.S. Department of Justice, the arrest and conviction record of arsonists ranks a poor last compared with the arrest record for other major crimes.

TABLE VIII

NUMBER OF INCENDIARY FIRES (1,000s)*

Classification	1970	%	1971	%	1972	%	1973	%	1974	%
Incendiary and suspicious fires	65.3	7	72.1	7	84.2	8	94.3	9	144.4	11
Unknown fires	162.0	16	166.2	17	154.2	15	150.5	15	159.2	13
Incendiary plus one-half unknown fires	146.3	15	155.2	16	161.3	15	169.6	16	224.0	18
Total fires	992.0		996.6		1050.2		1085.9		1270.0	

* Based on information from Gordon P. McKinnon and Keith Tower (Eds.), *Fire Protection Handbook*, 14th ed., 1976. Courtesy of National Fire Protection Association, Boston, Massachusetts.

TABLE IX

DOLLAR LOSS ($1,000,000s)*

Classification	1970	%	1971	%	1972	%	1973	%	1974	%
Incendiary fires plus suspicious	106.4	9	232.9	10	285.6	12	320	13	563	17
Unknown fires	997.7	45	1002.9	44	992.7	41	1045.3	41	1237	38
Incendiary plus one-half unknown dollars	705.3	32	734.4	32	781.9	32	1181.0	36	1181.5	36
Total dollar loss	2209.2		2266.0		2,416.3		2537.2		3260.0	

* From Gordon P. McKinnon and Keith Tower (Eds.), *Fire Protection Handbook*, 14th ed., 1976. Courtesy of National Fire Protection Association, Boston, Massachusetts.

TABLE X
ARREST-CONVICTION RECORD*
Comparison of U.S. Arrest Rates, 1974

Offense	Number Per 100 Offenses Reported
Murder	98
Rape	48
Robbery	34
Aggravated assault	52
Burglary	17
Larceny	20
Motor vehicle theft	16
Arson	
Incendiary and suspicious	9
Incendiary and suspicious plus one-half unknown cause	3

* From LEAA, *Survey and Assessment of Arson and Arson Investigation,* 1976. Courtesy of Law Enforcement Assistance Administration, Washington, D.C.

TABLE XI
DISPOSITION OF INDEX CRIMES, 1974*

Disposition	Number Per 100 Offenses Reported		
	Juveniles	Adults	Total
Arrests	9.56	11.64	21.20
Convictions	1.13	5.18	6.31
Sentences			
Incarcerations	0.008	3.17	3.18
Prison	—	0.66	0.66
Jail	—	2.51	2.51
Juvenile corrections	0.008	—	0.008
Probation	1.12	1.47	2.59

* From LEAA, *Survey and Assessment of Arson and Arson Investigation,* 1976. Courtesy of Law Enforcement Assistance Administration, Washington, D.C.

Table X shows that motor vehicle theft offenses result in a 16 percent arrest record. The incendiary and suspicious fires arrest rate is only 9 percent. Assuming the conclusion previously mentioned is correct and adding one-half of the unknown causes, the arrest rate drops to 3 percent because the crime was unidentified and no arrest was made.

However, these are only the arrests. Continuing our study, at-

TABLE XII

DISPOSITION OF ARSON CASES, 1974*

Disposition	Number Per 100 Incendiary Or Suspicious Fires		
	Juveniles	Adults	Total
Arrests	5.31	3.73	9.04
Convictions	0.63	1.24	1.87
Sentences			
Incarcerations	0.004	0.70	0.70
Prison	—	0.10	0.10
Jail	—	0.58	0.58
Juvenile corrections	0.004	0.01	0.02
Probation	0.62	0.54	1.16

* From LEAA, *Survey and Assessment of Arson and Arson Investigation,* 1976. Courtesy of Law Enforcement Assistance Administration, Washington, D.C.

tention is drawn to Table XI. This table, taken from the LEAA Survey and Assessment of Arson and Arson Investigation, indicates that for all crimes listed in Table X, an arrest rate of only 21 percent is followed by a conviction rate of only 6.3 percent.

Comparison of arson arrests with arson convictions (Table XII) presents an even more dismal figure. For incendiary or suspicious fires reported, the arrest record drops to 9 percent with less than 2 percent convictions. Of these, less than 1 percent result in jail sentences; one-half are put on probation. One of the sad commentaries on our present-day society is that 59 percent of the arrested suspects are juveniles. Small wonder that the professional arsonist finds the crime so tempting when the odds of conviction are so low.

Definitions

Webster's Second College Edition of the *New World Dictionary* defines *incendiary* as "having to do with the willful destruction of property by fire, causing or designed to cause fires, as certain substances, bombs, etc."

Arson is defined in the same source as "the crime of purposely setting fire to another's building or property, or to one's own, as to collect insurance." Most modern arson laws include bombing or an explosion, with or without an accompanying or incidental fire.

For purposes of this chapter, arson fires are divided into three separate classifications.

Criminal Arson Fires

Criminal arson fires are those in which the perpetrator has made no effort to conceal the fact that the fire was deliberately set but has relied upon the fire to destroy all evidence of any *accelerants, trailers, plants* and *timers* used to start the fire. These are the type of fires in which, if the arsonist failed to make positive ignition of the fire, the entire setup is found intact before the fire occurred.

Accidental Arson Fires

Fires intentionally set to simulate an accidental fire are called accidental arson fires. For example, gas may be deliberately allowed to accumulate in a building where it may be ignited and exploded in a number of accidental ways, such as a spark, arcing of a switch or striking a match. In such fires, which appear to be accidental, it is difficult to provide evidence or witnesses to prove arson. If proof beyond reasonable doubt cannot be provided, a conviction of arson cannot be obtained.

Fraud Fires

A fraud fire is one in which the arsonist has deliberately set a fire in order to collect insurance. A fraud fire may be either an accidental arson fire or a criminal fire. The distinction between the fraud fire and the other two classifications is the manner in which the case is litigated.

If substantial evidence is available for a criminal indictment against the perpetrator of an accidental arson fire, it can be reclassified as a criminal arson fire. A criminal arson fire can be deliberately set to defraud an insurance company. If in so doing the building is burned, then the crime is arson. In order to prove arson, substantial evidence is required to obtain an indictment. In many cases, evidence is insufficient to ensure a conviction or even to obtain an indictment. However, there may be sufficient grounds for the insurance company to withhold payment of the claim on the grounds of fraud. Since the basis of the defense of fraud against the insured is less stringent than in a criminal case, the arsonist can be deprived of profit and thereby suffer at least to that extent.

Fraud fires include only those cases where arson has been proven to the degree that litigation requires. In the event that the fire investigator cannot build a solid criminal case, he has the alternative of persuading the insurance company to defend the case against the arsonist and presenting his evidence to the insurance company.

ARSON LAWS

The legislative bodies of cities, counties, states and the federal government are constantly changing the statutes by which we live. Recent publicity given to the arson problem has resulted in a concentrated effort to improve the effectiveness of current arson legislation. Therefore, the fire or arson investigator is counselled to be thoroughly familiar with current and proposed legislation and its effect. Since arson laws vary from state to state, only the broadest concepts will be covered. The broadest of these concepts is the *corpus delicti*.

Corpus Delicti

The *corpus delicti* is the "body of a crime." It is the substantiated evidence that a crime has been committed. In arson, it is the burned or charred remains of the structure which was destroyed by the fire. In some jurisdictions, the law states that some portion of the building must display destruction of the wood fibers. Other jurisdictions require only that a charring has occurred. Some states will not accept smoke damage as proof of arson although, for purposes of an insurance policy, the results are quite often the same as if the building had been totally destroyed.

Most modern arson laws include an explosion as an arson attempt to destroy a building. The explosion need not be accompanied by charring, smoke damage or other indications of destruction by fire; the explosion itself is proof of arson.

Another element of the corpus delicti is the intent. Most jurisdictions define arson as a "willful and malicious act," making it necessary to prove that the burning was intentional rather than

TABLE XIII

MOTIVES OF CONVICTED ARSONISTS, NEW YORK CITY, 1974

Motive	Adults (%)	Juveniles (%)
Revenge	47	5
Pyromania	30	14
Malicious mischief (vandalism)	10	80
Crime concealment	9	2
Insurance fraud	4	0

* From LEAA, *Survey and Assessment of Arson and Arson Investigation,* 1976. Courtesy of Law Enforcement Assistance Administration, Washington, D.C.

accidental. The presumption in every fire is that the cause of the fire was accidental. It is the burden of the prosecutor in an arson case to overcome this presumption.

Motive is not considered an element of the crime. Although not a legal requirement, both judge and jury feel more comfortable when a logical motive is brought forth as a part of the evidence.

Burden of Proof

As in any criminal case, the burden of proof rests with the prosecution. In a criminal case, the burden of proof is much greater than in a civil case. The prosecution must show evidence *beyond a reasonable doubt* of the involvement and guilt of the accused.

In a civil case, the only requirement is that the *preponderance of the evidence* should point to the involvement and guilt of the accused. Therefore, should the arson investigator lack proof for a criminal case, the act of fraud can sometimes be grounds for civil litigation to the disadvantage of the arsonist.

Motives

As with most statistics on arson, data regarding the motives of convicted arsonists are lacking. The few studies which have been made confirm the fact that most professional arsonists are successful in evading detection, arrest and conviction (Table XIII).

It should be pointed out that Table XIII includes the motives of convicted arsonists in 1974. Statisticians and investigators now believe that arson fraud fires account for 21 percent of the number of fire insurance claims and 45 percent of the dollar loss. Comparing this statistic with the available statistics on arrests and convictions, it is obvious that much work must be done to improve the investigation and, consequently, the arrest and conviction record.

Search and Seizure

Modern changes in legislation regarding search and seizure of evidence have compounded the arson investigator's problems. No such problem at present exists for the fire investigator seeking evidence of the cause of a fire when no crime has been committed nor evidence of a crime found. However, the freedom enjoyed by the civil fire investigator may cause complications for the arson investigator. Should the civil investigator discover evidence of a crime, he has a duty and obligation to notify the nearest authorities. If an arson investigator removes evidence uncovered by a civil investigator without following the proper procedures, the evidence may not be admissible.

The obvious answer is for the civil investigator to conduct an investigation on the same level of legal requirements as the arson investigator so that evidence discovered by the civil investigator will be admissible. The civil investigator should become familiar with the problems encountered in search and seizure so that he does not pollute the evidence and jeopardize a case.

In a civil case, no such conflict exists and the fire investigator has the right to enter private property as long as a claim has been filed by the insured. The purpose of his investigation is to determine that a valid claim has been filed. If evidence of arson is discovered in the process, the fire investigator must then immediately notify proper authorities so that a criminal investigation can be conducted, based upon the evidence discovered by the civil fire investigator.

INVESTIGATION OF ARSON FIRES

Every fire should be investigated until the possibility of arson has been eliminated. Therefore, when the fire investigator arrives on the premises, his first duty is to seek out indications of arson. The general method of investigating a fire was covered in Chapter

4. Following these principles, the fire investigator will soon see the dilemma. He must eliminate arson as a cause, but the elimination of an arson fire involves the elimination of an accidental fire which could have been deliberately set.

The Paradox

The most difficult problem facing a fire or arson investigator is the determination of whether an accidental fire was deliberately set and made to appear as though it were an accidental fire. If the arsonist does a thorough job, there will be little evidence to convince the fire investigator that arson was committed. One general principle can be applied, however. This is the principle of consistency.

As an example, one of the most daring swindles in the annals of crime was perpetrated by a farsighted swindler who went to the extent of fabricating records in the archives of Spain to establish an heir in the state of Arizona. He carefully forged records in a 100-year-old manuscript deeding his wife 12,000,000 acres. She was thus established as the direct descendant of record. The fraud was perfectly executed with the exception of a glaring error. He used an ink which was not invented until 100 years after the date on the record which he forged. This oversight cost him his fortune.

For the fire investigator, the broad principle to be applied here is *consistency*. Had the ink been consistent with the age of the paper, the fraud might have been successful. This principle should be applied to the investigation of any accidental fire.

For example, if a twenty-year-old gas valve malfunctions and has not been touched prior to the accident, it would be inconsistent to find Teflon® tape on a valve fitting. The presence of Teflon tape (which came into existence in the early 1970s) would prove that some work had been done on the valve since its installation.

An example of inconsistency: The wires of a building were overheated by the normal loads placed on the circuits, and as a result, fire broke out in many places in the walls and conduit of the building. The building was destroyed. The initial assumption was that the contractor had built the building with inadequate wiring and was therefore responsible for the fire. The question nagging the fire investigator was why normal loads would suddenly cause a fire after the wiring had successfully carried the load since the building was built. No evidence was present of any significant change in the circuit loads between the time of installation and the time of the fire. Logically, therefore, a fire should have started one of the first times the normal full load occurred.

Analysis of the routine information gathered about the fire, such as the age of the building, the normal and maximum electrical circuit loads, health of the owner, financial status of the business, etc. revealed an inconsistency. The answer was that the arsonist had replaced the original wiring in the building with lighter wire which soon overloaded the circuit and caused an *accidental* fire. Had the wiring been installed by the contractor, the fire would have occurred shortly after the full normal loads were applied. Unfortunately for the arsonist, the laws of physics make this an irrefutable conclusion.

In order to be eliminated as a possible arson fire, the following factors should be considered:

1. Time of the fire.
2. Whereabouts of the owner at the time of the fire.
3. Did the owner have the technical knowledge necessary to perpetrate the crime?
4. Did the owner have access to the tools and equipment necessary to perpetrate the crime?
5. Were the movements of the owner at the time of the crime unusual?
6. Did the owner make any unusual moves immediately preceding the fire to establish an alibi? If so, check with family, neighbors and friends to see if the trip was planned for some time before the fire or if it was an unexpected development.
7. Check into the relationship between the owner, relatives and friends. Is there any reason to suspect revenge?
8. Does the owner have a record of setting fires?
9. Has anyone in his family a record of pyromania?
10. Has the owner's business recently taken a turn for the worse?

11. Has the owner increased his insurance recently?
12. Is there any reason to suspect that the owner could have perpetrated the crime?
13. Did he remove valuables?

Before concluding that arson has been committed, put yourself in the arsonist's shoes and imagine how you would "torch" the building and establish an alibi. Then, check for inconsistencies in your hypothesis. Next, assume that you were retained by the suspect's attorney to defend the case. What steps would you take to destroy the plaintiff's case?

If, after subjecting yourself to these extremely vital tests, you can conclude that the accidental fire is a deliberately set fire, then remind yourself of the creed of the arson investigator: "I am a truth-seeker, not a case-maker; it is more important to protect the innocent than to convict the guilty."

INITIAL EXAMINATION

It is necessary to alter equipment or some part of the physical plant to create an *accidental* fire. For example, if the "culprit" is a gas valve, it may need to be opened to cause it to fail. In order to open the gas valve, it would be necessary to unscrew the screws attaching the covers to the valve body. Therefore, the importance of first checking screws to see if there has been recent tampering cannot be stressed enough. This requires a detailed microscopic examination. A detailed examination of all the pipe connections must be made to determine whether the piping has been tampered with recently. If this is the case, the details concerning work completed on the piping system should be investigated thoroughly.

Almost as important for investigating an *accidental* arson fire is to look for the unusual. Any item which is out of place or is not in its customary position should be examined and explained. While these same principles apply in the investigation of a confirmed arson case, they apply even more in the case of *accidental* arson.

One fact in an accidental arson fire is that the fire takes deliberate planning and foresight. If the perpetrator is not experienced or knowledgeable in starting a fire and in disguising its origin as accidental, he may need assistance or some education to accomplish his purpose. He may go to the local library, trade school or other sources to obtain the expertise, knowledge, tools and equipment necessary to accomplish his end. Inquiries regarding the suspect's activities in the months or weeks preceding the fire may yield fruitful results.

While the suspect may not have the necessary expertise himself, it is possible that among his relatives, friends or business associates, an accomplice may be available who can supply the know-how and equipment; collaboration should be checked.

Inquiries into the suspect's profession, work history, education, training, military experience, hobbies, avocation, arrest and conviction record, history of pyromania or fire-setting, sports and a general background of his knowledge pertaining to the expertise necessary to perpetrate the crime should be obtained. This information can be garnered from his high school and college records, neighbors, police records, priest, rabbi, pastor, girlfriend, etc. If special tools were required, it is possible that the tools may have left markings on equipment that can be identified by means of a microscopic examination and comparison.

The professional arsonist is primarily concerned with performing a satisfactory job for his client and avoidance of detection. Quite often these two objectives go hand in hand.

For example, if the owner of a clothing store decides to "sell his stock to the insurance company" and has no insurance on the rented building, the arsonist might start a fire which would produce voluminous smoke to effectively render any clothing a total loss in the eyes of the insurance company but would not harm the building to any great extent. The advantage of this would be to accomplish the goal and make the fire appear as though it were an accidental fire.

Forced Entry

Signs of forced entry include crowbar marks on the windowsill or door jamb. You should check with the fire service or police to see whether the marks were made by them in gaining entry into the building.

Another sign of forced entry is broken glass. Normally, when a fire occurs, heat and pressure will cause windowpanes to explode outward, spreading glass ten to twenty feet outward from the building. There may or may not have been an explosion. Finding windowpane glass ten feet from a building does not necessarily infer that an explosion occurred. Other evidence must be present to substantiate the explosion.

If glass is found on the floor of the building, a detailed examination of the glass should be made to determine whether it was broken before or after the fire. If the glass was broken after the fire started, smoke film would be deposited on the glass. Firemen, in an attempt to ventilate or gain access to the building after the fire started, could have broken the glass. However, if the glass was broken before the fire, it would fall on the inside. The inside of the glass would fall on the floor face down. If the staining was caused by the fire, it would be protected. Therefore, if no staining was found and the breakage occurred before the fire, the only staining possible would be of the glass as it lay on the floor. If the glass landed with the outside up, then additional staining could possibly have occurred during the fire, as in the case of embers falling on top of the glass.

Direct Evidence of Arson

Every fire should be investigated under the assumption that arson could have been a cause. If arson can be eliminated, less effort will be required in the investigation. Therefore, examination of the scene for direct evidence of arson should be the first step.

The most obvious evidence of arson is the burn pattern itself. This is particularly true where the arsonist has used an accelerant to engage the entire building in fire very quickly but where the fire department was notified in time to stop total destruction of the building. Then the *corpus delicti,* or the evidence of the crime, can be detected from the remains.

Accelerants

An immediate search for accelerants should begin near the origin, since volatiles rapidly disappear in a well-ventilated building. A *Sniffer* should be used for the detection of accelerants. If hydrocarbons are detected, samples of the wood, ground or any other material should be tested for the identification of accelerants.

Multiple Origins

The presence of multiple origins is a strong indication of an arson fire. In order to verify multiple origins, carefully measure the char depths. Photographs and samples of accelerants should be taken at the low point of all suspected origins. In addition, photographs, sketches and measurements of the rooms, showing the location of equipment, furniture and combustible materials, will prove useful. The presence of windows and their relative positions (open or closed) should be noted.

Delay Mechanisms

The timing device used by even ancient arsonists to start a fire which would provide time for escape was a candle. A candle was set vertically in combustible materials (excelsior, wood shavings and newspapers soaked in kerosene). The burning rate of the candle could be accurately measured so that an arsonist had sufficient time to make his escape and establish an alibi.

Another common delay mechanism is the use of grains which swell when exposed to water. Grains such as rice, wheat and barley are placed in an open container. As the grain absorbs the water and swells, it raises the water level in the container. As the water level raises, a float makes electrical contact to touch off an ignition device. The delay can be varied by the size and shape of the container and the quantity of water and grain combination. The two variables (water and grain) can be accurately timed in advance, giving the arsonist a reliable timing device.

One ingenious method involves the use of phosphorus as a timing device. This depends upon the removal of water which prevents the phosphorus from spontaneously reacting with air. Such methods involve wooden or combustible containers filled with water and phosphorus. A steady drip or leak is started at the low point of the container. Once the water is removed, the phosphorus will ignite the container and in turn ignite other combustibles. The timing mechanism is regulated by the quantity of water and the size of the leak.

This same principle can be used with other chemicals. The arsonist determines the oxidant and the reactant and uses physical means to keep the two separated long enough for his escape.

The use of electricity as a delay device is limited only by the arsonist's knowledge of electronics. Electric clock motors, the size of a thimble, can provide a very reliable and accurate timing device.

The electric clock can be powered by normal household current or small hearing aid batteries. The batteries or household current also supply the energy to ignite the "set."

Electricity has been used indirectly by connecting the "set" to a telephone or an answering service. When the telephone rings, an electric circuit is closed and the ignition sequence is started. To avoid the possibility that another caller may activate the device prematurely, the telephone can be set so that it must ring in excess of twenty or more times for the timing mechanism to activate. This gives the arsonist time to catch a plane and make a long-distance call from a neighboring state when he wishes to activate the ignition mechanism.

Batteries are unusual delay devices because of their loss of power as they cool. A short time delay can be provided to an arsonist by freezing the battery and inserting it into the circuit before he leaves. After the battery has thawed out and is restored to full strength, it provides enough current and voltage to activate a timing mechanism.

Another technique using the same principle is to use a battery as a source of power to hold a solenoid closed. In this case, the battery power is used to hold the gas valve closed. Connected in series with the solenoid on the gas valve is a secondary circuit normally utilizing house current, which will supply a source of ignition. The power from the battery holds a switch in the open position on the ignition circuit. The strength of the battery holds these two switches against mechanical springs. As the battery power is consumed, the battery grows weaker and weaker until the solenoid spring overcomes the weakening current from the battery, flooding the area with a flammable gas mixture. A short time later, the dwindling battery will fail to hold the ignition switch in the open position. The ignition switch will close, causing ignition and an explosion and/or fire, destroying the "set." Variables are the power consumption of the two solenoids and the power supply of the battery. These can be varied to furnish the arsonist with any time delay required.

Another unusual use of electricity involved an arsonist who filled a light bulb with gasoline and inserted it into a timed electrical circuit. A hole was drilled in the top of a 100 watt bulb and the glass envelope filled with gasoline. The hole was sealed with a piece of tape. The bulb was then inserted base down in a table lamp with the lamp connected to a clock timer set for one hour. After the timer closed the circuit, the device failed because the arsonist had forgotten to turn on the lamp. As a result, when the timer closed the circuit, the light bulb did not explode and set the intended fire. The "set" was discovered, and the entire scheme became apparent.

The unusual is to be looked for–something that is out of place, something that has no reason for its location at the time. For example, a long length of zip cord leading from one wall outlet to another one, connected in series so that all wall outlets were joined together in a direct short circuit, would be suspect.

Plants, Trailers and Timers

Plants, trailers or timers may be found anywhere in a building. The arsonist uses them to accomplish two purposes. The first is to spread the fire throughout a building in a very short time. The second is to give him time to leave the premises before the fire starts. Any physical means which will accomplish this end should be searched for diligently. The fire investigator should not confine himself only to looking for kerosene-soaked papers, gasoline or candles burning down to a box of matches. These are the obvious. (In fact, they are so obvious that they sometimes may be overlooked.) The fire investigator should look for the unusual or anything out of place.

One type of arson plant involves breaking a light bulb without breaking the filament. This is done by heating it and immersing it in water. The exposed filament is then wrapped in a gasoline-soaked rag and used as a *plant* connected to a *trailer* to spread the fire throughout the house. Usually it is connected to a Molotov cocktail filled with gasoline or any accelerant.

The light bulb is screwed into a lamp and plugged into a timing device which will turn on the circuit at a set time. The arsonist then leaves the *timer* to close the circuit. Once the circuit is closed, the circuit shorts across the filaments, ignites the gasoline and starts the fire.

The Ventilated Fire

Another quick clue to an arson fire is the *ventilated* fire. Examination of the *fire triangle* reveals that voluminous amounts of oxygen are required to sustain a fire of any magnitude. In order to provide oxygen, an arsonist will guarantee that his fire has access to plenty of air. This is accomplished by leaving windows, skylights or doors partially or fully open, depending upon the fire requirements. He will also turn the furnace fan on if it is summer or may even go to the extent of wiring the fan circuit closed in the wintertime to ensure that the fan will continue running when the thermostat is satisfied. Therefore, the thermostat should be examined for an abnormally high setting. The fan controls should be examined to determine the status of the fan at the time of the fire.

The condition of all doors, windows and skylights should be inspected and anything unusual should be noted. Open windows found in the middle of winter would certainly be suspicious. However, in the middle of summer open windows may not cast too much suspicion on the building unless it is air conditioned.

Contents of the Building

When an owner decides to burn a building, he will often remove personal objects with sentimental value. For example, he may remove family pictures, plaques, awards, loving cups or other keepsakes. This same philosophy applies to householders who burn their homes for purposes of insurance fraud. They may remove their guns, stamp collection, beer can collection and, particularly, their pets. They will even empty the freezer and refrigerator in anticipation of a fire.

Quite often the fire will occur when the entire family and pets have left for the weekend. This in itself is not unexpected or unusual. However, if they took the family jewels and keepsakes with them, it can become circumstantial evidence of arson.

Figure 22. A refrigerator or freezer is an efficiently insulated container. The contents are seldom affected by fires. Quite often, an arsonist will empty a refrigerator in anticipation of a fire. The refrigerator or freezer should be checked for normal contents.

Negative Evidence

The preceding evidence has been direct, positive evidence supported by physical evidence or eyewitnesses. Negative evidence must be considered before arson can be proven beyond a reasonable doubt. This is the elimination of all natural and accidental causes. Therefore, we have returned once again to the purpose of this book, the elimination of all possible causes. Once all other possible causes have been eliminated, the only cause remaining has to be the true cause of the fire—*arson*.

Following this philosophy to its logical conclusion, if it can be successfully shown that no other cause except arson exists, and it can be shown that someone had the motive and opportunity, grounds exist for an indictment.

To eliminate all accidental and natural causes, we have to depend on circumstantial evidence. For example, in order to prove conclusively that a fire was not started by an electrical short, it is necessary to show that all of the circuit breakers were not tripped. If they were tripped, it must be proven that they were functioning properly and no short circuits were found which were not a result of the fire.

Circumstantial Evidence

Circumstantial evidence is defined as "indirect evidence" by *Black's Law Dictionary,* Revised Edition (West Publishing Company, St. Paul, Minnesota).

> *Indirect evidence* is that which only tends to establish the issue by proof of various facts sustaining, by their consistency, the hypothesis claimed. It consists of both inferences and presumptions. It refers to evidence of facts or circumstances from which the existence or nonexistence of fact in issue may be inferred. It has been defined in People vs Palmer, 1887, New York State, Rep, 817, 820.

In the eyes of the court, circumstantial evidence is not regarded as inferior to direct evidence. Quite the contrary, in fact. At times, an eyewitness may be proven totally incorrect by circumstantial evidence. This is not because the person was lying but because his or her powers of observation and recall were not reliable.

Most fire investigations result in the accumulation of circumstantial evidence. Examples of circumstantial evidence in an arson case would include doors, windows and skylights which were left open, accelerants distributed throughout the building, multiple origins, removal of all personal items from a building and a timer set for twelve hours.

A fire investigation revealed that a building owner and his family planned to take a twelve-hour trip which would place them 600 miles from the scene of the fire at the time it broke out. However, the evidence was discovered before the timer started the fire. All of the aforementioned facts are circumstantial evidence. They are facts from which it can be inferred that the owner either set the plant and timer and opened the windows, doors and skylights, or had hired someone to do it. This chain of circumstantial evidence can be very effective in convincing a jury that the owner had prior knowledge of the act, particularly in the absence of any forced entry to the building.

Scientific Aids Peculiar to Arson Cases

Since arson cases involve the willful performance of an unlawful act, it is quite often helpful to submit a suspect to a *lie detector test* or *polygraph*. This device is by no means foolproof and in some jurisdictions is not admissible as evidence. Perhaps its greatest use is the psychological effect it has on a guilty party. The polygraph is a scientific instrument. The results depend upon the operator and the manner in which the test is conducted; it is not 100 percent reliable. Certain types of emotionally conditioned people will respond unfavorably to questions with misleading results.

The polygraph is a device which records blood pressure, pulse, respiration and the electrodermal response. Professional standards for the machines and operators have been established by the International Society for the Detection of Deception. Before undertaking any examination of witnesses, both the machine and operator should be qualified under the standards set by this self-governing society. Insist upon two separate tests. Before accepting the conclusions, be sure there were no contradictions in the two results.

Another testing device is the Psychological Stress Evaluator (PSE), which measures stress in the voice. This technique was developed in Vietnam to measure voice characteristics. A micro-muscular tremor is usually present in the voice. However, when tension is present in the body and mind, it is absent. Psychopathic liars would show no change in stress since they have no moral code; thus, for these few individuals the machine is useless. The input requirement is a voice recorded on a compatible tape. The recording can be taken over the telephone or in person, either discreetly or indiscreetly. If taken indiscreetly, the test may not be admissible in court because it can be considered an invasion of privacy.

Compared with the polygraph, the advantage is that the original data (recording of questions) can be recorded in any environment with a tape machine. The polygraph is taken in a structured environment where the person is seated in a chair and literally strapped to the machine with a multitude of wires and connections.

Both the Psychological Stress Evaluator and the polygraph alone cannot be used to build a case. However, they are useful tools in eliminating suspects and speeding up an investigation.

DEVELOPMENT OF AN ARSON CASE

Once it has been established that the fire was not accidental, the arson investigator must proceed in an orderly manner.

The first step is to establish that a crime has been committed by the violation of a local, county, state or federal statute. The arson investigator should become familiar with the statutes so that he can better proceed on the basis of the evidence required to prove violation of a statute.

For example, in some states it is not considered a crime to burn your automobile, boat or snowmobile. However, it may be illegal in that same state to defraud an insurance company of the coverage provided under a fire insurance contract. Under these circumstances, a charge of arson against the owner may be thrown out of court while the charge of fraud by arson would result in a conviction.

Having established a statutory violation, the task of the investigator is to prove that a crime was committed. First, it must be established that property was damaged or destroyed by fire.

Photographs showing the extent of damage are usually sufficient when accompanied by the testimony of the investigator in establishing the corpus delicti.

Evidence from the corpus delicti, which shows how and where the fire was set, establishes the fact that the fire was of incendiary origin. It is usually good practice, and in some courts mandatory, to eliminate all other possible causes through the opinions of experts. These would include automobile mechanics who could testify that no mechanical malfunctions were found in an automobile fire. It might also include an electrical technician who could testify that the shorting of electrical wiring occurred as a result of a fire rather than the cause of the fire.

Expert witnesses come from all occupations. There are no specific educational or experience requirements. The court will evaluate each individual's qualifications to determine that he is truly an "expert" in the area of litigation, whether he is a mechanic, doctor, lawyer, engineer, etc.

The presence of accelerants, the burn pattern and evidence of unusually high temperatures would be sufficient to show that an accelerant had been used and confirm the opinion of the arson investigator.

The next step is to show that the owner had the motive and opportunity to commit the act of arson. Under some jurisdictions, this is not relevant until the owner files for insurance monies. Under certain circumstances, the owner can be trapped by the policy's wording. The usual policy states that the owner shall make a sworn statement as to the circumstances of the fire. If these statements are found to be false, the provisions against false swearing would relieve the insurance company of the obligation to pay the insured amount.

For criminal cases, the same opportunity and motive problems exist and can be solved through diligent investigation, questioning, interrogation and hard work. The motives include the need for money, owner dissatisfaction with the building or vehicle, arson to hide another crime or mechanical or structural failure of the building or vehicle which would result in enormous repair bills for the owner. Any other motives which might come to light would be valid reasons for the owner to burn his property.

Opportunity is learned by interrogation regarding the owner's whereabouts at the time of the fire or the time leading up to the fire. It is also helpful to investigate the persons who established the alibi of the owner, as they may be accomplices in the arson attempt.

In motor vehicle fires, it would also be prudent to examine the front end for damage. It is possible that a hit-and-run driver torched his car to destroy evidence of the crime. Evidence of this accident would include indentations on the front grillwork, head-lights or fenders where a body had been struck. Should this be a factor in the investigation, proceed as if the fire had not occurred. Fibers from the hit-and-run victim's clothing, although burned by the fire, could still be clinging to the metal where they were caught during the accident.

If tire prints at the scene are available, the contact surfaces of the tires with the ground will provide ample tread for identification of the pattern of the tires. Although the vehicle may be totally destroyed, these four pads, where the weight of the car presses against the road, will be intact. These provide a basis for comparison with tracks found at the accident scene.

Emotional problems involving the owner may also provide a motive. The marital status of the owner should be investigated, as well as his financial position. Either of these can provide a motive for arson. All witnesses to the fire should be carefully interrogated to determine any circumstances which might give cause to believe that arson had been committed.

Interrogation

Interviewing is the act of extracting information from witnesses who may be unaware that they have seen and heard incidents which can lead to a successful solution of a problem. *Interrogation* is the questioning of a person suspected of a crime. The goal is to extract admissions from the suspect which may reveal the truth regarding his involvement. The statement of the suspect is checked by the interrogator for discrepancies and verification, or refutation by the physical evidence.

In arson cases where the evidence of arson has been completely destroyed, the case can be solved only through intensive questioning and interrogation. This has been proven time and time again by experienced arson investigators. One estimate attributed 98 percent of cases solved to interrogation. The techniques of interrogation are beyond the scope of this book; an excellent reference is *The Practical Psychology of Police Interrogation* by Hugh C. McDonald.

CHAPTER 6

Electrical Fires

I T IS RARE INDEED to find a building in the United States that is not equipped with electricity. By simply running a cord from a convenience outlet, we have at our command a source of energy which can supply heat or power at the flick of a switch. It is the very presence of electrical energy which leads to the high percentage of fires attributed to this cause. According to the National Fire Protection Association, electrical fires account for 13 percent of all fires reported. This is one of the largest categories of fire causes in the United States. It is not surprising when the use and abuse of electricity is taken into consideration.

The household demand for electricity is overwhelming. It has caused the electrical industry to increase the household electrical supply from 60 ampere service to 100 ampere service in a decade.

Such a preponderance of electrical fires imposes a burden upon the fire investigator to become familiar with electricity, how it is handled and distributed, and its physical characteristics in normal use.

PHYSICAL CHARACTERISTICS OF ELECTRICITY

Electricity is defined as the flow of electrons due to a difference in energy potential between two points on a conductor. The analogy of electricity to water is a convenient method of explaining the flow of electricity. In using the analogy, the electrical conductor is considered as the equivalent of a pipe carrying water. In order to have the water move, a pressure differential must exist between any two points on the pipeline.

Voltage

In the case of an electrical conductor, the differential is called the electromotive force or electric potential between two points. The normal household pressure in an electrical system is called *voltage* and corresponds to the pressure in a pipeline which moves the water.

Current

The equivalent of the quantity of water flowing through a pipe, in the case of electricity, is the flow of free electrons moving along the surface of the conductor, between the conductor and the insulation or, in the case of bare wire, along the surface of the bare wire. The flow rate of electrons is measured in electrons per second or coulombs per second. The number of electrons per coulomb is 6.3×10^{18} (6.3 billion billion) electrons. Rather than handle such a large number, electrical calculations are simplified by the *ampere*. By definition, one ampere is the current existing when the flow rate is one coulomb per second. This is the equivalent of gallons per minute in a hydraulic system.

Resistance

The flow in a water pipe is resisted by friction in the pipe or by some other form of resistance which creates friction or requires energy. The devices which require energy to perform useful work are the equivalent of *resistance* in an electrical circuit. The resistance determines the amount of current permitted to flow through a *resistor,* such as a light bulb filament, heater coil or motor windings.

Ohm's Law

The relationship between voltage, current and resistance is stated in Ohm's Law:

$$E = IR$$
$$\text{Where } E = \text{voltage, in volts}$$
$$I = \text{current in amperes}$$
$$R = \text{resistance in ohms}$$

The voltage supplied is usually constant at the designed voltage for the particular installation. The line is designed to carry a maximum current at a constant voltage. The safety factor against overheating of the conductor (the line) is based upon the time and the percent overload. This will be shown in characteristic protective devices later in this chapter.

The voltage is maintained constant by the power company in charge of transmission or distribution of the power. It is regulated within stringent limits, and the duty of the power company is to maintain a constant voltage at the rated capacity of the line. Higher or lower voltages can cause problems.

The current, or amperage, is a function of the resistance of the line and the demand placed upon the line by consumers. To carry out the analogy of the water supply, the water pressure (voltage) is always present, but no water is used until someone turns on a faucet. The "turning on" of the faucet in the electrical analogy is the closing of a circuit so that amperes, corresponding to gallons per minute in water, are permitted to flow.

The resistance and current are interdependent. A high resistance requires very little current, or, to put it into the analogy, if an obstruction to the flow of the water is inserted into the line, very little water would get through per minute, although the pressure would be the same. In electricity, a high resistance obstructs the flow of current, although the voltage remains constant.

When substituting numbers into the equation $E = IR$, as for example with a resistance of 1,000 ohms at 120 volts, we can solve for the current, I, by transposing the basic equation so that $I = E \div R$.

Substituting the numerical values:

$$I = 120 \text{ volts} \div 1,000 \text{ ohms}$$
$$I = 0.12 \text{ amperes}$$

Now let us assume that the resistor broke down (shorted out) and instead of offering 1,000 ohms resistance, it only offered 1 ohm resistance.

Substituting these values into the same equation we find the following:

$$I = 120 \text{ volts} \div 1 \text{ ohm}$$
$$I = 120 \text{ amps}$$

Since the normal household service is only designed for 100 amps, there is a 20 percent overload in the current. If the resistance was in a household circuit regulated by a circuit breaker designed for 20 amps, then the 20 amp circuit breaker should immediately break the circuit. If it fails to break the circuit, then the next circuit breaker upstream (towards the source of power) should break the circuit as the current exceeds its rated capacity of 100 amps.

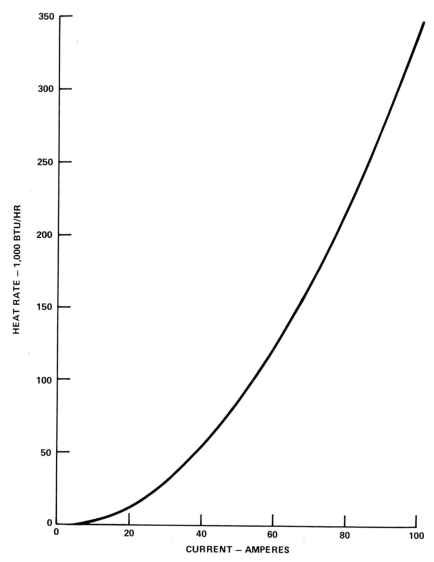

Figure 23. Heat rate versus current. The heat generated by an increasing current is an exponential function. The heat rises directly proportional to the square of the applied current.

Resistance versus Temperature

The one area where the analogy between hydraulics and electricity is inconvenient is in the explanation of the generation of heat through friction. It is difficult to imagine heat being generated by water flowing through a pipe. The fact is that it does indeed generate heat the same as any friction, such as a piston rubbing against the cylinder walls or a shaft rubbing against a bearing. However, the heat is dissipated by the water.

The equation for expressing the power dissipated as heat in electrical concept is as follows:

$$P = 0.2931 \ I^2R$$
$$\text{Where} \ P = BTU/hour$$
$$I = \text{amps}$$
$$R = \text{ohms}$$

Then:

$$\frac{BTU}{Hr.} = 0.2931 \ I^2R$$

An examination of this equation reveals the inherent danger in any electrical circuit. Normal changes in the current are to be expected as the resistance in a line changes. However, as pointed out previously, lowering the resistance increases the current. But as can be seen in the above equation, the heat produced is not directly proportional to the current but is proportional to the *square* of the current. The resulting heat rise of the increased current passing through the circuit can raise the temperature of the conductor to the point where it can start a fire. This is illustrated in Figure 23, which shows that a 100 ampere current can produce 350,000 BTU/hr. (The average residential furnace will produce 100,000 BTU/hr.) This points up the potential fire hazard inherent in an electrical system.

CHARACTERISTICS OF AN ELECTRICAL FIRE

The most obvious characteristic of an electrical fire is the presence of electrical current. While this may seem elementary, it is very surprising to consider the number of fires attributed to electrical causes where, although electrical wires were present, no current was flowing through the wires at the time of the fire. It is essential to any electrical fire that the current must supply the energy required for ignition and raise the temperature of the material ignited to the ignition point.

The mere presence of these two characteristics does not guarantee that the fire is of electrical origin. The fire investigator is faced with the problem of cause and effect. In the case of an electrical fire, the problem is determining whether a short caused the fire or the fire caused the short. This problem cannot always be resolved from the physical evidence but can often be determined through the process of elimination and by careful study of the circumstances leading up to the fire.

The origin of the fire must be carefully delineated and all other possibilities considered before attributing the fire to electrical causes. The possibility of a lightning fire, using the electrical system as a conductor, should not be overlooked. This is one of the easiest causes to eliminate and should be considered first.

Origin

The origin of the fire should be searched carefully for possible indicators of the cause of the fire. The origin of an electrical fire should contain conductors such as copper or aluminum wires, conduit, switch panels, electrical distribution boxes or other metal appurtenances. These would include electrical appliances such as radios, television sets, tape recorders or any other electrical appliance. The conductor need not be electrical in nature. A circuit can ground out to a metal conductor and arc to another conductor (such as a liquid), causing a fire.

Low voltage systems, such as doorbell chime circuits, telephone circuits, thermostats or other control circuits, usually do not carry enough power to cause a fire directly and therefore would not be considered a direct cause of a fire. However, a malfunction in such a control circuit may result in overheating of an appliance and would be considered the cause of the fire, although it did not of itself have the energy required for ignition. Arcs from low voltage systems can provide sparks which may ignite flammable vapors.

Characteristically, an electrical origin will have signs of arcing, such as holes burned in the conduit, the electrical panel or electrical appliance involved. The interior of appliances at the origin should be examined carefully to determine whether the heat which damaged the interior came from within the appliance itself or from an exterior source.

If the appliance itself was the origin of the fire, the interior burning will leave a distinct area of destruction from internal heat. This is particularly true if the cabinet is made of metal or some other fire-resistant material. This aspect will be covered in greater detail in Chapter 9.

Multiple Origins

Electricity is one of the few causes of fire in which multiple origins will be found. Multiple origins occur due to overheating a circuit. For example, let us assume that a circuit originates in the basement of a home. The circuit consists of two No. 12 AWG wires encased in insulation. The insulated wire goes from the distribution panel in the basement, along the floor of the basement, up through the walls to the third floor to service ten electrical receptacle baseboard outlets. Due to an oversight, one of the residents left a number of appliances plugged into the single circuit which drew excessive amperage for the fuse size controlling the circuit. The owner, tired of replacing fuses, placed a strip of copper between the bottom contact of the fuse and the screw threads of the fuse body. This effectively eliminated the control that the fuse had supplied over the circuit. When the current drawn by the appliances exceeded the rated capacity of the wire, the I^2R factor took effect.

The wires, subjected to a much higher current than their rating, overheated and proceeded to heat the insulation. The insulation in turn reached an ignition temperature and ignited. The entire length of wire was being heated by the current so that the insulation in the basement wiring ignited; the insulation on the wire in the walls between the basement and third story started a ventilation-controlled fire.

Since the fire in the basement was exposed to a better source of ventilation, it controlled the entire fire. Sometime later, the fire on the first and second floors broke through the walls. If much time had elapsed between the start and extinguishment of the fire, the fact that the first and second story fires originated in the walls may have been hidden. Should this occur, the rapidity of the fire's spread may falsely lead the fire investigator to the conclusion that the fire was of multiple origin and therefore had to be an arson fire.

CAUSES OF ELECTRICAL FIRES

According to a study in the *IAEI News,* the following causes of fires are given in descending order of occurrence:

1. Miscellaneous wires, devices and appliances 52.2%
2. Electrical origin (no other details reported and others not specified) 16.6%
3. Heating appliances and incandescent lights 10.5%
4. Cords 6.6%
5. Television sets 5.9%
6. Motor windings (not integral with appliance) 5.5%
7. Transformers 1.6%
8. Radios (tape recorders, reproducing equipment) .. 0.7%
9. Christmas decorations 0.4%

It is obvious from the above data that the major cause of electrical fires is found in wires involved in the distribution or use of the electrical energy.

Overloaded Circuits

The most common cause of an overloaded circuit is the misuse of an extension cord. An extension cord is flexible and insulated. It normally is a lighter wire than the average household circuit. It is common to find a No. 18 wire in a light extension cord with numerous other extension cords connected to it. Attached to these extension cords will be many appliances which overload the light cord, or *zip cord* as it is commonly termed. The overloaded zip cord will heat to the point where the insulation melts and a fire commences. The zip cord will be found at the origin of the fire, although the appliances responsible for overheating the cord may be at distant points or even in other rooms, depending upon the length of the additional cords and the location of the appliances. The entire length of the "culprit" cord will have been over-heated, although the subsequent cords may not. Such overheating is unlikely to trip a circuit breaker or blow a fuse, since the current required to melt the insulation on the zip cord may be less than the current required to blow the fuse or trip the circuit breaker.

Short Circuits

A short circuit is a closed circuit wherein the resistance offered to the circuit approaches zero. By reducing the size of the resistance in the equation $E = IR$, it can be seen that in order to hold the voltage constant, the current must increase. For example, should the resistance drop to 0.1 ohms, the theoretical current would be 1200 amperes:

$$I = E \div R$$
$$I = 120 \text{ volts} \div 0.1 \text{ ohms} = 1200 \text{ amps}$$

In actuality, however, the current carried by a circuit is limited to the maximum current available. The current is a function of the generating capacity of the power company supplying the energy and is regulated by the power distribution system with its accompanying safety devices. Considering the I^2R factor, it is obvious that a current of this magnitude would raise havoc with the electrical system of any house. Therefore, numerous protective devices preclude the possibility that 1200 amps would ever reach the short circuit for any length of time.

The amount of current reaching the short circuit point depends upon the action of the fuses or circuit breakers controlling the circuits. The control of the duration of a short circuit is a function of the characteristics of the fuse or circuit breaker. The control devices (fuses or circuit breakers) are usually located at the distribution panel and should be tested to make sure they functioned properly.

When a short circuit does occur, it will create a tremendous amount of heat at the point of shorting, which may result in melting the conductor and/or the contact point. These marks are distinctive and easily identified.

Unfortunately, a fire originating from some other cause can de-

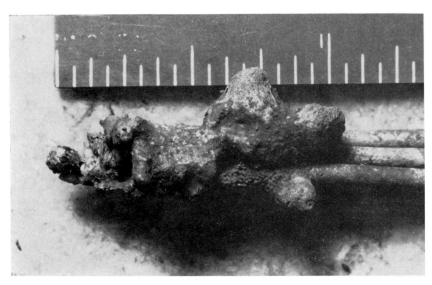

Figure 24. A direct short will raise the temperature at the point of arcing above the melting point of the conductor. This will form beads on the arced ends of the conductor. The melting point of copper is above temperatures normally found in a fire unless directly exposed to exceptional temperatures by gas jets or other high temperature sources.

Figure 25. The melting of a conductor is a function of the current and voltage available. This conductor was melted by arcing of low voltage and small current.

stroy the insulation on the wire and cause a short, which may mislead the fire investigator to conclude that the short was the cause of the fire.

This problem can sometimes be resolved by other factors involved in the case. For example, in the case of a short circuit which was suspected to be the cause of a fire, it was found that the electric clock on the same circuit had stopped at 8:25. Investigation revealed that the fuse had blown on the circuit (the distribution panel located in the basement was not involved in the fire). Verification with the tenant indicated that the clock was accurate and running at the time of the fire. However, eyewitnesses indicated that the fire had been observed at 8:10, some fifteen minutes before the clock stopped. Since it is assumed that the clock stopped when the fuse blew, it is obvious that the clock had been running for fifteen minutes after the fire had started. The conclusion is that the short circuit could not have started the fire. It should be remembered that fires will usually cause short circuits but can very seldom cause overloads.

INVESTIGATION

The basic fire investigation techniques are outlined in Chapter 4. The origin of the fire should be carefully determined, and it should be noted whether there is more than one origin. In the case of a multiple electrical origin fire, the same principles apply to each electrical origin. There must be an electrical circuit, control or appliance which was carrying an electric current at the time of the fire.

The condition of all electrical wires and equipment should be noted and photographed before removal. The status of all circuitry within the appliances and electrical apparatus should be determined as closely as possible before touching or moving any of the material. The reason for this is that fire destroys the insulation but may not necessarily cause conductors to contact each other or create shorts.

The status of each circuit should be checked immediately to determine whether the circuits are open or closed. This testing should be carried back to the distribution panel. The circuit breakers or fuses controlling the circuit should be checked to determine whether they are tripped. They should also be examined for continuity to determine whether the circuit is open or closed, regardless of the position of the operating handle. The operating lever can be wedged in the closed circuit position, but the circuit breaker will open under excessive current loads or short circuits, irrespective of the lever position.

Any sign of arcing should be identified by measuring and establishing the physical location with respect to the point of origin. The shorted area should be photographed, and the relationship between the shorted area and the origin of the fire should be shown from two different angles in the photographs.

It is customary for the fire service or a representative of the power company to disconnect the electrical power to a burning building. The time of disconnection should be determined to verify the time shown on electric clocks. This will eliminate any circuits whose clocks show times corresponding to the disconnect time.

After the area of origin has been determined, the circuitry should be sketched to establish a record of the wires as they were found. A scale drawing should be made to show the relative positions of all appliances and equipment connected in the circuit. The location of all switches and their status should be recorded. After all of the circuits have been checked for continuity, the sizes of the wires involved in the circuitry should be determined.

Determination of Switch Status

It is desirable for the fire investigator to determine whether a circuit was open or closed at the time of the fire. This can be easily determined by examining the fuses and switches which control the circuit. The switches, if not destroyed by the fire, will characteristically be smoke stained and clearly indicate the position of the switch at the time of the fire.

As shown in Figure 26, the exposed portion of the switch lever will be blackened by soot and the heat of the fire. The protected portion of the switch will be unstained due to the protection of the plate cover and the physical position of the switch.

The position of the switch lever will not indicate whether the circuit was open or closed. To make that determination, the circuit

Figure 26. The position of light switches during a fire can be determined by examination of the lever. The lever at the left was in the "off" position during the fire, as indicated by the heavy smoke staining on the top surface of the "off" lever. The switch at the right had also been in the "off" position but had been turned up to reveal the unstained surface, indicating it had been moved after the fire.

itself must be checked, as in the case of a two- or three-way circuit. The other switch involved will determine whether the circuit was open or closed.

The status of a circuit breaker can be determined by examining the surface of the switch exposed to the fire. If the circuit breaker toggle is in the "tripped" or "off" position, the toggle should be photographed and then moved to examine the surface of the "on" label. If the surface reveals smoke or fire damage, it can be assumed that the circuit breaker was in the "on" position before and during the fire and that the fire shorted out the circuit, causing the switch to trip.

Comparison of the "on" portion of the switch with that of the other tripped switches will indicate whether other circuits were on or off and give a good basis for comparison.

Wire Sizes

The National Electrical Code has established a color coding method for identifying the various wire sizes. This system is ade-

TABLE XIV

CHARACTERISTICS OF ELECTRICAL WIRES*

Size AWG	No. Wires	Stranded Diameter Each Wire Inches	Bare Diameter Inches	Ampacity Copper (30°C; 86°F)	Ampacity Aluminum (30°C; 86°F)
18	Solid	0.0403	0.0403	21	
16	Solid	0.0508	0.0508	22	
14	Solid	0.0641	0.0641	15	
12	Solid	0.0808	0.0808	20	15
10	Solid	0.1019	0.1019	30	25
8	Solid	0.1285	0.1285	40	30
6	7	0.0612	0.1840	55	40
4	7	0.0772	0.2320	70	55
3	7	0.0867	0.2600	80	65
2	7	0.0974	0.2920	95	75
1	19	0.0664	0.3320	110	85
1/0	19	0.0745	0.3720	125	100
2/0	19	0.0837	0.4180	145	115
3/0	19	0.0940	0.4700	165	130
4/0	19	0.1055	0.5280	195	155

* Adapted from NFPA, *National Electrical Code,* 1975. Courtesy of National Fire Protection Association, Boston, Massachusetts.

quate up until the time the identifying color has been destroyed by heat and fire. Once this has occurred, the only means of identifying the wire is by measuring its diameter.

Table XIV gives the size for American Wire Gauge Cables normally used in electrical circuits. The columns labeled *Ampacity* list the current-carrying capacity of the copper wire at 86°F, as established by the National Electrical Code. The circuit loads can then be calculated by determining the number of appliances on the circuit and examination of the fuse or circuit breaker to determine whether the circuit was overloaded.

Underdesigned Circuits

Local ordinances, as well as the National Electrical Code, should be checked to determine the permissible outlets or fixtures allowed on a circuit for a given wire size. An electrician or electrical engineer who is experienced in fire investigations should be consulted to check for proper circuitry design. If the building is old, the circuitry may have been installed when less stringent codes were in effect. Changes may have been made in the circuit by un-qualified persons. In such cases, increased loads may exceed the limits of properly designed circuits.

Short Circuits

A short circuit will usually leave signs of arcing, melted wires or melted fixtures. The ends of the copper wires will be burned off, and small droplets of molten copper will be found solidified on the ends of the wire (see Figs. 24 and 25).

The melting temperature of copper is 1,980°F. It is unusual for a fire to attain this temperature until the fire has burned for three hours (see Fig. 10) unless accelerants are present. In either of these cases, wholesale melting of copper will be accompanied by melting of other metals below that temperature, and the distinction between melting from arcing versus that from the heat of the fire can be determined.

The arced portion of the wire should be carefully photographed, and its position relative to the fire origin should be accurately measured in all three dimensions. If arcing was the cause of the fire, it should be at the *low point* of the fire. Subsequent burning below the point of origin should be less than that above, with the following exceptions.

In heavy arcing, such as occurs at electrical service entrances, molten metal will splatter from the point of arcing and ignite combustible substances located some distance from the actual point of arcing. Since the heaviest currents are found in the distribution panel, currents approaching 100 amps can occur before circuit breakers or fuses are blown, interrupting the circuit.

It should also be remembered that a short circuit can occur in an overloaded circuit. The short circuit is usually a result of the breakdown of insulation and subsequent physical contact between current-carrying wire and another conductor.

Current-controlling Devices

The two most common current-controlling devices are the circuit breaker and the fuse. A Fustat® is a patented device combining the features of a fuse and circuit breaker. All of these devices are commonly used in even the most complex installation as protective devices.

Figure 27. This mobile home fuse box consists of eight screw-type fuses and two plug-type fuses at the top of the box. The plug-type fuses supply power to the range and fuse panel. Should a massive short occur, the main fuse can be blown before the smaller fuse blows, giving double protection to the circuit.

Fuses

A fuse is a *weak link* in an electrical system. It commonly contains a fusible link (from whence it derives its name) which, when exposed to high temperatures imposed by the I^2R factor, heats to above its melting point. The melting of the fusible link breaks the circuit and protects the system until the link is manually replaced. The normal household fuse is a screw-type fuse with a glassine window through which the fusible link can be examined. If the fuse has blown due to a short circuit, the glassine window will usually be blackened because of the oxidation of the fusible link caused by the high temperatures involved. The blackened surface will be on the interior of the glassine window. The window can also be stained on the exterior by soot and smoke from the fire. However, this can be removed and the interior of the fuse examined.

If the fuse was blown due to an overloaded circuit, the melting point is usually reached at a much slower rate than in the case of

Figure 28   Figure 29

Figure 28. This fuse was melted by the continuing application of 15 amps for 10 minutes. Note the lack of smoke staining and the small portion of the fuse which has been destroyed.

Figure 29. This fuse, with the glass cover removed, shows the result of a direct short. The black staining on the porcelain body of the fuse also obscured the glassine window shown in Figure 28, necessitating its removal. Note the blackened surface and the amount of the fuse strip destroyed by arcing.

the short circuit. Therefore, the fuse will merely melt as the I²R factor gradually heats the fuse to its melting point.

Unfortunately, the melting point of the fusible link is between 600° and 800°F. This temperature is encountered in most fires, and if the fuse is exposed to this temperature, the fuse link will melt, interrupting the circuit. The investigator may not be able to determine whether the fuse was blown by shorting, by electrical overload or by the heat of the fire impinging upon the fuse. One method of making this determination is to compare all of the fusible links.

Each fuse should be examined. The relationship between the fire and the various circuits should be determined to see whether any of the circuits were unaffected by the fire. These circuits would normally have intact fuses and untripped circuit breakers. If the fuse has melted from an overload, tab ends will usually remain where the fuse link had been. If the fuse was melted by the heat of the fire, the tab ends will not be as obvious. It should be remembered that if the heat of the fire caused melting of one fuse, it would be most unusual if the same heat did not melt all of the fuses, assuming that the fuses are all the same size.

TYPE *S* FUSES. Type *S* Fuses represent a classification of fuses incorporating a time delay mechanism to overcome the inherent problem of short-term overloads. These overloads would suffice to melt a fusible link fuse of the Edison base screw type, although they would not overheat the conductor or electrical system to cause a fire. The effect is achieved by incorporating a heater element which must be heated by the I²R factor before the fusible link reaches its melting temperature, thereby opening the circuit. The Type *S* base also prevents increasing the fuse size in a given circuit. The base of each Type *S* fuse is unique to its given ampere rating so that different fuses are not interchangeable. The Type *S* fuse protects the circuit against the high currents required for the starting torque on electric refrigerator motors, etc.

FUSE IDENTIFICATION. The 15 ampere Edison base screw-type fuses are identified by a hexagonal window. For fuses of larger amperage, the window is circular. For Type *S* fuses, all of the bases are of different diameters and are not interchangeable.

Larger fuses of the cartridge type have the rating and name of the manufacturer on the metal portion of the fuse for identification if the label is destroyed.

Circuit Breakers

Circuit breakers are mechanical devices which eliminate the inconvenience of the one-shot screw-type fuse and the Type *S* fuse by incorporating the advantages of both into one reusable control.

The circuit breaker is placed into a distribution panel, and the circuits are connected directly to the circuit breaker. It consists of three elements:

1. The control lever
2. A heating element
3. An electromagnet

The control lever is arranged so that wedging it in the *on* position will not prevent the circuit breaker from opening in the event of either an overcurrent or short circuit. With an overcurrent, the bimetal heating strip expands unequally until the design temperature is reached, based upon the rating of the circuit breaker. At that time, the bimetal strip disengages the catch and a spring opens the contact points, breaking the circuit.

The electromagnetic coil works on the same principle. When a short circuit exceeds the current design limit, the electromagnet disengages the catch, releasing the contact point and breaking the circuit. The greatest advantage of the circuit breaker is the convenience it affords. It can be reset if a circuit is opened for any reason. The circuit breaker is *not* intended to be used as a control switch. It can be damaged if used in this manner.

CAPACITY OF CIRCUIT BREAKERS AND FUSES. The rated capacity of a fuse, fusestat or circuit breaker is the maximum continuous current that the device will carry without interrupting the circuit. For example, a 20 amp circuit breaker will carry a 20 ampere current continuously without interrupting the current. In practice, a fuse, fusestat or any current-regulating device will carry considerably more current than its rating before the circuit is broken.

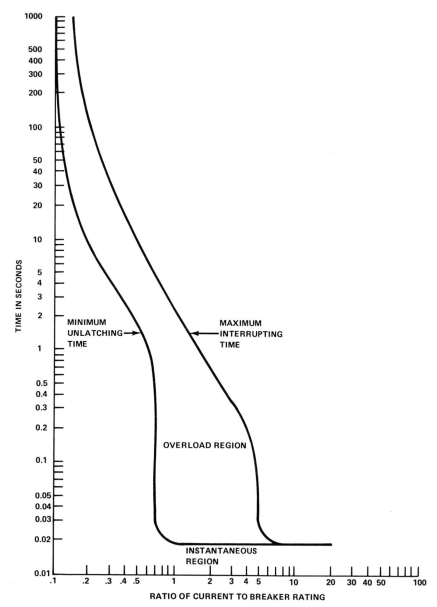

Figure 30. Typical reaction times of circuit breakers due to shorts or overloads.

Figure 30 is a composite graph showing a typical curve for circuit breakers. It should be noted in the instantaneous (short circuit) region that the maximum time required, regardless of the rating, is 0.02 seconds. These higher ratings are the interruption time for a direct short which, of course, introduces the maximum current. In the overload region, it can be seen that a circuit breaker can carry 50 percent overload for over fifteen minutes without tripping. It can also carry a 200 percent overload for 100 seconds without tripping. Under certain circumstances, these times and overloads can cause fires under conditions where the ambient temperature at the circuit is much higher than at the circuit breaker.

For example, if the circuit breaker is located outside in the dead of winter with a temperature of −20°F, the heat requirements for a bimetal strip to trip would be much greater than for the same strip heated in an ambient indoor temperature of 70°F. Under extremely cold conditions, these factors should not be overlooked.

Bad Connections

When a conductor is connected to a switch or appliance, an electrical connection offering a minimum of resistance to the flow of current must be made. These connections take many forms, such as soldered, screwed, bolted and compression connections to ensure solid contact between the conductor and the connection at the electrical device. It is at these connections that a high resistance area can be developed which increases the heat output of the current.

As indicated before, when a high resistance area develops in the circuit, the heat output increases. This same current flows through the entire circuit and will generate heat as a function of the I^2R factor. When the current passes through the resistance area, the heat output is increased at that point. If it is raised high enough, it can ignite any combustible material near the bad connection.

The fire investigator should examine all connections to ensure that they are tight, with the proper amount of overlap in screwed connections, as illustrated in Figures 33 and 34. Compression connectors should be snug but not so tight that the wires have

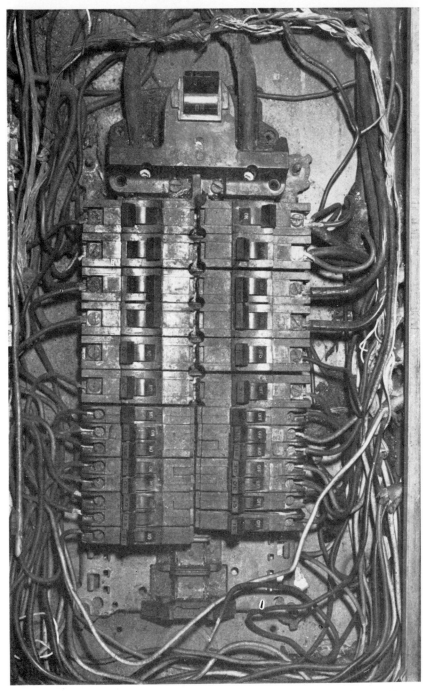

Figure 31. This typical circuit breaker panel shows the panel immediately following a fire. The tripped circuits can be detected by the position of the levers. The panel should be photographed before any switches are moved to check relative positions. The circuit breakers should be checked to determine whether the circuits are opened or closed before moving any levers.

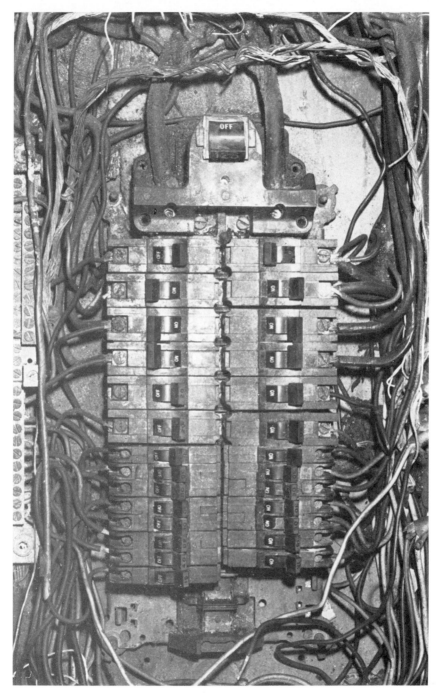

Figure 32. After the status of the circuits has been determined, the switches should be thrown in the opposite position to verify that the switches have not been moved since the fire. Moving a switch to its opposite position will reveal a protected indicator, as shown in this photograph.

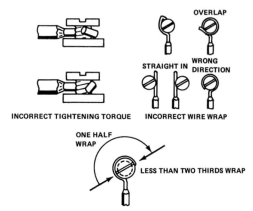

Figure 33. Incorrect methods of terminating aluminum wire at wire-binding screw terminals of receptacles and snap switches. From *Hazard and Analysis, Aluminum Wiring*, 1975. Courtesy of U. S. Consumer Product Safety Commission.

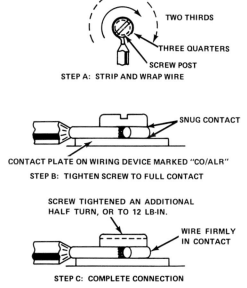

Figure 34. Correct method of terminating aluminum wire at wire-binding screw terminals of receptacles and snap switches. From *Hazard Analysis, Aluminum Wiring*, 1975. Courtesy of U. S. Consumer Product Safety Commission.

failed structurally. The shearing of the wires at a compression connection would create a high resistance area which could lead to a fire.

Improperly spliced wires can be another source of fire and are characterized by shorting and arcing at the splice. Bad connections can be made with copper as well as aluminum wiring. The bad connection is the result of poor workmanship or the improper use of materials or methods. However, the effect of improper connections or workmanship on aluminum is more pronounced than it is on copper conductors.

Aluminum Conductors

Aluminum is a particularly useful material when used as an electrical conductor. Its light weight, good conduction characteristics and competitive cost makes it a desirable material for this use. It is used extensively and almost exclusively for transmission lines and has been used extensively in distribution systems as well as in industrial and residential applications.

Two physical characteristics of aluminum may cause problems to the user of the product. The first is its high coefficient of expansion; the second is its affinity for oxygen.

Aluminum's high coefficient of expansion causes problems in several ways. Changes in the weather cause changes in the length because the metal expands and contracts more than steel or copper (Table XV). Expansion is due not only to ambient temperatures but also to thermal-electrical causes such as overheating with high currents. Because currents create heat in a wire, it is necessary to use a larger wire when aluminum is substituted for copper. Table XIV from the National Electrical Code shows the ampacity of various wires. It clearly indicates that, as compared to copper, aluminum requires larger sized wire to carry equal current. This causes problems in the connection of the larger wire to appurtenances.

These problems result from the fact that the smaller the area of contact between two conductors, the higher the resistance and the higher the resultant temperature. This increase in temperature causes a corresponding increase in the expansion of the aluminum

TABLE XV

COEFFICIENT OF EXPANSION*

Material	Coefficient × 10⁻⁶
Aluminum	13.3
Copper	9.2
Steel	9.9

* From T. Baumeister and Lionel Marks, *Standard Handbook for Mechanical Engineers,* 7th ed., 1967. Courtesy of McGraw-Hill Book Company, New York.

conductor. It presses inordinately against its connector, resulting in stresses to the connector which can result in permanent deformation of both the connector and the wire.

The expansion of the aluminum may extrude the wire from the connection. As the connection cools, the aluminum shrinks, loosening and increasing the resistance of the connection. This occurs repeatedly and compounds the problem. The looser the connection, the higher the resistance and temperature and the greater the rate of expansion. The end result is failure of the connection and overheating of the conductor which may lead to a fire.

The second problem encountered with the use of aluminum is its affinity for oxygen. The minute that aluminum is exposed to air, it oxidizes, which increases its resistance. The problem is solved by using chemical deoxidizers or by tightening the connections to break the aluminum oxide and make a good connection by virtue of the torque applied to the connecting mechanism.

A minor factor concerning aluminum is the possibility of electrolysis. When two unlike metals are immersed in an electrolyte, a current will flow. Therefore, whenever aluminum is used as a connector, a suitable connection must be made to eliminate the possibility of electrolysis in the presence of an electrolyte. This is solved by using special connections between aluminum and copper wires. If aluminum wire is encountered in a fire, the investigator should look for evidence of bad connections and lack of sufficient slack in the wire. This same problem can also occur with copper or steel wire.

Occupants should be questioned for clues indicative of bad connections, regardless of whether the wiring was copper or aluminum. Clues to look for include the following:

1. A warm light switch or receptacle face plate. This would be an indication that the switch or receptacle was subjected to overheating due to a faulty connection. If this were reported, the suspect switch or receptacle should be carefully examined to determine whether it was the cause of the fire.
2. Strange or distinctive odor of the smell of burning plastic near a receptacle or switch.
3. Flickering of lights not caused by appliances or other external events. These would be indicative of a broken wire which was arcing, melting and rewelding until it reached a point where the arcing was severe enough to cause a fire.

Metallurgical Examination of Electrical Wire

In an electrical fire, the cause of the fire is overheating of the electrical wire through an overcurrent or short which raises the temperature of the wire to the point where the surrounding combustibles are ignited. These combustibles can be the insulation on the wire itself or anything near enough to be affected by the heat given off by the wire.

As a result of the heating of the wire, the wire may be subjected to metallurgical changes which are readily apparent from microscopic examination of the interior structure of the wire.

Effect of Heat on Copper Wire

When copper wire is heated close to the melting point, several microstructural changes occur. The first metallurgical change is grain growth of the copper crystalline structure. This grain growth is a function of the peak temperature reached and the length of time that the elevated temperature was maintained (see Fig. 36).

Another change occurs when oxygen is adsorbed by the copper near the surface of the wire. Oxidation occurs, changing the boundary structure of the grains. This may leave voids between the enlarged grains.

If the wire is heated above the melting point of the copper, *dendrites* are formed as the molten metal resolidifies. Without

Figure 35. This photograph shows the internal structure of a normal copper conductor at ×200 magnification. The lines parallel to the surface of the wire (at the top of the photograph) are striations made by drawing the wire through the forming die.

Figure 36. In this ×200 photomicrograph, the copper conductor was torch heated to 1000° F and water cooled. Comparing the grain size with that of Figure 35 indicates the grain growth due to heating.

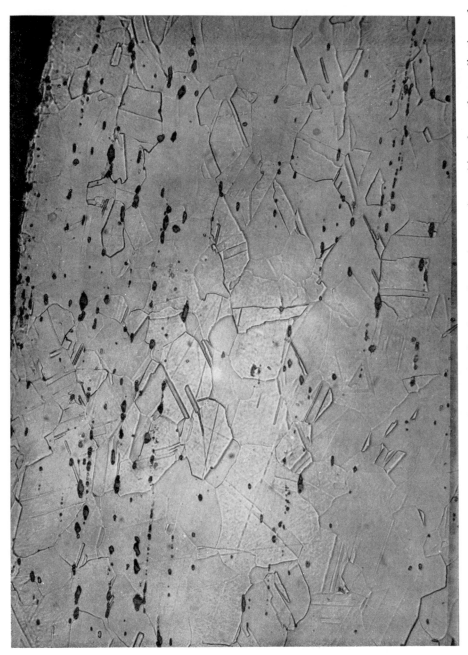

Figure 37. This ×200 photomicrograph of a copper wire shows the grain growth resulting from the application of 150 amps for 30 seconds and then air cooled.

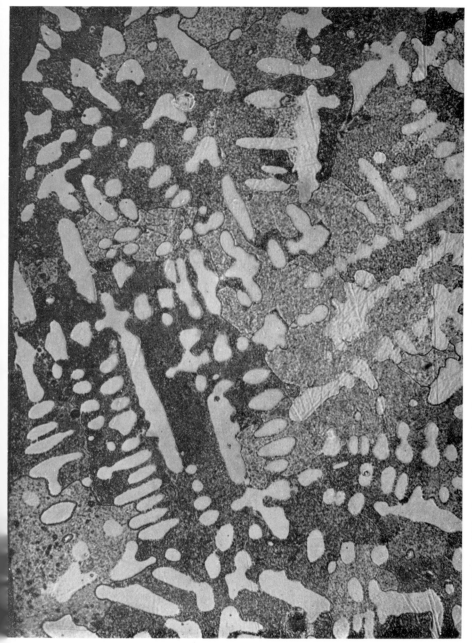

Figure 38. This ×200 photomicrograph of copper wire shows the effect of applying 115 amps for 1.5 minutes and subsequent air cooling. The heat melted the wire, which resolidified, forming dendrites (the light-colored areas).

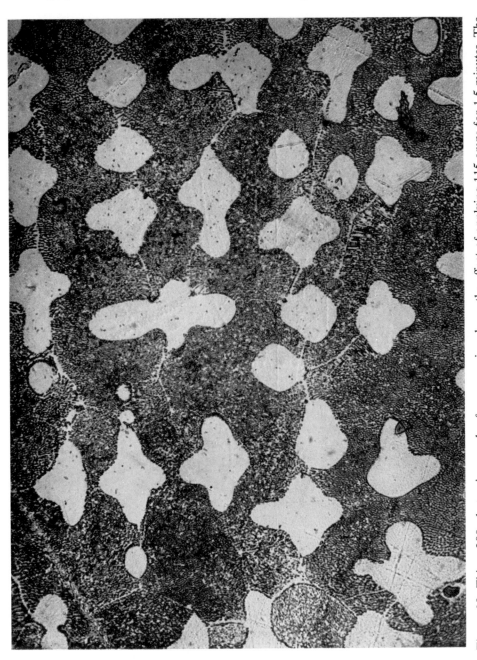

Figure 39. This ×200 photomicrograph of copper wire shows the effect of applying 115 amps for 1.5 minutes. The wire, which was melted, and then reheated to the melting point and air cooled, forming dendrites.

melting, no dendrites are formed. A dendrite is an irregularly shaped mass of the resolidified copper which resembles the branching of a tree. These are illustrated in Figures 38 and 39.

Copper for electrical use is 99.90 percent pure. It must be maintained at this purity because small changes in the purity result in great changes in the conductivity of the copper. This would be detrimental to its use as a conductor. Therefore, the chemical requirements of copper wire are strictly adhered to in the manufacturing process.

Metallurgical Differentiation of Heat Source

Unfortunately for the fire investigator, it is impossible to determine whether the metallurgical changes in a wire occurred from internal heating by an overcurrent or from external heating by a fire. The problem lies in the three variables involved. The first variable is the maximum temperature to which the wire is raised. Grain growth is proportional to the final temperature reached. Whether the wire is air cooled or water quenched by the fireman's hose has no effect on the final grain size.

The second variable is the duration of the temperature. A minor temperature increase with a long duration can increase grain size to the equivalent size of a higher temperature over a shorter time period. Grain growth also appears to be larger at the surface of the wire than at the core, regardless of whether the wire is heated internally or externally.

The change in grain size in the interior of the wire or on its surface has no significance as to the heating source, nor does the fact that the external heat was radiant or convective.

The formation of dendrites occurs regardless of the source of heat which melted the wire. The dendrites are subject to more growth the longer the melting point is maintained and the longer the remelted mass is maintained above the melting point. Again, the source of heat is immaterial.

The third variable is the amount of oxygen available to the surface of the wire. In an electrical fire, the insulation is closer to the source of heat, which is the surface of the electrical wire. It, of course, is pyrolyzed first, and through ignition of the insulation, surrounding combustibles are ignited. This in turn speeds up the destruction of the insulation and the reduction of the insulation to ash. Until an external blow or force acts upon the insulation, it will invariably remain upon the surface of the copper wire. In an external fire, the insulation is burned from the outside inward, thus acting as a thermal insulation to the wire and slowing the heating process. However, the rate of oxidation of the surface of the wire is the same regardless of the cause (external or internal heat) once the insulation has been destroyed.

Effect of Subsequent Fire on Fires of Electrical Origin

Another factor which compounds the problem for the fire investigator is that a given electrical fire can destroy all evidence of the electrical nature of the fire.

For example, a fire can start through shorting and arcing of an electrical wire. Let us assume that the fire has started and in the meantime the electrical current has been interrupted by the reaction of an upstream circuit breaker. The electrical wire, melted by the arcing, could be easily identified if the fire was extinguished immediately. Unfortunately, this is not always the case.

Let us assume that the fire continues to burn and heat is generated which *remelts* the beaded ends of the arced wires. Should this occur, all evidence of the original cause of the fire will have been destroyed by the reheating and remelting of the ends of the copper wires. If the electrical conductors were aluminum, its lower melting point would compound the problem. Temperatures reaching the melting point of aluminum are much more likely than temperatures reaching the melting point of copper.

Evidence

The evidence to be sought, protected and examined is generally limited to the electrical wires, connections and devices involved in the fire. Samples of the wire at the origin should be obtained. If testing will be done on the burned-out wires, samples of unburned wire, preferably from the same circuit, should be taken. This eliminates any possibility of variation between manufacturers of the copper wire and the insulation. These control samples should be large enough so that destructive tests can be run and the physical characteristics determined.

If it appears that a surge of electricity may have occurred, the atmospheric conditions should be checked for the possibility of lightning; the transformer and distribution circuits should be examined where power enters the building.

Statements

All witnesses to the fire should be questioned as to what they saw, heard and smelled. Questions regarding electrical fires should be addressed to the person who turned off the electrical energy at the fire, particularly regarding the time it was turned off, the method and disconnect location.

A check with eyewitnesses will reveal whether the lights were on when the fire was first observed and at what time the lights went off. Determine whether the lights went off simultaneously, light by light or circuit by circuit.

The sound of electric motors operating (furnace fan, air conditioner, refrigerator, freezer, etc.) should be recorded regarding time and location. If any sparking, arcing or other electrical disturbance was noted, or if the smell of ozone, burning insulation or electrical material was experienced during the fire, it should be recorded.

CHAPTER 7

Gas Fires

THIS CHAPTER COVERS gas fires caused by leaks or control valve malfunctions. Since a gas fire may or may not accompany an explosion, explosions are explained in Chapter 8. Malfunction of appliances and controls is covered in Chapter 9.

The NFPA attributes 1 percent of fires to *Gas Fires and Explosions.* This classification is defined as those fires and explosions involving escaping gas. Excluded from this category are heating and cooking appliances, and welding and cutting torches.

It is the exceptional home in the United States that does not have some form of controlled energy such as electricity or gas. Very few businesses can operate without some form of energy. This energy is usually found in the form of electricity, natural gas or a combination of propane or butane. This means that nearly every residence, commercial or industrial building contains electricity and/or gas which are possible causes of fires. To determine the actual cause of a fire, the possibility of a gas fire must be eliminated or confirmed. In order to understand how gas can cause a fire, more detailed explanations of the nature and characteristics of gas are in order.

CHARACTERISTICS OF GAS

The gas most familiar to everyone is air. We live at the bottom of a sea of air. Air is actually a mixture of gases, as shown in Table XVI.

Air is 78 percent nitrogen, which is lighter than air; 21 percent oxygen, which is heavier than air; 0.03 percent carbon dioxide, which is 1½ times the weight of air. Noting the difference between the lightest gas (nitrogen) and the heaviest (carbon dioxide), it becomes apparent that air is a mixture of gases with an average specific gravity of 1.

The other obvious fact is that some characteristic of the gases

TABLE XVI

COMPOSITION OF DRY AIR BY VOLUME*

Substance	Percent	Specific Gravity (Air = 1)
Nitrogen	78.03	0.9718
Oxygen	20.99	1.1053
Argon	0.94	1.379
Carbon dioxide	0.033	1.5282

* From American Gas Association, *Gas Engineers Handbook,* ed. by C. George Segeler, 1965. Courtesy of Industrial Press, Inc., New York.

prevents the heavier gas (carbon dioxide) from settling to the bottom of the sea of air in which we live. Were this to occur, we would have all suffocated long ago.

Two factors prevent the settling out or the *stratification* of the various gases. One is the presence of air currents which cause mechanical mixing of the gases and keep them in a uniform state. The other is the molecular diffusion which results in a continual mixing of the gases due to the inherent energy of motion each molecule of gas possesses.

Specific Gravity of a Gas Mixture

As a result of molecular diffusion and mechanical mixing of the gas, the resultant gas mixture has a specific gravity of its own, based upon the specific gravities and percentages of the component gases.

The specific gravity of the mixture is calculated as follows:

$$SG_m = SG_a \times \%_a + SG_b \times \%_B \ldots SG_n \times \%_n$$

Where SG_m = the specific gravity of the mixture,
SG_a = the specific gravity of component a where air = 1
$\%_a$ = the percent by volume of the mixture of a, etc.

For example, specific gravity of a 9.5% propane mixture (SG propane = 1.52) can be determined as follows:

$$\begin{matrix} \text{(air)} & \text{(propane)} \\ SG_m = 1 \times 0.905 + 1.52 \times 0.095 = 1.05 \end{matrix}$$

Considering air as a mixture, and realizing that diffusion and mixing keep the relatively large percentages of oxygen and nitrogen in a uniform mixture, it is easily understood why and how a flammable gas, when released into a room, will rapidly form a flammable mixture.

Mechanical Mixing

Under ideal conditions, with still air and still propane, their interface would mix at the rate of approximately four square feet per hour. This natural diffusion, based upon the molecular activity of the gases, would not enable a furnace to operate if it were dependent upon the natural diffusion rate to provide a flammable mixture in the combustion chamber.

Instead of relying upon the molecular action of gas to form a flammable mixture, the furnace designer uses the physical mixing characteristics of a fluid gas stream to form a flammable mixture. The raw propane exits through an orifice which regulates the proper amount, enters a mixing chamber where it is mixed with primary air and is then passed through the ports of the burner into the combustion chamber. The mixing of the gas in this process forms a flammable mixture in a matter of seconds rather than the hours that would be required for molecular diffusion. The same principle applies to the mixing of propane or any other flammable gas with air to form a flammable mixture. Invariably it is accomplished by mechanical mixing rather than molecular diffusion.

Gas Migration

Passages for gas from one room to another depend upon a difference in pressure or a *draft*. Once a differential pressure has been established between two areas, the air and gas will flow from the high pressure area to the low pressure area through any available opening, even a porous concrete block wall. Gas has also been known to pass through a keyhole, the cracks around a door, soil (for a considerable distance) and porous membranes.

Another seemingly unlikely passage for gas is through an electrical conduit. The conduit, although it appears to be full of wire, can carry gas between rooms, floors and, in rare cases, buildings.

This occurs when the size of the conduit is large enough and the pressure differential great enough to force the gas through.

Circulation of Air

Since we cannot rely upon diffusion to explain the formation of an explosive mixture of gas and air, we must rely upon the forces causing air circulation in a building to provide the mixing of the gas into a flammable or explosive mixture. To do this, the characeristics of the building and rooms in relation to the infiltration and circulation rate of air must be examined.

Infiltration

Infiltration is the leakage of air through and around doors, windows and exterior walls of a building. Any cracks in the walls, ceilings or floors will permit the passage of air and gas through the building at a rate dependent upon the wind and the exterior and interior temperatures and pressures of the building. The chimneys and vents of water heaters, furnaces, clothes dryers and other appliances all provide passageways for air to move in and out of the building. Anywhere that air can go, gas can go.

When air is moving out of a building, it tends to decrease the pressure within the building. An equivalent amount of air will attempt to enter the building to compensate for the exhausted air and equalize the pressure.

In addition to these natural ventilation sources, windows and doors can be left ajar. Windows or vents can be deliberately opened to provide ventilation in the case of arson. All of these factors can affect the mixing rate of gas entering a room. If the room were completely air-tight, the gas would be displacing air within the room and at the same time mixing and diffusing with the air in the room.

This mixing will be affected by the possible outlet of gas through vents or openings which may be available. Some of these outlets would be the vents for other appliances. For example, an electric dryer may have just been turned off. The residual heat inside would still send a warm air current up its stack, continuing the draft already existing until the temperatures of the stack and dryer have cooled to the point that the natural draft no longer carries gases out of the room. This would set up a convective current between the source of the leak and the exiting gas, drawing the leaking gas into the exiting gas stream.

Wind Forces

These natural drafts, or *stack effects,* can be nullified by downdrafts which reverse the flow and drive the gas mixture into the furnace room. This will add turbulence and more gas to the mixture by reducing the flow of fresh air to the room, which would normally dilute the gas mixture. This would also prevent the removal of gas already mixed.

Thermal Drafts

Normally, gas entering the building is delivered through utility pipes. During the wintertime this gas is cooler than the air within the building. However, the flow rate of the gas is such that the gas will soon absorb heat from the building as it travels to the appliance. It is usually warmed to ambient or room temperature by the time it reaches an appliance.

Assuming that a furnace is operating properly, it would be warm. The gas may thus be even warmer than the ambient room temperature. Therefore, consideration should be given to the gas temperature as it left the furnace or as it left the leak and entered the room. Assuming that the gas is at room temperature, it would follow the convective pattern of air within the room and rapidly mix with air to form a flammable mixture.

Any changes in the ventilation or circulating system of the room, such as the starting of the ventilating fan on the furnace, would affect the mixing of gases in the room, as well as change the temperature of the gases within the room by introducing either colder or warmer air into the room.

Anyone opening or closing doors within the building, which may let colder or warmer air into the building, would also affect the mixing of gases in the room. This should be noted, particularly in the wintertime. The mere movement of a person's body through a room will set up enough convective currents to substantially affect the mixing rate of the gases.

Calculations of Infiltration

The American Society of Heating, Refrigerating, Ventilating and Air Conditioning Handbook of Fundamentals, gives the rate of air changes as shown in Table XVII. As can be seen, approximately two changes of air take place per hour in rooms with windows or door openings to the exterior. A room with no windows or exterior doors could experience a complete change of air within two hours. An average room with dimensions of 8 × 10 × 12 ft. would contain a volume of 960 cubic feet. Assuming a 2.5 percent volume of propane gas, this means that the leakage rate would have to be 24 cubic feet per hour to form an explosive mixture to completely fill the room. In view of these figures, it is no wonder that most explosions occur with less than 25 percent of the possible volume of gas in the room.

Environmental Influences

Consideration must be given to the environment in which a leak occurs. The mere presence of a leak is not conclusive evidence that it is the cause of the fire. It is possible that the leak was a *result* rather than the cause of a fire. Considerations to be taken into account include the flow rate of the leak in terms of cubic feet per hour, the specific gravity of the gas, the condition of the room in which the gas source is located, the existing temperature inside and outside, the existing convective currents or artificially induced ventilation, furnaces or other natural man-

TABLE XVII

AIR CHANGES TAKING PLACE UNDER AVERAGE CONDITIONS IN RESIDENCES, EXCLUSIVE OF AIR PROVIDED FOR VENTILATION*

Kind of Room or Building	*Number of Air Changes Taking Place Per Hour*
Rooms with no windows or exterior doors	0.5
Rooms with windows or exterior doors on one side	1
Rooms with windows or exterior doors on two sides	1.5
Rooms with windows or exterior doors on three sides	2
Entrance halls	2

* Reprinted from *ASHRAE Handbook of Fundamentals,* 1972.

made circulatory devices which could change the mixing rate of the gas and air.

For example, if the gas is natural gas and the pilot light has gone out in a well-ventilated furnace, in the absence of a down-draft the gas mixture would be vented out the chimney. It would diffuse harmlessly into the atmosphere until the flow of gas had cooled the furnace and stack to the point that a draft could no longer be sustained. The gas mixture would then accumulate in the furnace unless a downdraft occurred and carried the gas mixture back into the room.

ODORANTS

The gas industry has long recognized the inherent danger of a gas leak. To provide a means of early leak detection, the industry adds odorants to their products for identification of a leak.

Neither natural gas nor the liquid propane (LP) family of gases has a natural odor. It is necessary to add an odorant to these gases so that the consumer can detect their presence. According to the NFPA Standard 58, the odorant should be characterized by a distinctive odor so that the presence of gas can be detected at less than one fifth of the lower limit of flammability of the gas. The requirements are met by the use of one pound of ethyl mercaptan or one pound of thiopane per 10,000 gallons of LP gas, or the equivalent.

Properties of Odorants

An odorant should be chemically stable with a high odor intensity under all conditions. It should have a distinctive odor and be nontoxic and nonirritating in combination with the gas or with the products of combustion. It should be easily introduced into the system at a low cost per gallon.

Commercially available odorants are generally a compound containing sulfur, which provides the distinctive odor. Table XVIII is a listing of the more readily available commercial odorants and their characteristics.

The effectiveness of the odorant is measured by its ability to be detected by the average person under normal conditions. The dosage is such that it can be detected in proportions as low as one

TABLE XVIII

CHEMICAL AND PHYSICAL PROPERTIES OF ODORANTS*

Chemical Description	Molecular Weight	Specific Gravity		Sulfur Content by Weight	Sulfur Contribution at 1 lb./MMcf
		Gas (air = 1)	Liquid (water = 1)		
Tertiary butyl mercaptan	90	3.14	0.799	35.5%	0.25 gr./100 cf
Isopropyl mercaptan	76	2.65	0.820	42.0%	0.295 gr./100 cf
Normal propyl mercaptan	76	2.65	0.836	42.0%	0.295 gr./100 cf
Secondary butyl mercaptan	90	3.14	0.830	35.5%	0.25 gr./100 cf
Thiophane	88	3.07	0.999	36.4%	0.255 gr./100 cf
Dimethyl sulfide	62	2.16	0.85	51.6%	0.632 gr./100 cf
Ethyl mercaptan	62	2.16	0.84	51.6%	0.362 gr./100 cf
Diethyl sulfide	90	3.14	0.84	35.5%	0.25 gr./100 cf
Methyl ethyl sulfide	76	2.65	0.84	42.0%	0.295 gr./100 cf

* Courtesy of NGO Chemical div. Helmerich & Payne, Inc.

fifth of the amount recommended by the manufacturers under ideal test conditions. This means that the presence of the gas can be detected long before the lower flammability limit of the gas is reached.

This does not automatically eliminate any danger of an explosion, but it does allow the consumer early detection of small leaks before sufficient gas has accumulated to cause a major explosion.

Danger exists where failure of major appliance controls allow considerable amounts of gas to escape in a short period of time with no one present in the immediate area to detect the gas. Detection depends upon a human nose in the vicinity to identify the odor as a gas.

Odorization Failure

Being a chemical in small quantities, the odorant can be affected by other chemicals. It may be chemically unstable and break down or react with materials in the pipeline or the walls of the pipeline itself. The odorant may also be physically adsorbed or absorbed by the gas lines or containers prior to the leak. If the odorant is neutralized, it loses its effectiveness and gas may accu-

mulate until an explosive mixture is formed and a source of ignition found.

The mercaptans, the most active of the various odorants, are
the least stable and have the highest odor strength. They have an
affinity for oxygen and form disulfides, which are much more
stable. The disulfides also have the least strength of odor. If
enough oxygen is present in the gas system, the mercaptans can
be completely oxidized, destroying their effectiveness.

Soil Adsorption

When a leak develops in a buried pipeline, the gas, particularly
if under great pressure, will find its way through the interstices of
the soil to the surface and into the atmosphere. The escaping gas
can travel considerable distances (several miles) if the cover over
the buried pipeline consists of asphaltic pavement, concrete or
frozen earth, which forms an impermeable barrier to the transmission of the gas. The odorant will be absorbed or adsorbed by
the soil. The degree of adsorption is a function of the velocity and
volume of gas, texture of the soil, depth of the gas line and concentration of the odorant. A trench dug for pipeline burial is
usually backfilled with granular material. This provides plenty of
adsorptive surfaces and interstices through which the gas may
pass. In effect, the soil acts as a filter to remove the odorant.

The Personal Factor

As mentioned earlier, the presence of an odorant does not
guarantee that it will prevent an explosion. The odorant is pas-
sive and depends upon the human nose or some type of detector
to observe and identify its presence. Not everyone is aware that
gas has an odorant added, nor does everyone know or recognize
the odor of gas despite efforts to publicize it.

Many gas companies have spent thousands of dollars on mail-
ing postcards which have been saturated with the odorant. The
customer is instructed to scratch the surface of the postcard,
which will release the odorant used in their gas system. This cam-
paign was carried on for several years, but many persons are still
unaware that an odorant is added. Others may have an olfactory

sense incapable of identifying the gas. The fire investigator should question persons who claim that they did not smell any gas regarding their olfactory senses and ability to smell.

Gas industry personnel who are exposed daily to the gas and have their clothing saturated with the odorant have difficulty in detecting it when associated with a leak.

Therefore, the mere presence of an odorant in the gas source does not necessarily guarantee that an odorant was either in the escaping gas that caused an explosion or was capable of being detected by the victim of the explosion. Further, the victim may not have been aware that an odorant was contained in the gas. Therefore, should the question of an odorant in the gas arise, the gas should be analyzed to verify that an odorant was supplied in conformance with NFPA Standard 58.

LEAKING GAS SYSTEMS

If the origin of the fire has been located near a gas-fired appliance, or if an explosion preceded the fire, the next step would be to test the gas system for leaks.

Temperatures attained in a fire have a definite effect on the structural components in a building. In a residence fire, temperatures can reach 1800°F. Should the piping, whether it be steel or copper, reach these temperatures, the gas contained in the pipe will also reach these temperatures, which will increase the pressure within the pipe.

The weakening of the pipe from intense heat may approach the point where the pipe will blow out to relieve internal pressure. This is quite common in copper gas pipe but not in steel pipe. However, leaks do develop at joints due to stress relieving and internal pressure.

The predominant effect on the steel or copper pipe is to weaken the pipe so that the weight of the pipe and/or appurtenances causes the pipe to sag, straining the joints and causing leaks.

Pipe Joints

A threaded pipe connection differs from a nut and bolt in that the pipe threads are usually tapered. This means that as the joint is tightened and more pressure is exerted on the joint, more stress will be exerted on the components. The nipple, or inner portion of the pipe, is in compression while the outer portion, or coupling, is in tension. These two equal and opposite forces press against each other to hold the joint sealing compound in place.

Steel Joints

In steel piping, the joint is usually accomplished by screwing a tapered thread fitting into a corresponding tapered coupling or fitting. The threads of the male portion of the joint are coated with a sealing compound and the threads are engaged. A pipe wrench is used to tighten the joint until sufficient pressure is obtained and a secured joint is made.

Under fire conditions, the temperature of the steel is raised. As the temperature of the steel increases, its tensile strength decreases to approximately 50 percent of its original strength, as shown in Figure 6.

In addition, the piping compound is decomposed by the heat and no longer serves its purpose. Increasing pressure from the gas within the pipe will find a path out through the decomposing pipe sealing compound.

At the same time, the temperature of the steel will start to relieve itself from its original stressed condition, relieving the tension and compression in the original joint. After a fire, the joint may no longer be *gas tight*. It will lose its sealing compound and its tensile and compressive strength. In all probability, it will leak. The question then arises as to whether the leak was in existence before the fire. Currently, there is no method to determine whether the joint leaked prior to the fire. A microscopic examination of the sealing compound may reveal paths through the joint where gas has eroded a path through the sealing compound. However, this does not clarify the question of a leak existing prior to the fire.

Cause of Leaking Joints

Leaking joints can be caused by building settlement. Settlement exposes pipes to forces which may crack joints or welds (Fig. 40). The leak may be underground, but by testing the ground for the

Figure 40. This gas line leak was caused by weld failure. The natural gas migrated from this leak into a home, where the flammable mixture found a source of ignition, causing an explosion which demolished the home.

presence of gas, the leak can be quickly located. The local gas company will assist in locating gas leaks in the ground.

Pipelines are broken during construction activities. Excavating equipment breaks lines, causing leaks. When this equipment hooks a pipeline, it usually causes failure of a joint. The joint may be located some distance from the point of contact with the pipeline. This leak may continue to exude gas until an ignition source is found, resulting in an explosion. If unexplainable gas is detected, the construction activities within a one-half mile radius should be checked for a ruptured line and gas source.

Copper Fittings and Joints

Brass fittings which are threaded and treated the same as steel piping will have the same characteristics except that all changes will occur at a lower temperature. The stress-relieving temperatures of brass are between 400° and 800°F. It is more probable that a threaded brass pipe connection will leak after a fire than one made of steel.

Copper pipe may be connected in four ways:
1. Soldered connections
2. Compression fittings
3. Flared fittings
4. Screwed fittings

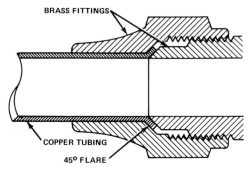

Figure 41. *Flared joint.* A flared joint is made by compressing the flared copper tubing between the two conical surfaces of the brass fitting, effectively forming a seal at the inside and outside contact with the flared tubing. Courtesy of The Weatherhead Company, Cleveland, Ohio.

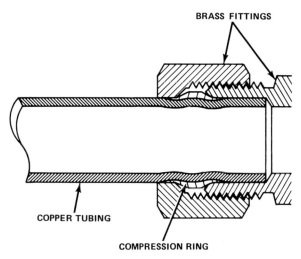

Figure 42. *Compression fitting.* A compression fitting is made by compressing the compression ring into intimate contact with the copper tubing and the mating surfaces of the brass fitting. Courtesy of The Weatherhead Company, Cleveland, Ohio.

Soldered connections are seldom used on gas piping but are more common on water pipes. The most common gas pipe connections are flared or compression fittings.

FLARED FITTINGS. In flared fittings, the end of a copper pipe

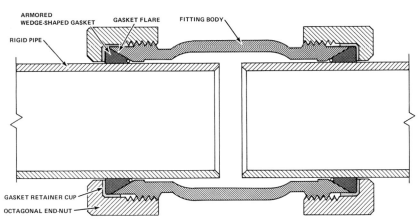

Figure 43. *Flexible compression joint.* In this joint, the seal is afforded by a confined wedge-shaped gasket which seals against the outer surface of the rigid plain end pipe and gasket flare in fitting body on both ends of the joint. This type of joint provides for angular deflection between pipe ends and repeated expansion and contraction between pipe ends without gasket slippage on pipe. There is no provision in the design of the product to resist longitudinal pullout forces. The end user is responsible for providing adequate pipe anchorage with the restraint offered by the soil or other means at the time of field installation. Courtesy of Dresser Manufacturing Division, Dresser Industries, Inc., Bradford, Pennsylvania.

is flared into a 45°angle to form a sloping shoulder. The sloping shoulder is clamped against a matched shoulder of the male portion of the brass fitting by a female portion which clamps the flared end between the male and female portions of the brass fitting. This forms a gas-tight joint by compressing the soft copper piping between the matched brass fittings. No foreign substance is required to seal this type of joint. However, if the end of the pipe is not squared before flaring and burrs are not removed, the matching faces will not form a perfect seal. This will leave gaps which will permit gas to escape. This should be found during testing and the joint refitted.

COMPRESSION JOINTS. A compression joint is formed by compressing a brass ring against copper tubing (Fig. 42). This forces the brass into the copper, forming a compression fit between the surfaces of the fitting and pipe.

Of the three types of joints, the flared joint is the most efficient and causes the least trouble in using copper tubing for gas.

Problems experienced with steel threaded joints also occur with brass or copper joints. The annealing or stress-relieving temperatures of the brass pieces are reached between 400° and 800°F. The longer the joint is exposed to high temperatures, the lower the temperature required to achieve the same stress-relieving results. When these temperatures are reached, the compression in the male end and tension in the female end are relieved so that the brass fittings relax and release the pressure on the surface of the copper tubing. The rise in gas pressure within the tubing will open an avenue of escape to relieve the pressure within the pipe. Therefore, the pipe will leak during testing.

FLEXIBLE COMPRESSION FITTINGS. To allow for limited degree of horizontal and angular pipe movement, a flexible compression joint is used, as shown in Figure 43. The joint is effectively sealed by compressing a confined flexible gasket against the outer diameter of the pipe and gasket flare in fitting body. This fitting provides a flexible connection between two plain end pipes. They are designed to absorb repeated expansion and contraction forces and movement so that the minimum of axial stress is transferred from one pipe length to adjacent pipe length, etc.

Gas Cocks

On gas cocks, particularly older models without springs to maintain constant torque, the nut which attaches the plug to the valve body may have been peened at the factory to prevent the nut from unscrewing. Should a service man decide that a valve is too stiff and loosen the nut, the peening will be upset. However, it will be obvious that the torque on the nut has been altered.

It could also be inferred that the valve had been disassembled and reassembled if a different type of lubricant is present. Therefore, the valve manufacturer should be contacted and the grease analyzed to ensure that lubrication is provided by the original factory grease.

When the nut is removed from the valve stem, a torque wrench should be used to determine the torque required to loosen the nut. Should this be less than factory specifications, the inference would be that someone had tampered with the valve between the time it was set at the factory and the time of the explosion. However, if the valve has been exposed to heat, it would reduce the factory torque setting and would be an invalid indication of the torque at the time of the explosion.

On valves equipped with a spring for constant tension, the torque on the nut is not critical as long as the spring is compressed. If the spring has been compressed, it will provide constant tension on the valve. If it has been compressed beyond the range of its compressibility, it would then be suspect. The tension on the spring can be tested with a torque wrench to determine if it meets factory specifications.

Regulators

Gas regulators are mechanical devices used to decrease pressure downstream from the regulator. They will normally give trouble-free service for years. However, like any mechanical device, they have their weaknesses.

Diaphragm Failure

One of the problems encountered with a gas regulator is diaphragm failure. The diaphragm is a flexible membrane providing movement for the regulating spring and lever. If the diaphragm develops a leak, it permits gas to travel to the atmospheric side of the membrane.

When regulators are installed inside a building, normal practice is to run a pipe from the atmospheric side of the membrane of the regulator to the exterior of the building to vent the gas. Without venting, a leaking diaphragm will allow gas to accumulate within the building and form an explosive mixture. Only a source of ignition would be required to cause an explosion.

If the regulator has not been destroyed by the fire, the faulty diaphragm is easily detected by testing the regulator. Caution should be taken in testing so that the test pressures are not in excess of normal pressures subjected to the regulator.

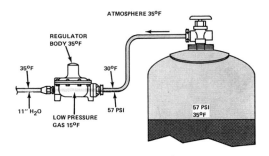

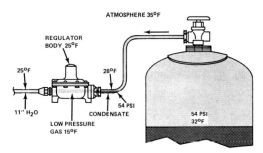

Figure 44. *Cause of regulator freeze-up.* As propane cools, it can carry less water. With an ambient temperature of 35°F, the gas gives off heat as it goes from the high pressure to the low pressure side, causing the temperature to drop to 30°F at the inlet and 15°F within the regulator. Should this continue for any length of time, as shown in the lower diagram, the endothermic action lowers the temperature of the gas as heat is lost in lowering the pressure. A condensate forms upstream from the regulator, and water is given up as the temperature decreases. This causes freezing at the orifice of the regulator. From *Plain Facts About Freezing Regulators,* Form MCK-1024. Courtesy of Fisher Controls Company, McKinney, Texas.

Freezing Regulators

Another problem which plagues regulators is the presence of water in the gas. A combination of atmospheric and internal pressures and temperatures of the LP gas will cause condensation and freezing at the point of discharge from the high pressure side to the lower pressure side of the regulator. This is graphically illustrated in Figures 44 and 45.

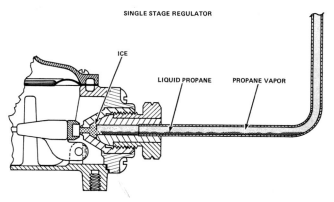

Figure 45. *Results of regulator freeze-up.* The condensate at the inlet of the regulator forms ice, which will cause regulator freeze-up unless the action of the regulator can break up the ice jam. In subfreezing weather, the ice can back up into the supply line, as shown in Figure 46, or the regulator can fail, as illustrated in Figure 47. From *Plain Facts About Freezing Regulators,* Form MCK-1024. Courtesy of Fisher Controls Company, McKinney, Texas.

If this occurs, ice will form in the regulator pressure chamber. The formation of ice at the inlet to the regulator causes failure of the regulator to reduce the pressure, permitting excess pressures into the low pressure system. This causes overfiring in the appliance and can cause a fire by an overheated appliance.

Out-of-gas Fires

An easily detectable cause of gas fires is the fire resulting from refilling an LP gas tank which has run empty. The NFPA has established procedures to follow when a service company supplies gas to a customer, particularly when the customer has "run out of gas." The inherent danger lies in appliances without 100 percent shutoff controls. When an appliance so equipped runs out of gas, the pilot light is automatically extinguished through lack of adequate fuel.

Should the pilot light or the main gas control malfunction when additional fuel is added and the control valve has not shut off the flow of gas, the flow will immediately start with the addition of LP gas at the tank. It is then only a matter of time until a flam-

Figure 46. When sufficient water is condensed from propane or butane, it can form at the orifice and back up toward the tank. This ice, in subzero weather, will remain even after the gas line and pressure have been released. The ice can be found by breaking the connections to reveal the presence of water in the propane.

mable mixture is formed and a source of ignition is found. Often it has been the owner of the appliance who strikes a match to relight the appliance.

The procedure established by the NFPA includes turning off all gas shutoffs to the appliances and the main shutoff to the house before refilling the tank. The tank is then refilled, and the main valve to the house can be turned on. The pilot lights are immedi-

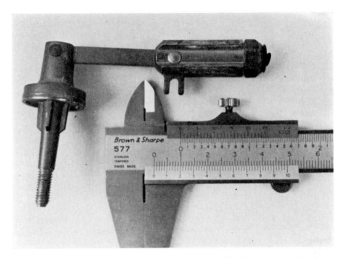

Figure 47. *Regulator failure.* This regulator failed to control the gas pressure when the seal (at the left end of the operating lever) stuck to the orifice, tearing the rubber seal so that it would not function properly. Testing of the regulator revealed a downstream pressure far in excess of normal. Examination of the operating lever revealed the cause of the malfunction.

ately relit, and the appliances are recycled to ensure that the controls are functioning properly, particularly with 100 percent shutoff controls.

It is easily deducible when an explosion or fire occurs from these circumstances. The customer runs out of gas, calls the local gas company, and the gas company fills the tank without turning off any gas cocks and without relighting the pilot lights. Failure of the service man to perform these necessary acts will usually result in an explosion if a valve malfunctions or if it does not have a 100 percent shutoff.

The worst explosions and fires will occur when the main control of an appliance is malfunctioning and permits large amounts of gas to enter a building. Time required will be whatever time is necessary for the escaping gas to form an explosive mixture and find a source of ignition. These factors can be determined from the violence of the explosion, the direction of the destructive forces, the relative location of the appliances involved and a time-flow rate of the orifices involved. A few simple calculations, which will

Figure 48. *Over-rich explosion.* When an over-rich gas mixture explodes, the resultant rolling flame front may scorch or ignite combustible materials, as witnessed by this scorched pipe insulation.

be covered later, are all that is required to determine whether a pilot light or main burner leak was responsible.

EXPLOSIONS

Overrich Explosions

A gas fire is usually preceded by an explosion. The explosion may be so minor that it does not leave evidence that an explosion occurred.

Enough gas could escape from a furnace burner to enable ignition to cause a minor explosion. The explosion occurs at the interface between the gas and the air. The balance of the gas in the room would be in an overrich condition. This means that the initial explosion would cause turbulence. This in turn would rapidly mix the gas with available air to cause a rolling fire which would heat any combustible substance near it. However, no violent explosion would occur. If this occurs and the explosion is not

Figure 49. *Over-rich explosion.* Turbulence caused by an over-rich gas explosion may ignite fires in unusual places, as seen in the arm of this sofa.

violent enough to rupture the gas lines supplying the furnace, the fire in the furnace could continue to burn until the external fire destroyed the electrical controls, shutting down the furnace.

Lean Explosions

On rare occasions, a lean explosion will rupture a gas line and undiluted gas will emanate from the ruptured end of the line, supplying large amounts of fuel to the fire. Should this occur, it may go unnoticed by firemen arriving on the scene. The firemen should be questioned to determine whether the fire diminished in intensity after the gas supply was shut off. This will indicate whether the fuel had a measurable effect on the fire. It will also be a clue in determining whether the furnace or appliances were being supplied with fuel.

The possibility exists that the tank ran out of gas during the fire. Check the remaining fuel content of the LP gas tank at the time of your inspection. It should also be determined who shut the tank off, when this was done and its effect on the fire.

The local gas company can provide information on the date of the last refill and the approximate rate of gas consumption by the appliances involved. Results based upon the quantity of gas consumed are unreliable. This is due to the fact that the BTU content of gas is high and the gauges on the LP gas tank are not accurate indicators of the remaining gas in the tank.

LEAK DETECTION

Once it has been established that gas is a possible cause of a fire, the gas line supplying fuel to the building should be checked. Since the gas line may have been exposed to the fire, it is dangerous to use gas as a testing medium. However, it may be necessary to use gas to test a connection near the entrance of the gas line to the building. This can easily be done by applying a test solution, such as a diluted solution of a liquid shampoo or a commercial leak detector. Commercial leak detectors have the added advantage of a low freezing point which enables them to be used in cold weather.

A convenient location to break the gas line is at a union. Before breaking the union, a soap bubble solution should be applied to the joint and the gas turned on briefly to find out if the joint is leaking. Once it has been established that the joint is not leaking, the joint can be broken and an air source attached for testing the balance of the line.

A convenient method of testing a gas line is to use an automobile tire pump, gauge and adapter (Fig. 50). A suitable gauge can be attached to the connections so that the original pressure in the line can be duplicated. Care should be taken not to exceed the original working pressure of the line, or it may open an otherwise gas-tight joint.

If the line will not hold air, it will have to be tested for any gross breaks in the line as a result of the fire. Should broken pipes be found, the break should be examined to determine if it was the cause of the fire. The broken pipes can then be capped and

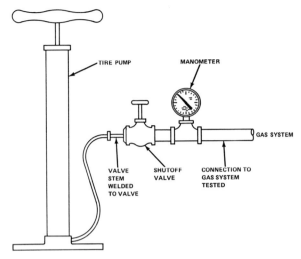

Figure 50. Leak-testing equipment including tire pump, shutoff valve and pressure gauge. From Earl A. Clifford, *Practical Guide to LP Gas Utilization,* 1969. Courtesy of Harbrace Publications, Inc., Duluth, Minnesota.

Figure 51. Testing for a gas leak can be dangerous if propane is used. A portable tank of breathable air can be substituted for the gas source in testing systems where extensive damage has occurred.

plugged and the line again checked. Portions of the line can be tested beyond the break so that the entire line is subjected to a pressure check. Photographs of the leaks should be taken showing the bubbles and relative amount of leakage as well as the leak location. A time test can be run to quantify the leak, using a chart similar to that shown in Figure 53.

According to the NFPA, the maximum gas pressure allowed inside a building should not exceed the pressures required to ensure safe operation of the equipment, plus an allowance for normal pressure drop. Most buildings are restricted to 11 inches water column (W.C.) for propane gas and 7 inches W.C. for natural gas. If appliances are located on the line and the control valves were leaking, the shutoff valve (if any) can be tested for leaks. If it is not leaking, it can be turned to the "off" position, isolating the leaking control valve. The leaking control valve should be tested at a later date. One leak does not preclude the presence of other leaks. The first leak may have been the result, not the cause, of the fire. The fire investigator should continue his examination and testing of the entire gas system, thus eliminating the possibility that another portion had a leak which started the fire.

This process will also establish whether the fire caused any leaks, as well as the extent of leaks. Once the origin of the fire has been established and all piping has been inspected, the appliances located furthest from the origin can be rapidly tested and eliminated as a cause of the fire. This will give an indication of the intensity of the fire and degree of destruction of control valves that are located closer to the origin of the fire.

A pressure gauge attached to the system will indicate a leak by a loss in pressure as test air escapes from the system. The rate at which the needle on the pressure gauge drops will reveal the leak rate. Testing should be continuous until all leaks are found and identified. An opinion should be formed regarding whether they were in existence before the fire.

Natural Gas Leaks

NFPA Bulletin 54 describes the methods in which gases enter a house from a broken pipeline. Natural gas subjected to 10

Figure 52. Sewers which contain no traps, or traps from which water has evaporated, become natural pathways for sewer gas to migrate into a basement.

pounds per square inch of pressure tends to rise from the pipeline and find its way through granular material used to backfill a trench. If the ground is covered by an impermeable material, the gas will follow this underside until it finds an avenue of escape. The gas can also be contained by frost, frozen earth, a blanket of ice and snow or silt deposited by flood waters.

All of these will contain the gas and force it to follow the underside of its containment. The gas will follow the line of least resistance, which is usually the course of the gas line in either direction. It will usually go uphill until it finds a path into a building where the sewer or gas lines enter. This point of penetration is usually not a gas-tight hole, due to the need to provide movement of the gas line due to expansion and contraction forces.

At this time, the gas, which has not had any great opportunity to mix with air, will enter the basement and mix with air as it enters. This may immediately form a flammable mixture which will change the density or specific gravity of the mixture. The flow of the flammable mixture will be controlled by convective forces in the building. The source of ignition is the only remaining factor required to cause an explosion and fire.

LP Gas Leaks

Since LP gas is heavier than air, it is usually found in containers located outside of a building and sitting on pads on the ground. LP gas containers are seldom buried in the ground. Therefore, should a leak develop at the tank, the gas would fall to the ground, mixing as it traveled, and follow the contours of the ground very much like water. Should the gas find an opening into a basement, it would fall into the basement and mix as it entered. Once it attained access to outside air, the wind or natural drafts would govern its rate of mechanical mixing with air to form a flammable mixture. Then, all that is required is a source of ignition to cause an explosion and fire.

CONTROL MALFUNCTIONS

The most common cause of gas fires is a leak in the gas system caused by a malfunctioning control. Controls may be classified in two categories. *Safety shutoff controls* regulate the flow of gas to the main burner. This is usually regulated by a thermostat which controls the opening and closing of the main valve supplying gas to the burner. A pilot light is normally used to supply ignition to the main burners, but on older valves, *no safety shutoff* is provided for *pilot light failure*. Without this feature, if the pilot light is out and the thermostat calls for heat, the malfunctioning control would pour gas into the combustion chamber, but no source of ignition would be available. The gas pouring into the combustion chamber would eventually overflow into the room until it found a source of ignition, causing an explosion and/or fire.

The other type of appliance control is a *100 percent shutoff control valve*. This means that in the event of a flame-out on the pilot light, the valve would automatically shut off the flow of gas to both the main burner and the pilot light. Therefore, should the thermostat call for heat while the pilot light is out, the automatic control would prevent a flow of gas to the main burner and the pilot light.

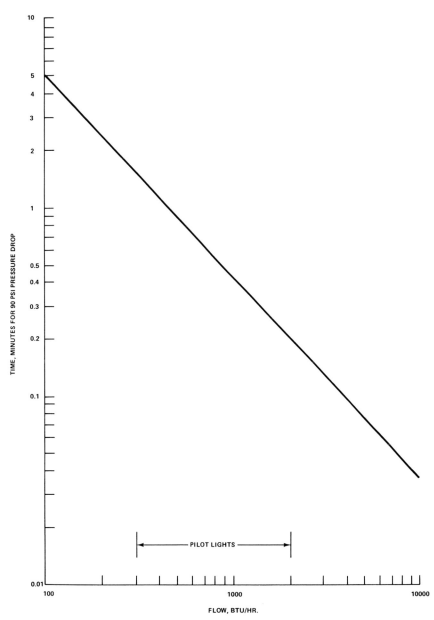

Figure 53. *Flow Rate Chart.* A simple flow rate chart can be constructed by timing pressure drops through various orifice sizes. This chart can then be used to compare the rate of leaks with the use of test equipment such as shown in Figures 50, 51 and 54.

Figure 54. To test the flame characteristics of burners, the appropriate gas must be used. This can be done by attaching a single-stage regulator and the appropriate valves to a small gas bottle and then to the appliance being tested.

Pilot Light Failures

The pilot light on an appliance is a small flame which burns continuously and supplies ignition to the main burner. The pilot lights burn various quantities of gas, depending upon the size of the appliance and the requirements of the pilot light.

In recent years, the gas appliance industry has manufactured gas ranges without 100 percent shutoff on the pilot light. The size of the orifice was reduced in output so that the rate of diffusion of the gas by mechanical mixing and molecular diffusion would not be capable of producing a hazardous mixture. Figure 53 shows the relative flow of propane in terms of BTUs per hour. It should be noted that below the range of 500 BTUs per hour, no shutoff is required, as indicated in NFPA Code 54 (National Fuel Gas Code).

Determination of Flow Rate

After a leak has been detected, it is vitally important that a determination of its flow rate be made. Figure 54 is a photograph of apparatus which may be of assistance in making this calculation. In testing the system, a gauge should be used so that the working pressure is not exceeded. This gauge can be used to make a rough determination of the flow rate.

This is easily done by using a stopwatch and a set of orifices. Once the gas line has been checked and all leaks have been eliminated, the orifices can be added onto the line and the pressure raised to working pressure. Then the orifice is opened and the time required to bleed the line is timed with a stopwatch. Knowing the capacity of the orifice will give the time required for gas to leak in terms of BTUs or cubic feet per hour. By changing the orifice, a chart, such as Figure 53, can be drawn against the time it takes to bleed the system. This should be done until the time becomes so short that it is immeasurable, usually an indication of an open line or a broken line with full access to the atmosphere. By isolating each leak in the piping system, the time can be compared with the leaking time for the various orifices and an approximate flow rate can be determined.

Figure 54 shows another method involving the same technique. It consists of a propane tank connected to a high pressure regulator with a gauge. This passes into a lower stage regulator, also equipped with a gauge. The same technique is used, and a calibrated curve is set up for the larger dial as well as the smaller dial. The addition of the second gauge and the much higher pressures involved in the high pressure storage tank give the system a larger capacity for measuring flow rates.

One word of caution: In setting up the test system, the source of propane should be isolated from the system to ensure against leaks. If this is not accomplished, it is conceivable that a leak in the shutoff valve from the propane supply could exactly equal the leak in the system so that the gauge would show a level reading for a substantial time period. An acceptable test period to assure that no leak exists in a system would be fifteen minutes without any drop in the pressure gauge reading.

Test Media

When testing extensive systems, particularly where the possibility of another explosion exists, air should be used as a test medium to minimize this danger. Pipes being tested may contain substantial quantities of the flammable gas which they carried before the fire. Therefore, even if air is being inserted into the line, until all lines are purged of their original gas, danger of an explosion still exists.

When using air as a test medium, the test pressure for air should be adjusted to the equivalent pressures supplied by the original gases.

For example, if a pressure regulator and air are being used for the test medium, you would obtain a reading of 16.8 inches W.C. downstream from a low pressure regulator which would ordinarily register 11 inches W.C. if propane were being used. This is because the pressure exerted by a gas is a function of its specific gravity. A heavier gas would register a lower pressure than a lighter gas working against the same spring.

To obtain the equivalent test pressure required to simulate the original effect on the regulator exerted by the flammable gas, it is necessary to multiply the normal operating pressure by the specific gravity of the gas. The temperature and pressure of the test medium should be recorded so that a comparison of the conditions existing at the time of the fire can be made. If substantial differences in temperatures are found between testing and the time of the fire, the test pressures can be adjusted to correspond to the pressures existing before or after the fire, which would compensate for the difference in temperatures and pressures.

For example, if a natural gas system is being tested with air, the normal gas pressure (7 inches W.C.) would be multiplied by the specific gravity of the natural gas to obtain an equivalent air pressure of 3.9 inches W.C.

Table XIX lists the equivalent test pressures for three common gas fuels. For example, if air is run through a pressure regulator set for 11 inches W.C. for propane and a gauge connected to the downstream side of the regulator, the gauge would read 16.8 inches W.C.; this would be the equivalent of using propane as a test medium. Do not readjust the regulator to deliver 11 inches W.C. If propane is used as the test medium, the gauge will register 11 inches W.C.

TABLE XIX

EQUIVALENT AIR TEST PRESSURES*

	Air	Propane	Butane	Natural Gas
Specific gravity ·················	1.00	1.52	2.02	0.56
Normal pressure ·············	—	11″ W.C. 10 psi	11″ W.C. 10 psi	7″ W.C. 10 psi
Equivalent air pressure ········	—	16.8″ W.C. 15.2 psi	22.2″ W.C. 20.2 psi	3.9″ W.C. 5.6 psi

* From *Rego LP Gas Serviceman's Manual*. Rego, Chicago, Ill.

The same principle applies to the volumes of gases. When air is used as the test medium, the quantity of air should be multiplied by the factors shown in Table XX to obtain the equivalent flows of the three common gases. For example, if the flow of a pilot burner is measured at 0.56 cubic feet per hour and you wish to determine the amount of leaking propane through the pilot orifice, multiply by 0.81 and obtain 0.45 cubic feet per hour flowing through the orifice.

Test Conditions

The test conditions should duplicate conditions existing at the time of the leak which preceded the fire. The flow of gas is a function of the pressure and temperature. Both of these factors play an important role in any testing procedure and should be given careful consideration.

Test Pressures

Before applying a testing medium at any pressure, the pressure existing prior to the fire should be determined and duplicated. Various piping handbooks, codes and gas manuals give testing procedures to be used for the detection of leaks. However, these are preventive measures taken to discover the existence of a leak to be eliminated. The pressures used in this type of testing are usually far in excess of the normal working pressure.

It is possible that if an excessive working pressure is used, this excess pressure could cause a new leak. These test pressures used to determine the presence of a leak may be five times the normal working pressure. Codes which recommend test pressures in excess of working pressures should not be used as a criteria for establishing test pressures for determining the presence of leaks

TABLE XX

EQUIVALENT AIR TEST VOLUMES*

Air Flows	Equivalent Gas Flows	
1 CFH†	0.81 CFH	Propane
1 CFH	0.71 CFH	Butane
1 CFH	1.29 CFH	Natural Gas

* From *Rego LP Gas Serviceman's Manual*. Rego, Chicago, Ill.
† CFH—Cubic Feet per Hour.

in a fire investigation. Under no circumstances should the normal working pressures ever be exceeded in a fire investigation test.

Care should be taken in applying pressure on the upstream side of a regulator in the event that the regulator is inoperable. Should this occur, the upstream pressure would be passed downstream and exceed the working pressure. If this happens and a control valve is subjected to excess pressure, the gaskets could rupture and cause new leaks. It is best to isolate the control valves until the line has been completely tested for leaks.

TESTING THE CONTROL VALVE

Control valves are designed for specific pressures and uses. If this pressure is exceeded during testing, the gaskets might fail and the springs which normally close the cutoff valves could open, giving an indication that the valve is malfunctioning. Therefore, it is vitally important that the pressures existing at the time of the fire, or prior to the fire, be the pressures used in testing, *not* factory test pressures.

External Examination

Before testing the control valve, its external characteristics should be examined and noted. The possibility of damage by heat should be evaluated. Damage to the valve should be determined regarding its exposure to radiant, convective or conductive heat. Any foreign material which has adhered to the valve should be observed and classified. These substances could include water from the firemen's hoses, dry chemical from a fire extinguisher or debris which fell on the valve as a result of the fire. These substances should be identified and the effect upon the valve recorded.

If possible, the valve should be tested in its original location to detect external leaks. If none are detected, the valve should be photographed in place and removed to a laboratory for testing the internal function and leaks.

Before removal of the valve, an examination of the screws, gaskets, control knobs, levers and external wiring should be carefully made and documented. The examination should detect any possible tampering with the valve since its installation. Signs of recent work on the valve would include replacement of screws, gaskets or wiring with new materials. These are readily identified if the valve has not been subjected to heavy trauma from the fire. The age of the patina on the screws, gaskets and external wiring should be the same as that of the remainder of the valves and surrounding controls. Comparison of these will usually reveal any recent work completed on the controls associated with the apparatus.

Testing for External Leaks

In order to test the control valve, it is necessary to connect the valve to a constant supply of air at the proper pressure. The pressure should duplicate the pressure existing at the time of the fire as closely as possible.

Although the valve has been tested in place at the fire scene, it should be retested by closing all outlets from the valve to detect any external leaks that might have developed from removal to the laboratory. If a new external leak is found, it should be duly recorded and the cause of the leak ascertained to learn why it was not apparent at the fire scene. This testing should be completed without touching any controls and with as little disturbance as possible in closing off the valve outlets.

With the control levers of the valve in their original position, the next step is to check the automatic shutoffs, if any, to see

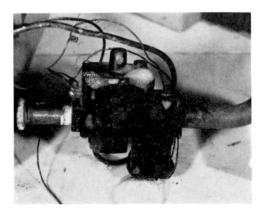

Figure 55. Testing control valves is done by applying a soapy solution to the joints where leaks could occur. This includes the various gaskets on a control valve. This photograph shows leaks caused by total destruction of the gaskets on a control valve.

whether they functioned. For normal residential and some industrial applications, as many as three outlets could be located on the valve body. These would include a bleeder for the gas-controlled valves, a safety pilot line and a main burner line. Assuming that a leak is found, the source can be isolated by closing all three outlets and opening them one at a time. This will reveal which line is leaking. In valves which have been exposed to high temperatures, it is quite possible that all three valve seals have been destroyed by the fire and all three will leak.

BLEEDER TUBE LEAKS. The mere presence of a leak does not automatically identify the origin and source of the fire. A good example of this is a bleeder tube leak. Should a bleeder tube leak at its maximum rate, it would be unlikely that it could have been the cause of the fire. The volume of gas emitted from a bleeder tube is so small that the natural convective currents would dissipate the gas to a nonflammable limit faster than the gas could accumulate. Only in rare circumstances could enough gas accumulate from a bleeder tube to provide enough energy to start a fire. These circumstances would include a tightly sealed room with combustion air obtained from the outside confines of the room. However, this possibility should never be overlooked, no matter how remote the probabilities.

PILOT LINE LEAKS. Pilot orifices seldom pass more than 1,000 BTUs per hour or less than ½ cubic foot of propane per hour. In sealed rooms, or rooms where a very low rate of air change occurs, the leaks from a pilot line accumulate sufficient quantities of gas to cause an explosion and fire.

The rate of accumulation is a function of the convective currents in the room and the rate of air exchanges, as well as the rate of flow of the gas from the pilot burner. Pilot burners normally pass air through an orifice which immediately forms a flammable mixture in the range of 5 percent. Once the combustion chamber of the furnace has cooled and a natural draft will no longer carry the flammable mixture from the chimney, the flammable mixture can accumulate within the combustion chamber. The mixture can back out of the entrance where the primary and secondary air was drawn into the combustion chamber. This can occur only if the combustion air is taken from the room which contains the furnace. If the combustion air is brought in from the outside, the gas mixture is confined to that area unless a leak exists between the combustion chamber and the furnace room.

While it is not impossible for a pilot light to provide the necessary gas for explosions, the probability is low and much less if another source of combustion is contained within the same room. For example, if a water heater is located in the furnace room using the same type of gas as the furnace, it may take combustion air from the same area as the furnace. If the furnace is off, it no longer uses combustion air. This means that a reversal of combustion air flow can occur in the furnace. Leaking gas from the furnace, mixed with the combustion air, can be drawn into the combustion chamber of the water heater where it will be ignited, causing a flashback to the furnace which may reignite the pilot light if the explosion is not too violent. However, if the explosion is violent it could ignite combustible materials as the flammable mixture is ignited by the pilot light or main burner flame of the water heater.

In the absence of a constant source of ignition, such as a gas water heater, an electric water heater can also supply the necessary source of ignition for the flammable mixture from the furnace or pilot light. An electric water heater will strike an arc as circuits are closed to supply heat to the water heater element. This can be the necessary ignition spark for the flammable mixture emitted by the furnace.

Although the probabilities are greater that a pilot burner leak can cause a fire or explosion, the great majority of gas control malfunctions are caused by the gas from the main burner. Normally it would take from six to twenty-six hours to fill a basement with a flammable mixture from a pilot burner, but it would take only a short time for a main burner on a furnace to supply enough flammable mixture to blow the house apart.

Leak-limiting Orifices

Section 1.3.1.13, *Venting of Gas Appliance Pressure Regulators,* Volume 4, 1976 Edition of the National Fire Codes, states that the gas appliance pressure regulators which require atmospheric pressure on one side of their diaphragm for the proper function of the diaphragm shall be equipped with a vent pipe leading outdoors, or into the combustion chamber to a constantly burning pilot light, unless constructed or equipped with a vent which limits the escape of gas through the vent opening in the event of diaphragm failure. The allowable flow rate should not be greater than those indicated in Table XXI.

Table XXI reveals that the National Fire Protection Association permits 1 cubic foot of undiluted LP gas per hour to flow into an area, regardless of its size or airtightness. It also permits 2½ cubic feet per hour of natural gas to flow into an area of unknown configuration. The presumption is that the gas will diffuse faster than it can accumulate. Practically, it will never reach the lower limit of flammability.

These flow rates would be 2500 BTUs per hour for propane, 3280 BTUs per hour for butane, and 2600 BTUs per hour for natural gas.

Main Burner Leaks

When the main burner on a central gas furnace leaks, the quantity of gas can range from 18,000 BTUs per hour (7.2 cubic feet per hour) to 100,000 BTUs per hour (40 cubic feet per hour). At these flow rates, the minimum amount of propane required to demolish a typical residence is quickly available.

TABLE XXI

MAXIMUM FLOW RATES OF LEAK LIMITING ORIFICES*

Gas Pressure Regulator	Test Inlet Pressure Inches W.C.	Maximum Gas Flow Rate, CFH†	
		Nat. Mfd., Mixed, LP Gas-Air	Undiluted LP Gases
Listed class I	10.5	2.5	1.0
Listed class II	21.0	2.5	1.0
Unlisted	21.0	2.5	1.0

* From *National Fuel Gas Code*, ANSI Z223.1/NFPA No. 54-1974, Table 1-3. Courtesy of the American Gas Association, Cleveland, Ohio, and National Fire Protection Association, Boston, Massachusetts.
† CFH—Cubic Feet per Hour.

A leak through the main burner orifice will mix with primary air going through the mixing chamber and secondary air in the combustion chamber. At that point, the flammable mixture can exit either through the chimney, if a draft is in operation, or through the entry for the combustion air. In so doing, the flammable mixture can be redrawn through the intake area for primary air. The primary air now consists of the flammable mixture that is again mixed with 100% propane coming from the orifice. This process can continue until an overrich mixture is formed by the continuing enrichment of the air. Not all of the air is recirculated; each cycle will spill some of the premixed air into the furnace room.

This enrichment accounts for the numerous fires which occur as the mixture of propane and air is emitted from the furnace, emanating into a room in decreasing percentages of propane. Once this mixture is ignited at the lower limits, the resulting explosion will ignite the overrich mixture, causing burning at the interface of the overrich mixture and the leaner air front.

Simulating the Furnace Control Failure

If a leak is found in the control and isolated, it should be dismantled to determine what physical malfunction caused the leak. If no leak is found, it is necessary to check the control for a malfunction. A step-by-step record should be made of the procedure so that the controls can be retested to verify the malfunction.

To check the furnace control, the history of the control prior to the fire should be obtained regarding its use or misuse. Questioning should be directed to the person who last started the furnace or apparatus as to the procedure followed; it should be duplicated in simulating the starting of the furnace. Testing should be done with air. Electrical controls, such as thermionic generators, flame sensors and other electronic devices, should be tested by duplicating the currents necessary to activate the controls.

Examination of the electronic controls is also part of the procedure. An ammeter can measure the *hold-in* and *dropout* milliamperages of the electronic controls. This will establish whether the application of the proper amount of current is operating the valves or relays which open or close the safety, bleeder or main burner valves.

A step-by-step procedure following the manufacturer's recommendations and the actual procedure of the operator should be performed. Any deviations from the recommended or specified currents, resistances or amperages should be recorded, as well as the reaction of the control valve in terms of gas flow.

Each of the three systems (bleeder, pilot and main burner lines) should be tested independently, simulating the necessary electrical or electronic steps required to open the valves. If a malfunction is found, the next step is to dismantle the valve to determine the cause of the malfunction.

Random Malfunctions

If no malfunction of the valve is found, it does not eliminate the control valve as the cause of the explosion or fire. Random malfunctions (one-time occurrences) do happen. These cannot be duplicated in the laboratory. A common one-time malfunction of the valve is caused by buildup of foreign material on the valve seats which, when the valve is closed, permits gas to leak. The resultant explosion will jar the valve and valve seal sufficiently to disrupt the foreign material, causing the valve to effectively close.

Other causes of random malfunctions include a valve stuck in the "on" position. This occasionally occurs when the valve surfaces, which are normally lubricated, become dry and cause the valve to stick. Quite often the presence of this type of sticking can be detected by internal examination of the valve.

Internal Examination

Once it has been determined necessary to examine the interior of the valve, drawings or blueprints of the valve should be obtained for study. This will enable the investigator to determine the physical and mechanical operation of the valve so that it may be successfully opened. It also is an aid in determining possible causes of malfunction and the areas most likely to malfunction.

If possible, a duplicate valve should be obtained and a step-by-step procedure for dismantling it should be established. Only then should an attempt be made to dismantle the suspect valve. Before dismantling the valve, the screws, gaskets and external wiring should be examined for recent work performed on it, and if so, that it meets factory specifications. Then, following the outlined step-by-step procedure, an examination can be made. Photographs should be taken during each step, and a careful inventory of essential parts should be taken and checked against the factory specifications.

Any foreign material should be photographed and carefully removed for identification. These substances can build up on the lips or edges of the rubber, neoprene or plastic seals commonly used for valve faces and prevent the valve from sealing tightly.

The electrical connections on solenoids, electromagnets or other electrical connections should be carefully examined for tight contacts. The surfaces of switches or switch contacts should be inspected to determine their function and operational ability. These can be visually checked by applying proper voltages to the control and observing its function. Once the valve has been opened and thoroughly examined, it should be reassembled in the reverse procedure followed in the opening of the valve. After closing the valve, it should be retested to see that it is functioning or whether the breakdown and reassembly has caused a malfunction.

GAS MIXING WITHIN THE FURNACE

Under conditions of normal combustion and aspiration of gas in a furnace, the gas is mixed with combustion air. The air is mixed in a mixing chamber and exits through the ports, where it is ignited by the pilot light or previous combustion. A stable burning flame establishes itself at the tip of the burner so that the velocity of the incoming gas is equal to the burning velocity of the flame and all gas is consumed at that point or in the combustion chamber. The heated gas rises and exits through the normal combustion heat exchange chamber and out through the flue.

When gas failure occurs, gas enters the combustion chamber and absorbs heat from the heat exchanger walls. This cools the heat exchanger and combustion chamber. It slows the natural draft until air entering the combustion chamber is as cool as the exterior air. The draft in the chimney or flue is stopped by the increased density and drop in temperature of the gas, perhaps assisted by a downdraft. At this point, the gas dynamics of the system change radically. If free communication exists between the combustion chamber and outside air, the gas mixture in the combustion chamber can enter the room.

It is necessary to separate the two types of combustion air by function. *Primary* combustion air is mixed with the gas as it ejects from the orifice into the mixing chamber. At that point, primary combustion air is drawn into the mixing chamber and the mixed gas, which is within the flammable range of the gas, is emitted from the ports.

Secondary combustion air is required for complete combustion of the gas and is normally mixed with the burning gas as it emits from the burner port.

Once the draft has stopped, secondary air flowing into the combustion chamber may slow and reverse its direction. As the gas mixture emitted from the ports fills the combustion chamber, it builds up pressure, causing back pressure on the normal flow of the secondary combustion air. This allows a flow of the gas mixture, which is available within the flammable range, to circulate back to the intake of the primary air and reenter the combustion chamber. This mixture, already within the flammable range, is recirculated through the mixing chamber and further enriched. During this same time it spills mixed gas into the room where the secondary combustion air originated. This process can continue until an overrich mixture exists which precludes an explosion within the combustion chamber. In the meantime, however, a mixture of gas and air in the flammable range, and being enriched, is being forced into the room through the intake provided for combustion air.

Stratification

It is erroneous to conclude that since LP gases are heavier than air, if an LP pipe broke, raw gas would flow like water into the room. If this were true, 100% gas would be found at the bottom of the room (too rich to burn), and 0% gas at the top of the room, with the explosive limit of the gas existing within several feet of the floor. However, mechanical mixing prevents absolute *stratification* or *layering* of the LP gas.

It is impossible to introduce large quantities of LP gas into a room without inducing turbulence at the point where the gas leaves the pipe. This turbulence causes mechanical mixing, which increases the rate of diffusion by enlarging the interface between the raw gas and the air. The mixing rate is a function of ventilation, temperatures of the gas and room air temperature gradiant between the top and bottom of the room.

IGNITION

After the origin of the fire has been determined and a leak in the gas system or control has been found, the next step is to identify the source of ignition.

A flammable mixture, or raw gas, will not start a fire by itself; an ignition source is needed. The minimum energy requirement for ignition is 0.25 millijoules. It takes very little energy to ignite a flammable mixture. The spark from a light switch can supply sufficient energy to perform the act. It is usually impossible to isolate an absolute source of ignition, especially where more than one possible source is available.

For example, when a furnace control malfunctions and the furnace is located near an operating gas-fired water heater, it can be assumed that the water heater could be the source of ignition. Whether the water heater was operating on pilot or main burner, it would draw air from the furnace room (and thereby escaping gas from the furnace) for combustion. The only requirement is that the furnace and water heater have a common air source.

The source of combustion air for each appliance should be checked because quite often the source is not the room that contains the appliance but a stack or duct from the outside. This is particularly true in mobile homes.

Another source of ignition is open windings on electric motors with brushes which give off sparks. It must be proven that the motor was operating at the time of ignition. It is embarrassing for an investigator to state that, in his opinion, the electric motor was the source of ignition, only to have the opposing expert prove that the electrical cord supplying energy to the motor was unplugged.

In every gas fire or explosion, it is obvious that ignition occurred. However, the source of ignition should not be dismissed as obvious but should be determined with a high degree of probability. Each possible source should be proven by determining that the appliances were operating and were the most probable source of ignition.

The source of ignition does not necessarily need to be located in the area where the explosion occurred. For example, when a valve malfunctions and a furnace room finds no ready source of ignition, gas will continue to pour from the main burner. The furnace will fill and the gas will eventually migrate upstairs as a result of convective currents until a source of ignition is found. The ignition will light a trail of gas which will flash back immediately to the main source in the basement, causing an explosion and fire, due to the overrich condition.

CORROSION OF COPPER GAS LINES

Corrosion of copper gas lines occurs when a high sulfur content fuel or odorant reacts chemically with the walls of the copper gas lines. The sulfur in the gas, regardless of its source, reacts with the copper to form copper sulfide. The interior of most control valves will accumulate small quantities of sulfide. By tapping a copper gas line, the copper sulfide will loosen from the walls.

The gas control valve industry has known of this problem for many years. The size of the scale which builds up is large enough to plug a pilot orifice. To prevent this from occurring, the industry uses an aluminum pilot line since it does not react with the sulfur. For further precaution, the gas valve industry provides a filter in the valve upstream from the aluminum line leading to the pilot light. No such filter, however, is provided in the main stream of gas leading to the main orifice. The industry depends upon the construction of the valve and the shape of the main cavities in the control valve to act as a trap to collect the larger flakes of copper sulfide.

Sulfides accumulate in the copper pipeline and periodically break loose. These are carried into the main control valve by the gas stream and are sufficient to cause a substantial leak by preventing the valve from closing.

BLEVES

A BLEVE is an acronymn for a "Boiling Liquid-Expanding Vapor Explosion." BLEVES commonly occur when an LP gas container is exposed to excessive heat. The heat flux must be greater than the capacity of the relief valve. As a result, pressure within the tank rises until the metal, weakened from the high temperatures, fails. This sudden release of internal pressure of the vaporized gas has a reaction similar to that of a rocket. The expanding gases pushing from the interior of the halves act as a rocket; the container ends may be propelled for considerable distances as the pressure is released.

Ignition of the vapor is immediate, although the liquid phase

Figure 56. When a propane tank car is subjected to external heat, the resultant pressures will often blow out the end of the tank car. The larger portion will take off like a rocket, sometimes propelling the tank car 400 feet, as shown in this photograph. The car originated on the other side of the railroad embankment after being derailed and exposed to a fire caused by other fuel tanks burning.

of the LP gas may fall to the ground as a cold spray, due to the refrigeration effect of the expanding gas.

The effect of a BLEVE on railroad tank cars, which carry up to 30,000 gallons of propane, are spectacular since the long end of the tank car resembles a rocket in size, shape and weight. Figure 56 is an illustration of the tremendous force generated by a BLEVE.

CHAPTER 8

Explosions

THE DEFINITION of an explosion is difficult to formulate, as it is hard to differentiate between fast-burning fires and explosions. An explosion is normally considered to be any oxidation exceeding a given rate. The difference between explosions and detonations is based upon the rate and pressures produced.

The clues to look for in an explosion are any physical evidence of forces exerted on structural components of the building. This would include broken glass located some distance from the building. The glass could have been shattered from the fire; however, if an explosion preceded the fire, the glass will be found lying outside of the window some distance from the building. The significant clue will be that if the explosion was the start of the fire, no smoke staining will be found on the glass. The glass will be perfectly clear, broken and will be on the outside of the building. The distance will depend upon the magnitude of the explosion.

In a small explosion, the glass will be the only substance found any distance from the building. If the explosion occurred as a result of the fire, smoke staining on the glass will occur. Therefore, the first clue to look for is whether the glass is clear and the distance it traveled. If clear glass is located a considerable distance away, then undoubtedly an explosion preceded the fire and other clues should be examined.

The next clue is evidence of wall movement in the building, depending upon the explosive medium. Natural gas, being lighter than air, would form a flammable mixture closer to the ceiling, or in the upper levels of the building. LP gas will gravitate toward the basement, forming an explosive mixture above the basement floor. The basement windows could be blown out. If the gas has been in the building for some time, it could dissipate and a more violent explosion could occur upstairs due to the over-rich density of gas in the basement. The overrich density of this gas will be accompanied by a rolling fire front as the overrich gas mixes with

224

available air and burns rapidly rather than exploding.

If the gas mixture is lean, no fire may occur because the fuel will have been consumed by the explosion. The velocity of the gas will be greater in a lower limit explosion than in an upper limit explosion.

If other explosive substances are found, such as any member of the dynamite family, an explosives expert should be called. Bombs are beyond the scope of this book; the concern here is with accidental explosions.

Having determined that an explosion did occur, the next step is to determine the origin of the explosion. The primary objective is the determination and identification of explosive materials present. For example, in an all-electric building, no gas would be available. Therefore, the explosion would have been caused by another explosive material.

The point from which the explosive source originated has to be established. If the explosion was violent, this can easily be accomplished, particularly if it was in a basement. The windows can be checked. If all windows are blown out and the glass is blown for an equal distance, then it is most likely that the explosion occurred there. If no upstairs windows are broken, look for pieces of furniture or objects located in the basement to determine whether they have moved and the direction of the force which moved the objects.

The next step is to determine what in that area could have caused the explosion. For example, an open gas can could have permitted the evaporation of gasoline, forming an explosive mixture with enough energy to blow out the basement windows.

Other sources of explosives could be propane cylinders. Evaporation could occur from gas tanks for which the cap had been left unsealed. Paint thinners are another source of explosive mixtures, especially lacquer thinners. Any of these could have been the source of an explosion.

Some of the more exotic causes of explosions are listed under incendiary fires and will not be repeated here. They consist of substances which by themselves are relatively harmless but when combined form a hazardous mixture which will explode when subjected to excessive heat or when triggered by an explosive cap.

DEFINITIONS

In Chapter 3 it was pointed out that oxidation occurs at varying rates. Under controlled conditions, the burning velocity of a gas, for example, is a function of the air-fuel ratio and the pressure at which the gas is burning. Under normal conditions, the velocity of the burning gas as it escapes the port of a burner is around 1 mph, or 1.5 feet per second. In a free-burning fire, the burning velocity of the flame can vary from less than 1 mph to tornadic winds which are not uncommon in fire storms.

As the chemical reaction of combustion occurs, it has been shown that the reaction accelerates exponentially with the temperature of the fire. In an explosion, the rate reaches unprecedented velocities.

In order to understand this phenomenon, let us consider a typical water heater burning at the rate of 30,000 BTUs per hour, using propane as a fuel. The burning velocity of the flame front is approximately 1.5 feet per second. In one hour, the propane burner has consumed 30,000 BTUs.

Let us assume that the burner was unlit and for one hour poured a flammable mixture into a room without ventilation so that the entire contents of the room contained 30,000 BTUs of a flammable air-propane mixture.

Now let us introduce a source of ignition to this air-propane mixture. The source of ignition can only occur at one point in the room. An example would be an electric motor which started up, providing an arc which ignited the mixture. The ignition of the propane would propagate from the point of ignition to all points of the room within a fraction of a second. In other words, all of the energy contained within that flammable air mixture will be converted into heat in the space of a microsecond. It will be accompanied by expansion of the gas contained in that room. As the burning occurs, pressure will build up within the room until all of the gas has been converted into a hot flaming mixture. The sudden increase in pressure would exert a force against the confines of the room. This resultant force and pressure rise, due to the instantaneous burning of the gas, is termed an *explosion*.

Let us assume that the propane burner, instead of leaking premixed gas, emits pure propane from a broken propane line. As the raw gas is ejected from the broken pipeline, some of the gas would mix with air, forming a flammable mixture. At the interface of the mixture with room air, the gas would form a lean mixture. After an hour or so, a variable mixture would be found at the lower limit of flammability. Closer to the broken line, it would exceed the upper limit of flammability and the mixture near the break would be too rich to burn.

Now let us assume that the leading edge of the gaseous cloud reached a source of ignition. The lean mixture would not burn, but the lower limit mixture would immediately ignite, causing an explosion. This would result in turbulence of the air in the room and mixing of the overrich gas with air in the room. The resulting dilution of the overrich mixture would continue to feed the exploding mixture so that a *rolling explosion* would occur. This explosion may last several seconds as the remaining overrich mixture is involved by turbulent mixing of the gas and air.

In the first case, the duration of the explosion and length of time the hot gases were exposed to the atmosphere may not be long enough to ignite anything but the most highly combustible material (see Fig. 49).

In the second case, however, the duration of the explosion and the rolling flame front would be longer and have a higher heat content. This would ignite more material and be more apt to sustain a fire.

Let us assume that the same explosion occurred outdoors. Without any resistance, the expanding gas would push against surrounding air, compressing it and making more room for the expansion of gases. Without walls to obstruct gas movement, the pressure rise of the gas would be significantly less than would be found in a room where pressure was confined.

In the case of confined expansion of the gas, a window may break or a door be blown open, which permits the expanding gas to escape. This lowers the pressure within the room by *venting* the explosion, thus lowering the maximum pressures experienced by the confines of the room.

Although we have previously referred to the *flammable* limits of gas, the term is synonymous with the *explosive* limits of the gas.

The only factors which determine whether an *explosion* or a *burning* of the gas occurs are the amount of gas available to the source of ignition and the confinement.

Before the advent of automatic pilots on gas ranges, burners were lit with matches. It was not uncommon to turn on the burner and light the match a second or two later. A minor explosion would result as the gas was ignited a foot or so above the top of the burner and the flame flashed back to the escaping gas from the burner, leaving a properly lit burner.

Deflagration

Deflagration refers to the rapid burning of a substance in an uncontrolled situation. The main difference between deflagration and explosion is the slower burning rate experienced in a deflagration. In an explosion, the rate of burning is fast enough so that an audible sound accompanies it. In a deflagration, the velocity of the burning is such that no audible sound, such as a muffled explosion or bang, is heard. The burning rate is subsonic as the chemical reaction progresses.

Detonation

Anyone who has pumped up a tire using a hand pump can testify that the compression of air is accompanied by the generation of heat. When a substance reaches its auto-ignition temperature, it will ignite spontaneously. In an explosion, the expansion and burning of gases occurs in such a manner that the pressure is increased on the remaining unburned, unreacted gases. As this pressure is exerted on the gas, its temperature increases until it exceeds the auto-ignition point. When this occurs, the gas auto-ignites, causing an explosion unrelated to the chain reaction of the original flame front. This compression of the gas to above the auto-ignition temperature is termed *detonation*. The speed at which the reaction develops the shock wave and temperature rise is supersonic and can occur in practically any explosion. It most frequently occurs in structures where the length is much greater than the diameter, such as pipes, tunnels and mines.

Structural Ruptures

Another type of explosion encountered is the rupture of a confining structure, such as a boiler. When the material contained within the boiler explodes, it exerts pressure upon the walls of the boiler itself. When the shell of the boiler ruptures, the escaping gas exerts pressure upon the room containing the boiler and, again, the pressure is confined and seeks to relieve itself. A secondary container rupture (the room containing the boiler) can occur. Both of these ruptures are termed *explosions*.

The secondary explosion is occasioned by the continued expansion of the gas even though it has been released from the initial container, the boiler. It exerts pressure against the room, and unless this pressure is relieved by venting or destruction of the room, it can continue to expand and destroy the entire structure.

The key to this phenomenon is the expansion of the gases. By contrast, the same tank can be exposed to the same internal pressures by a hydraulic fluid, such as water. Since water is relatively incompressible, the expansion against the walls is immediately relieved when rupture occurs.

CLASSIFICATION OF EXPLOSIONS

According to the NFPA, explosions are classified under *Gas Fires and Explosions.* This classification includes fires caused by gas explosions and explosions that involve gas that has escaped from piping, storage tanks, equipment or appliances. Excluded from this category are heating and cooking equipment as well as welding and cutting torches. This classification accounts for approximately 1 percent of all fires by cause.

Fireworks, Explosions

This classification accounts for less than one third of one percent of all fires. This category includes all explosions that do not fall into established classifications, such as those caused by uncontrolled chemical reactions or escaping gas. It also includes those where a fire and explosion were known to have occurred but the cause or origin has not been determined. It does not include explosions where the source and ignition media are known, such as the explosion of an oil-fired space heater. This would be classified under *Defective Oil-Fired Heating Equipment.*

By observing the classifications of these two categories, it is easy to see that quite a number of explosions are reported in other categories and are not reported as *explosions* per se. Therefore, the percentages regarding the relative significance of explosions as causes and origins of fires are misleading.

The reason for this is that an explosion is considered an effect rather than a cause. The effect of the explosion is the result of the rapid expansion of the burning gases, which are the ignited materials. The source of ignition can range from an overheated bearing to static electricity.

Gas Explosions

Gas explosions occur when the gas-air mixture lies within the flammable range and is exposed to a source of ignition. Normally, the flammability limits govern the ability of the mixture to explode. The only difference between an explosion and a burning of the gas is the rate at which it occurs. Under certain conditions, the limits of explosion, i.e. the faster rate, can be affected and change the pressures induced by the burning gases.

According to the NFPA, the maximum pressure exerted in deflagration of gas-air mixture is approximately eight times the initial pressure. This may increase to as much as twenty times in a fuel-oxygen system. Pressures from gas detonations are about twice those of deflagrations so that a fuel-oxygen mixture could yield forty times as much pressure. A fuel gas explosion may exert 50 to 100 psi against the walls of a structure. Since the average wooden frame home cannot resist pressures in excess of 1 psi, considerable damage will result from fuel gas explosions (see Figs. 59, 60, 61 and 62).

Dust Explosions

According to the NFPA, combustible dusts in air behave similarly to a flammable gas-air mixture. Normally, the dust-air velocity of the explosion will be subsonic, and therefore, detonation will not occur. The exception is that under certain circumstances in coal mines, coal dust clouds have detonated.

Figure 57. A gas leak within a furnace, caused by a malfunctioning valve, may result in an explosion. The gas within the furnace, however, will usually be in an over-rich condition. The resultant explosion will not damage the furnace but may move it from its original position, as indicated by the lines on the basement floor.

Figure 58. When an explosion occurs within a furnace, the results are as illustrated here. It is important to determine whether the explosion occurred within the combustion chamber or in the circulating air chamber.

A dust explosion occurs when suspended dust particles are ignited. This phenomenon is similar to a vapor explosion in that the concentration of dust must exceed a lower explosive limit before an explosion can occur. Although the average lower explosive limit is about 0.065 ounces of combustible dust per cubic foot, this average lower explosive limit varies with each type of material. Except for the common denominator of the lower explosive limit and the fact that the suspending medium is usually air, the reaction of dust explosions differs in many ways from the vapor explosion.

Dust is generated by the industrial processes in grain elevators, rolling mills, grinding operations, processing plants and many others. Fine dust is the result of many operations. Dust is held suspended in air due to its minute size but eventually will settle on horizontal surfaces, where it still presents a hazard.

Once the suspended particles have exceeded the lower explosive limit, they will explode by deflagration of the minute particles in the air. The violence of an explosion will stir up heavier concentrations and result in further explosions generated from the first explosion. The ignition for subsequent explosions will be supplied by the initial explosion.

Accumulated dust on shelves, ledges or other horizontal surfaces will be disturbed and mixed with air, causing additional explosions. Unlike vapors, the dust concentration has no upper explosive limit and deflagration will continue as long as material is available. There is no limit to the number of chain reaction explosions which can occur as long as material is available.

Dust is found everywhere in operations involving combustible substances. Much like the gasoline vapors encountered in refineries, the hazard is always present. The explosion hazard is controlled by eliminating sources of ignition, similar to techniques exercised in refineries.

Much work has been done by the Bureau of Mines in studying the phenomenon of dust explosions. For the last thirty years, they have conducted experiments to determine the hazards of dust explosions. They have developed three indexes which rate the explosion hazard involved in dust. They include *ignition sensitivity, explosion severity* and *index of explosibility*.

Ignition Sensitivity

Ignition sensitivity is equal to the ignition temperature times minimum energy times minimum concentration of Pittsburgh coal dust divided by the ignition temperature times the minimum energy times the minimum concentration of the sample dust.

Explosion Severity

The explosion severity is equal to the maximum explosive pressure times the maximum rate of pressure rise of the sample dust divided by maximum explosive pressure times the maximum rate of pressure rise of Pittsburgh coal dust.

Index of Explosibility

The index of explosibility is equal to the product of the ignition sensitivity times the explosion severity.

The indexes are dimensionless quantities and are arbitrarily set at 1 for a dust equivalent to Standard Pittsburgh coal. The purpose of these three indexes is to give a relative comparison of the hazard involved. By checking the explosibility index, the fire investigator can make a judgment regarding whether it is probable that a dust explosion occurred.

Table XXII is compiled from an extensive list of substances having an explosibility index of 1 or greater. An explosibility index of 1 to 10 is classified as a strong explosion by the Bureau of Mines Index; an index above 10 is classified as a severe explosion.

TABLE XXII

EXPLOSION CHARACTERISTICS OF VARIOUS SUBSTANCES

Gases	*Maximum Pressure psi*	*Maximum Rate of Rise psi/sec*	*Concentration % By Volume*
Acetone	83	2,000	6
Acetylene	150	12,000	13
Butane	97	2,300	5
Ethyl alcohol	99	2,300	12
Hexane	92	2,500	2.5
Hydrogen	101	11,000	35
Naphtha	94	2,500	2.5
Propane	96	2,500	5
Toluene	92	2,400	4

Dusts	*Maximum Pressure psi*	*Maximum Rate of Rise psi/sec*	*Concentration oz/cu ft*	*Explosibility Index*
Agricultural Products				
Alfalfa	66	1,100	0.100	0.1
Casein	66	1,000	0.045	0.6

Cinnamon	114	3,900	0.060	5.8
Coffee	44	500	0.085	0.1
Corn	95	6,000	0.045	8.4
Corn cob	110	5,000	0.030	12.2
Corn starch	115	9,000	0.040	35.6
Cottonseed	104	3,000	0.050	2.2
Flax shive	81	800	0.080	0.2
Grain, mixed	115	5,500	0.055	9.2
Grass seed	76	1,000	0.060	0.4
Gums	80	2,800	0.030	22.9
Hemp hurd	103	10,000	0.040	20.5
Malt, brewers	92	4,400	0.055	6.5
Milk, skim	83	2,100	0.050	1.4
Nut shells	106	4,700	0.030	13.8
Pectin	112	8,000	0.075	10.3
Pits, fruit	104	4,400	0.030	7.4
Potato starch	97	8,000	0.045	20.9
Rice	93	3,600	0.045	4.5
Safflower	84	2,900	0.055	5.2
Soy beans	99	6,500	0.035	7.5
Sugar	91	5,000	0.035	13.2
Wheat	103	3,600	0.055	2.5
Wheat flower	95	3,700	0.050	3.8
Wheat starch	105	8,500	0.025	49.8
Wheat straw	99	6,000	0.055	5.0

Metal Powders

Aluminum, atomized	73	20,000+	0.045	10
Aluminum—cobalt alloy	78	8,500	0.180	0.4
Aluminum—copper alloy	68	2,600	0.100	0.3
Aluminum—lithium alloy	96	3,700	0.100	0.6
Aluminum— magnesium alloy	86	10,000	0.020	10
Aluminum—nickel alloy	79	10,000	0.190	0.6
Aluminum—silicon alloy	74	7,500	0.040	3.6
Boron	90	2,400	0.100	0.8
Calcium silicide	73	13,000	0.060	2
Chromium	55	4,000	0.230	0.1
Coal, Pittsburg seam	83	2,300	0.055	1.0
Ferromanganese	47	4,200	0.130	0.4
Ferrotitanium	53	9,500	0.140	1.3
Iron, carbonyl	41	2,400	0.105	1.6
Magnesium	90	9,000	0.040	10
Manganese	48	2,800	0.125	0.1
Silicon	82	12,000	0.110	0.9
Tantalum	50	2,600	0.200	0.1
Thorium	48	3,300	0.075	10

Dusts	Maximum Pressure psi	Maximum Rate of Rise psi/sec	Concentration oz/cu ft	Explosibility Index
Thorium hydride	60	6,500	0.080	10
Tin	37	1,300	0.190	0.1
Titanium	70	5,500	0.045	10
Titanium hydride	96	12,000	0.070	6
Uranium	53	3,400	0.060	10
Uranium hydride	43	6,500	0.060	10
Zirconium	55	6,500	0.045	10
Zirconium alloy	43	300	—	—
Zirconium hydride	60	6,500	0.080	10
Plastics				
Acetal, linear	113	2,900	2.00	10
Acrylonitrile	85	2,600	1.00	10
Alkyl nitrosomethyl amide	174	8,000	2.00	10
Allyl alcohol	91	7,500	0.50	10
Cellulose	117	4,100	1.00	10
Cellulose acetate	135	4,000	1.00	10
Epoxy	94	5,000	1.00	10
Ethyl cellulose	120	5,500	1.00	10
Gilsonite	89	3,800	0.50	10
Gum DK	87	8,500	0.50	10
Lignin, hydrolized-wood-type fines	102	5,000	0.50	10
Methyl methacrylate	84	3,100	1.00	10
Nylon	95	3,600	1.00	10
Petrin acrylate monomer	236	16,000	2.00	10
Phenol formaldehyde	77	3,500	0.50	10
Phenol furfural	88	8,500	0.50	10
Polycarbonate	96	3,300	1.00	8.6
Polyethylene	80	5,500	0.50	10
Polypropylene	76	5,000	0.50	10
Polystyrene	77	5,000	0.50	10
Polyurethane foam	87	3,700	1.00	10
Polyvinyl butyral	84	2,000	0.50	10
Rubber, synthetic hard	93	3,100	0.50	10
Shellac	73	3,600	0.50	10
Styrene	96	1,100	1.00	10
Vinyl chloride	95	3,300	1.00	10
Wood flower	113	4,900	1.00	9.9

Vapor Explosions

Vapor explosions are caused by the vaporization of flammable liquids. The more volatile the liquid, the more probable that it formed the vapor which exploded. Liquids in this classification include ether, gasoline, alcohol, paint thinners, solvents and naptha.

These vapor mixtures will usually be heavier than air and will form a rich mixture close to the floor or the surface of the liquid. The rate of evaporation will be a function of the surface area exposed to the atmosphere. Therefore, a spilled container, where the liquid has covered a floor surface with a large quantity of a thin film of liquid, will create a much more explosive atmosphere than an open container.

The longer that a container remains open in an unventilated room, the more likely that the vapor will diffuse into the atmosphere and form a potential bomb. For example, a pint of gasoline in a normal 10 by 12 by 8 foot room could completely fill the room with an explosive mixture. The resultant explosion would demolish the room, whereas the gasoline, if spilled on the floor and immediately ignited, could be easily extinguished.

Decomposition Explosions

A *decomposition explosion* is caused by the breakdown of a compound into its various elements. The decomposition occurs with the release of tremendous amounts of heat and the expansion of tremendous volumes of gas. Chemically, the material is transformed from a solid to a gas with the emission of heat. In other words, the substance which had previously been a solid is converted into a gas almost instantaneously. The problem is that the volume of the solid occupies a greater physical space when converted to a gas. As the material tries to convert to the gaseous state, it expands to many times its original size. In this rapid rearrangement of molecules plus the addition of heat, the heat raises the temperature of the gas, which in turn increases the space required for the gas. At the same time, this activity increases the pressure exerted on the walls of the container, resulting in a pressure relief explosion.

Pressure Relief Explosions

Pressure containers are designed to resist the normal operating pressure of the contents plus a factor of safety. In addition, a safety relief valve is usually supplied so that excess pressures will open the valve and a limited amount of the contents will be released to lower the pressure within the container. However, if the pressure is increased faster than the pressure relief valve can lower the pressure, the pressure within the container can build up to the point where it exceeds the structural ability of the container to withstand the pressure and it will rupture. This sudden release of pressure within a container, due to structural failure, is a *pressure relief explosion.*

The contents of the container determines what effect the rupture of the container has on its surroundings. For example, if an LP gas tank explodes, the liquid contents of the gas will immediately try to assume the configuration of a gas and expand violently with the release of considerable pressure. Normally this pressure would not be sufficient to cause any damage. However, if a spark is encountered at the interface of the expanding gas liquid where a flammable mixture has been formed through turbulence, then this would ignite, causing a BLEVE, or a "Boiling Liquid-Evaporating Vapor Explosion."

If the container happens to be a water tank, it can still cause a fire. The reason that the water is becoming steam is usually due to failure of the heat controls to shut off the flow of heat to the heating element of the appliance. The heat in turn raises the temperature of the water until it is converted into steam at approximately 212°F. As the pressure of the steam increases, it also increases the boiling point of the liquid or water, but the steam is still being formed by water until it is quite possible that the pressure inside exceeds the structural capacity of the tank and the tank ruptures. The rupture releases the steam and deforms the metal in the tank and may extinguish the fire. Whatever caused the fire to get out of control may fail to control the escaping gas so that it escapes until it finds a source of ignition and a second explosion follows.

Acetylene-Oxygen Gas Welding Explosions

Any gas welding or cutting system which uses oxygen as a mixing agent with a flammable gas can lead to explosions. The explosion may be the result of running out of gas on the oxygen or acetylene tank. It can occur with either tank, and it does not matter whether the flammable gas is propane, acetylene, hydrogen or some other highly flammable substance. The important feature of this system is that oxygen be piped into a mixing chamber without a check valve to prevent the flow of the mixed gases into the container which is about out of gas and operating under lower pressure.

The sequence of events is as follows. The container starts to exhaust its gas supply, and the pressure drops towards zero. However, the container is full of either air or oxygen. In the meantime, the other container, containing a flammable gas, is operating at a much higher pressure (15 to 60 pounds per square inch, depending upon the type of burning, cutting or welding being done).

Under normal conditions, the two gases mix as they enter the mixing chamber and are expelled through the burning, cutting or welding tip of the torch. The restricted orifice of the cutting tip presents an obstacle to the flow of gases. The resultant back pressure can eventually exceed the pressure from the nearly empty supply tank. When this condition exists, the pressure from the full tank will force the mixed gases to reverse the flow of the supply line to the air or oxygen tank into the regulator valve on the supply tank. At that time, the flame will enter the hose from the cutting tip and, since the mixture is already flammable, the flame front will follow the mixture upstream to the regulator on the nearly empty tank where the confinement is less, and the pressures will build up to explosive pressures. An explosion will occur which normally blows off the face of the regulator, permitting the entrapped burning gas from the interior of the regulator and the line to escape to the atmosphere where it can and usually does start a fire. These fires are caused by a tank of nearly exhausted fuel so that the pressure differential in the line permits this chain of events to occur.

They can and have been prevented by the addition of check valves which prevent the flow of gas back toward the regulator. These are attached directly at the connection to the cutting torch handle just upstream from the mixing chamber of the torch. A quick examination will reveal whether these are present. If they are present and functioning properly, then this could not have been the cause of the explosion. Other sources of the explosion should be sought. The clue to this type of explosion is the lack of check valves and the blown-out face of the regulator on the nearly empty tank.

Low Velocity Explosions

A low velocity explosion is one in which the speed of the reaction and the shock wave developed by the expansion of the gases is slower than the speed of sound. The speed of sound is approximately 1,128 feet per second at sea level, or approximately 767 mph. The burning velocity of gases under controlled conditions varies from 1 to 15 feet per second. In uncontrolled burning, the velocity increases tremendously, due to simultaneous expansion of the gas.

Low velocity explosions are the result of the ignition of gas-air mixtures within the flammable range, flammable liquids which have been vaporized to form a vapor-air mixture within the flammable range, a mist or dust cloud of minute particles of a flammable substance or a mixture of chemicals—all providing the necessary oxygen to produce an exothermic chemical reaction.

In the previously mentioned explosions, the explosive material, or the source of energy, was distributed through the room so that the source of energy was the entire atmosphere. Another type of low velocity explosion is that caused by the concentration of energy at a point, such as a firecracker or a stick of dynamite. On a larger scale, the aforementioned boiler would still be a point explosion in relation to the boiler room or the entire building. Regardless of whether the explosion occurs at a point in a room or even an entire building, the pressures developed are a function of the rate of chemical reaction and the confinement of the structure.

If the original liquid container is still available and had leaked during the time preceding the fire, the container will not be ruptured due to expansion of the gases within the container, as the vent hole which permitted the liquid to evaporate or leak out would be sufficient to vent the buildup of pressure within the container.

If the container is ruptured, it can only mean that the container had been tightly sealed and the rupture occurred as a result of internal pressure which was prevented from escaping due to the tight seal. Rupture of a container is positive proof that the container was tightly sealed and contained either liquid or vapors (not necessarily flammable) before the fire.

In the event that the explosion occurred without a fire, it is quite possible to locate a stain on the floor where the flammable liquid had flowed and been left standing until it evaporated, forming the vapors which led to the explosion. Again, the source of ignition is a necessary factor to be considered in evaluating the source of the explosive material.

Chemical Explosions

A chemical explosion is usually a result of a chemical reaction which "got out of control." These are commonly found in processing plants or industrial installations where highly volatile chemicals or flammable materials contact each other, resulting in the generation of flammable gases or explosive reactions between the chemicals involved.

In these cases, the industrial process will need to be studied by an expert to determine what step in the process led to the generation of the gases which resulted in the explosion, the source of ignition which caused the explosion or the chemicals which generated the explosion.

Many processes include highly flammable gases which are controlled by eliminating a source of ignition. A good example of this is the petroleum industry, where tank cars are loaded and unloaded constantly with full knowledge that highly flammable gases, i.e. gasoline, are being manipulated. Refineries, loading docks and ordinary service stations deal with this commodity daily and, by eliminating a source of ignition, keep the number of explosions to a statistically satisfactory minimum.

Concentrated Explosions

Concentrated explosions are explosions which occur at a small point. These would include the firecracker, a stick of dynamite or small container-type explosions referred to earlier. The obvious clues to these fires are the containers themselves. For example, fragments of the firecracker or dynamite case or metal particles from a pint propane bottle might be found at the scene of a concentrated explosion. The explosion may be small enough so that damage is insignificant or may be large enough to cause considerable damage and still be a low velocity explosion.

Additional evidence would be the presence of ignition devices. Most pressurized containers are sensitive to temperatures exceeding 120°F. When this temperature is exceeded, the internal pressure in the container is raised to the point where it may exceed the structural integrity of the container. This causes an initial explosion as the tank ruptures, permitting the contents to flood the room as the compressed gas expands. Immediately upon leaving the container, the compressed gas vaporizes and mixes with air. This forms a flammable air-gas mixture which will immediately ignite, resulting in a secondary explosion. If this secondary explosion results in a fire, then a new problem arises.

This problem concerns whether the fire was caused by the secondary explosion or whether it was caused by the increase in temperature which heated the tank to the point of rupture. The presence of a ruptured pressurized can containing flammable liquid must always be investigated to determine which came first, the rupture of the tank or the fire.

Clues to this answer will sometimes lie in the origin of the fire. When a tank ruptures, or a flammable gas explodes, no well-defined origin will be apparent. Therefore, careful examination of the premises and the lack of a well-defined origin can sometimes provide clues regarding the true origin of the fire.

High Velocity Explosions

High velocity explosions usually involve a chemical reaction which does not involve combustion of hydrocarbons. Most low velocity explosions involve a highly increased burning rate and the oxidation of hydrocarbons. Most high velocity explosions are caused by combinations of nitrogen and oxygen.

A high velocity explosion is characterized by much sharper sounds, higher velocity burning (or the absence of burning) and the development of much higher pressures than experienced in the low velocity explosion. It includes the rupture of tanks where the pressures experienced are much greater than those encountered in low velocity explosions. It is termed *pressure relief* rather than tank ruptures to differentiate between the low and high velocity aspects of the phenomenon.

The detection of the cause of these explosions is best attacked from a metallurgical standpoint to determine whether the structural material of the tank which failed meets the required specifications of the manufacturer. In addition, the material should also meet the test standards of the American Society of Testing and Materials (ASTM), American National Standards Institute (ANSI), Occupational Safety and Health Adm. (OSHA) or any regulatory body governing the reliability of structures.

Failure of structural vessels can be caused by metal fatigue due to the length of time the material was subjected to internal stresses. The material can fail due to exposure or to stresses in excess of the manufacturer's specifications and design criteria. The material can also fail through stress corrosion due to the exposure of the material to unpredicted materials, which results in corrosion of the internal structure of the metal, leading to failure of the integrity of the structure.

Metallurgical tests, both destructive and nondestructive, can be run on most materials to determine whether the material failed at normal or subnormal stresses which the structure was designed to resist. The circumstances leading up to the failure of the structure should be carefully reviewed to discover whether the structure was subjected to greater stresses than it was designed to resist. Earthquakes, wind, floods or other natural phenomena should always be considered as possible causes of additional stress upon a structure.

EXPLOSIVES

Explosives are chemical compounds specifically designed, manufactured and distributed for the accomplishment of a controlled explosion. The explosion is completed through the chemical release of heat and gas, including the physical effect of creating high pressures as a result of converting a solid into a gas with the release of tremendous amount of heat. This causes expansion of the produced gas due to the heat effect. It includes explosions which could be classified as either high or low velocity explosions. The burning rate is so rapid that it is classified as an explosion rather than a deflagration, although from a chemical viewpoint the material is destroyed by combustion rather than decomposition.

Also included with these burning-type explosives are the propellants used to develop pressures in missiles. These include black powder, smokeless powder and liquid or solid rocket fuels.

Classification of Explosives

The Department of Transportation has divided explosives into three classes, with a separate classification for blasting agents. These are listed in descending order of sensitivity to explosions.

Class A Explosives

This class possesses highly sensitive detonation characteristics and includes dynamite, desensitized nitroglycerin, lead azide, mercury fulmanate, black powder, blasting caps, detonating primers and certain smokeless propellants.

Class B Explosives

This class has a highly flammable hazard rating and includes the propellant materials. They are considered less hazardous than Class A explosives.

Class C Explosives

These include manufactured articles which contain limited quantities of Class A or B explosives as one of their components. It includes detonating cords, explosive rivets or other manufactured production items. These explosives will not normally mass-detonate under fire conditions.

Blasting Agents

Blasting agents are manufactured from an explosive material but are not easily set off by impact as are the other three classes. Blasting agents will burn, but the reaction will not reach detonation velocities.

The NFPA defines a blasting agent as any material or mixture consisting of a fuel and oxidizer, intended for blasting, not otherwise classified as an explosive and in which none of the ingredients are classified as an explosive. This is providing that the finished product cannot be detonated by means of a No. 8 blasting cap when unconfined (a No. 8 blasting cap is one containing two grams of a mixture of 80 percent mercury fulmanate and 20 percent potassium chlorate, or a cap of equivalent strength).

For reasons of safety, the construction, mining and general industry have adopted the use of fuel oxidizer systems which do not contain explosives. The most common blasting agent is called *Anfo*. This is an acronym for Ammonium Nitrate Fuel Oil. The combination of ammonium nitrate and fuel oil makes a safe and very effective substitute for dynamite. Ammonium nitrate explodes by decomposition and causes the fuel oil to ignite through the combination of pressure and heat released by the ammonium nitrate. A strong detonator is required to initiate the explosion, but the additional safety found in the use of the substance makes it well worth the additional cost of detonation.

Anfo is not suited for wet applications, however, so water gels or slurries are used. These utilize ammonium nitrate and a powdered aluminum which is uninhibited by the presence of water.

Should the presence of an explosive be suspected, an explosives expert should be contacted to assist in identifying the explosive. When commercial blasting agents are used, a detonator is required. The detonator will commonly be in the form of a blasting cap. The detonating device for the blasting cap may either be a fuse or an electric circuit. Remnants of these may often be found blown some distance from the center of the blast. If a commercial fuse was used and portions of the unburned fuse are found, it can be readily identified by the manufacturer's markings. If electric caps are used, the debris and wiring should be secured for comparison purposes. These can be compared with wire found in the home of a suspected arsonist. The markings on the wire where connections were made will sometimes be identified by, for example, characteristics of the cut made by a pair of side cutters.

Types of Explosive Materials

Commercial explosives are generally designed for a specific purpose, such as road building, breaking rock in quarries, mining and demolition of buildings. Practically all of them depend upon a detonator cap to initiate the explosion. The detonator cap is not only a chemical requirement but also a safety device. If explosives were sensitive enough to explode without a detonator cap, the increased hazard would far outweigh their usefulness.

Military explosives are designed for an entirely different use. Commercial explosives are used within a relatively short time of their manufacture compared with explosives manufactured for military use. Military explosives must be stored economically for long periods and must explode and function reliably when needed. They too must depend upon a detonator which must be capable of rough handling and hard usage in the field.

Should the fire investigator become involved in a case where military explosives were involved, he should contact the Federal Bureau of Investigation, the U.S. Army Ordnance Department or the Military Provost Marshal having jurisdiction over the military aspects of the case. Military explosives will not be covered in this text. The types of explosives are divided according to their usage.

Primary High Explosives

Primary high explosives are those which have the highest pressure rise due to the fact that they detonate when subjected to high heat, shock or friction. Their ability to develop high pressures with small quantities of material make them highly useful as a detonator of the high explosives, which are not as sensitive to the activating mechanisms.

Primary high explosives include mercury fulmanate, lead styphnate and lead azide. The primary use of these is to initiate an explosion. They are seldom used exclusively as a blasting agent.

Secondary High Explosives

Secondary high explosives are those which require a detonator consisting of a primary high explosive. These explosives were developed as a safety measure for commercial use. In well-ventilated situations, secondary high explosives will burn without detonating. Their normal function, however, is to be placed in a confined situation so that detonation is accomplished by the use of a primary high explosive.

Secondary high explosives commonly used today include dynamite, nitroglycerin, TNT, RDX and PETN. Secondary high explosives differ from blasting agents in that secondary high explosives are cap sensitive and blasting agents will not detonate by the action of a No. 8 detonator when unconfined.

Low Explosives or Propellants

These consist of black powder, smokeless powder and rocket fuels. They characteristically burn rather than detonate, although under proper confinement, many materials can be made to detonate by the increase of pressure accompanied by increase in temperature to the auto-ignition point.

INVESTIGATION OF EXPLOSIONS

An explosion, as defined previously, can vary from a mild puff, such as occurs in the delayed ignition of a gas burner, to an atomic holocaust. The forces involved can vary from an insignificant push to the shattering of structural members of a building, as shown in Figures 59, 60 and 61. An explosion may or may not cause a fire.

Some jurisdictions in the United States may not classify an explosion as arson unless burning occurs. Under the old definition of arson, mere singeing would not be sufficient to form a corpus delicti; the structure would have to actually burn. Modern arson laws have overcome this problem and have included *explosions* along with *burning* in the definition of arson.

Regardless of whether or not explosions without fire are considered arson in a given jurisdiction, every effort should be made to investigate as many explosions as possible. By examining explosions where fires have not occurred, the investigator will soon learn to detect signs of an explosion where burning has accompanied an explosion. The investigator will also become familiar with the residues left by an explosion, particularly those left by specific explosive materials.

Once the investigator has become familiar with the physical and chemical characteristics of an explosion, his next step is to determine whether the fire caused the explosion or the explosion caused the fire.

Figure 59. Structural damage caused by explosions should be photographed from all four sides so that a visual estimation of the physical damage can be evaluated. (See Figs. 60 and 61.)

Investigation of an Explosion Without a Fire

The investigation of a fireless explosion is simplified by the fact that only one physical event has occurred. No problem is encountered in differentiating between the cause and effect. An explosion has occurred, and efforts can be concentrated on the explosive medium, the source of ignition and why and how the explosion occurred.

This investigation is similar to determining the origin and cause of a fire. The origin of the explosion should be determined in the same manner. Starting from the outside, the location of debris from the explosion should be carefully documented with a sketch and accurate measurements showing the location of the building components. The location of structural components should be photographed to show their position in relation to the general area, particularly the origin.

Figure 60. By photographing all four sides of the building, a better portrayal of the damage is presented. Compare with Figure 59 for the difference in structural damage.

Aerial photographs of the entire area (Fig. 61), including the most remote pieces of debris, are useful in illustrating the general area and the magnitude and direction of the forces. Both color and black-and-white photographs should be taken to guarantee that all debris can be identified. The resolution of black-and-white photographs is usually superior to color photographs in aerial photography.

The size, dimensions and weight of the largest pieces of debris should be recorded for an evaluation by an explosives expert and structural engineer. It may be necessary to run metallurgical tests to make this determination. Since these tests are usually of a destructive nature, sufficient grounds for suspecting mechanical failure should be established before any such testing is done.

The investigator should work his way inward toward the center of the forces which destroyed the structure. The movement of the

Figure 61. An aerial photograph dramatically displays the entire destruction of a building in one photograph. The same portrayal cannot be made from ground without the use of at least four separate photographs. (See Figs. 59 and 60.)

structural components would be away from the center of the explosion. By working in a circle, the detection of movement of the components will direct and guide the investigator to the origin of the explosion. Normally, this would be the area of most severe damage.

At the origin, it would not be uncommon to find shattered wooden studs, columns or joists. The closer to the origin, the stronger the forces and the more severe the shattering effect would be upon the structural components.

The Origin

Once the origin has been isolated, photographs should be taken before attempting to determine the cause of the explosion. The

Figure 62. If the scene of an explosion is visited long after it occurred, quite often repair work or additional demolition work may have been done which alters the appearance of the building considerably.

origin will usually be a room. Having determined that a room or area of a building is the apparent origin, the investigator must determine whether the room, or a point in the room, was the origin.

This is accomplished by a detailed examination of the shattering and movement of the room components. If they appear equal and no reason is found to assume that they should be different, due to the structural configuration, the investigator should search for a gas or vapor which may have filled the entire room, so that the origin of the explosion was the room itself.

If gas lines are present, or other indications of flammable liquids which could have vaporized and caused the explosion are found, the next step is to determine the source of ignition. The source of ignition could be located a considerable distance from the origin of the gas. For example, if a gas line was broken by construction work in the street, causing a leak in the house, the gas could migrate upstairs to an open stove pilot light and flash back to the basement before an explosion occurred.

In an explosion, it is not likely that the source of ignition can be determined beyond a shadow of a doubt. The best an investigator can do is to determine the most probable cause of ignition.

If no source of flammable vapors are found, a detailed study of the floor should be made in search of *cratering*. Cratering occurs when a high explosive is exploded. It will produce a shattering effect on concrete, wood or other structural materials in the immediate vicinity of the explosive detonation. Should cratering be found in an area 3 or 4 feet in diameter, the most probable cause of the crater and explosion would be a high explosive such as dynamite. A diligent search should be conducted for fragments of the detonating mechanism, such as fragments of the detonator cap, burned-out fuse wire or other mechanical devices used to detonate the explosive.

A detailed search of the walls, floors and ceiling may uncover signs of fragmentation from a pipe bomb. These would appear as jagged holes in the floor, ceiling, or walls made by bomb fragments after the bomb had exploded. If these are found, the investigator (unless he has jurisdiction) should contact the proper authorities, and bomb experts may be called to assist.

Since some explosives leave detectable residues, the origin should be carefully examined for foreign substances not normally found in the area. These residues can be chemically analyzed to determine the explosives used.

A detailed examination of the circumstances leading up to the explosion will reveal whether the explosion was an accidental or malicious event. This will determine whether the investigation becomes a criminal investigation of arson or remains a civil investigation of an accidental explosion.

Investigation of an Explosion With a Fire

When an explosion occurs and starts a fire, the evidence of the explosion may be destroyed by the fire. Although it has been exposed to the ravages of the fire, some evidence may remain and, in some cases, may not be damaged to any great extent. This is particularly true of metal fragments from a bomb or the metal parts of a timing mechanism. The residue from black powder may be consumed in the fire but may be identified chemically by its residue.

If gas lines are present, the lines should be tested for leaks and the probability that the leak was caused by the fire should be weighed. If the gas controls have been destroyed by the fire and it cannot be determined if a gas leak occurred, the probability of a bomb or high explosive should be eliminated by the lack of cratering and shattering of joists and by the general nature of the explosion. It is not uncommon to find shattered floor joists, studs or rafters in a natural gas or LP gas explosion, particularly in a low limit explosion where the velocities are much higher than in a high limit gas explosion. It is in these cases that the process of elimination is sorely tried as a means of arriving at a cause.

Before the investigation begins, the initial step is to determine whether the fire was caused by the explosion or the explosion was caused by the fire. Quite often, explosions may cause fires and additional explosions may result.

The fire investigator's approach should be to make a chronological determination as to what occurred. The first step is to identify anyone who heard the initial explosion and determine if evidence of a fire existed prior to the explosion. Following this procedure, the investigator should determine what explosion occurred after the fire and the time and approximate location where the explosion occurred. It is not uncommon to have multiple explosions occur as pressurized containers are overheated to the point where the vessel ruptures, causing secondary explosions.

The origin of the fire and explosion should be identified separately. The investigator should concentrate on one aspect at a time. The origin of the explosion should be determined first, and quite often it can be determined that the explosion was the origin of the fire. The origin of the explosion should be determined as outlined in the preceding material on the investigation of the origin of an explosion without a fire. Once the origin of the explosion has been determined, the origin of the fire should be traced. Should they be found in the same general area, the problem of whether the fire caused the explosion or the explosion caused the fire can be attacked.

Separate Origins

When separate origins are found for an explosion and fire, the cause of the fire should be determined in relation to the explosion. For example, an explosion may break a pipeline carrying flammable substances some distance from the origin of the explosion. This in turn may start a fire, which was caused by the explosion. One method of making this determination is to check the char depths on shattered timbers. If the fire occurred first, a time lapse between the fire and explosion would have given time for the structural members to become involved in the fire. In other words, the char depths on joists, studs, rafters or wall panels will have achieved a given depth due to the fire. After the explosion, the shattered surfaces will be exposed to fire, and the lesser char depth on the shattered structural members will be discernible due to the length of time that the other portions of the structural members have already been burning. Had the explosion preceded the fire, the char depth on the shattered portions of the structural members would be the same as on the unshattered portions.

If this method fails, careful examination of surfaces can be made to determine whether the explosion drove particles of soot or carbon out of the building on glass fragments. Had the fire been burning prior to the explosion, it would have deposited carbon or soot on the windowpanes at higher elevations. If the explosion broke the glass and blew out windows some distance from the building, the glass may have soot or carbon deposits from the fire. If the explosion preceded the fire, no soot deposits will be present on the glass. This test is of no probative value if explosions were known to have occurred after the fire was well underway. It is only significant when the fire had been burning for some time before the explosion occurred.

Explosions and Fires with a Common Origin

When the explosion and fire occur in the same general area, the two can usually be differentiated following the principles outlined above. In any fire, numerous small explosions are heard due to containers of volatile substances, such as gasoline cans, solvent containers, etc. These will be ruptured due to internal pressure which normally can be attributed to the fire. The exception would be an arson *plant* where the container was deliberately ruptured by the application of a small, controlled source of external heat to give the appearance of an accidental fire. A ruptured container indicates that the container had been sealed and that internal pressure had ruptured the container, due to the heat of the fire. The explosion in this case is usually a dull, muffled explosion accompanied by an intensive increase in the burning rate as the flammable contents are rapidly consumed in the fire. This is usually accompanied by a loud "whoosh." The presence of ruptured cans should be investigated by questioning occupants to reveal whether containers were normally stored on the premises and their contents. The presence of empty containers with loose or missing caps should also be investigated.

A loose cap on an unruptured container indicates that it had not been sealed prior to the fire. Had the container been sealed, even though it were empty, the increase in pressure, due to the heating of the air within the container, is usually sufficient to cause rupture as a result of expansion of the air within the container. Each container, even though empty, should be identified by the owner or occupant regarding its contents or lack of contents. The fire service personnel should be questioned concerning their observation of explosions during their entire operation and fire fighting activities.

Characteristics of Gas Explosions

Gas explosions involving methane, natural gas, propane or butane are characteristically low velocity explosions. When the mixture involved is at the lower explosive (flammable) limit, the velocities obtained are much higher than when the explosion involves a mixture at the upper limit or higher. An explosion at the lower end will usually not cause a fire, although some evidence of singeing will be present. The damage and destruction wrought by the force of the explosion is a function of the volume of the lower limit gas mixture.

Volume of Gas

If leaking gas completely fills a room with a flammable mixture, a greater explosion and more damage will result than in a room half filled with gas. The volume of the room and the volume of gas needed to provide a flammable mixture at the upper range can be calculated. The minimum amount of gas required to satisfy the lower limit can also be calculated. The probability that the gas was at either of these two limits can be best determined by consideration of the source of ignition. For example, if the room needed to be completely filled before gas would infiltrate or migrate toward a source of ignition, an upper limit explosion can be safely assumed.

The procedure for making this analysis is based upon the last known time that the gas leak did not exist, its flow rate and the time of the explosion. Refer to Chapter 7 for a more detailed procedure for making this determination regarding the timing of gas leaks.

CHAPTER 9

Appliance Malfunctions

A LTHOUGH NOT A SEPARATE classification by National Fire Protection Association statistics, fires caused by mechanical failure of control systems or the appliances themselves provide the subject of this chapter. By extracting the NFPA subclassifications which could be attributed to the failure of a control system or the breakdown of an appliance, it appears that 14 percent of the total number of fires could be included in this reclassification.

The need for controls is obvious in even the most elementary appliance, such as a gas-fired kitchen range. The top burners, where the majority of cooking is accomplished, have a rotating knob which opens a gas valve to control the flow of raw gas through a ported burner. A minimum amount of gas is required to sustain a small flame for keeping pots warm. For heating or cooking, a much higher flame is needed. The size of the flame is controlled by rotating the control knob of the valve. By rotating this knob, the valve is opened to permit a variable amount of gas to escape from the burner ports according to the need of the operator. The knob is a *primary control* of the flow of gas through the burner. It is a manually operated valve under full control of the operator.

The top burners may be equipped with other controls. The modern range is equipped with a pilot light or an electric igniter. When the gas valve is opened, ignition is guaranteed by the presence of a small flame from the pilot light, burning less than 500 BTUs per hour. An electric control system will automatically shut off the gas if the igniters fail. Failure of either of these sources of ignition would be noticed immediately by the operator, who would attempt to remedy the situation. The worst condition would be delayed ignition of the outrushing mixture resulting in flare-up of gas that could singe or burn the operator. Manufacturers of ranges make every effort to ensure ignition at the instant that gas emanates from the burner ports.

The modern gas oven in a kitchen range is an example of multi-purpose controls. The oven range has a dial which is normally graduated in degrees Fahrenheit. The operator sets the dial for the desired temperature. The oven is an insulated rectangular box with an insulated access door.

The dial can be of two different types. One may be a gas valve which regulates the quantity of gas flow into the oven similar to the gas control on the top burner. This gas valve has been calibrated at the factory to maintain the temperature registered on the dial, provided the oven door is closed the majority of the time. In other words, BTUs per hour released by the valve will raise the oven to the temperature set on the dial.

The other type of control is a thermostatically controlled gas valve. The gas valve is opened and ignition provided by either a pilot light or an electrical igniter. Instead of being a variable flame, the flow of gas into the oven is constant and set to provide the maximum temperature attained by the oven.

To maintain lower temperatures, a thermostat is provided to turn the gas on and off. Once the thermostat is satisfied, the flow of gas is stopped and the oven allowed to cool to the lower limit of the thermostat. After the oven has cooled, the lower limit switch closes and calls for heat. Heat is supplied again by the full flow of gas until the upper limit of the thermostat is reached and the cycle repeats. This on-and-off oven cycling provides an average uniform temperature in the range of that indicated by the dial.

The oven is protected from accidental overheating by a control called an *upper limit switch*. For example, if the thermostat stuck and gas continued to flow, it would soon overheat the contents of the oven and possibly cause a fire within the oven. If the insulative qualities of the oven were insufficient, the transfer of heat by conduction could ignite combustibles located near the oven. As a safety measure, an independent upper limit switch connected in parallel with the thermostat will cut off the flow of gas to the oven. It is obvious that without these controls, it would be difficult and would require constant manipulation to regulate the amount of heat required in cooking. Similar types of control problems exist in blast furnaces, steam boilers, steam generators and other gas-burning equipment. Gas has been used as an example, but the same principles apply to fuel oil, coal, electrical energy or any energy source.

FUNCTION OF CONTROLS

As indicated in the first above mentioned example, the manually operated top burner valve required an operator to change the valve setting. If left unattended, the valve would continue feeding gas to the burner as long as gas pressure and volume were available. In the second case, however, the oven performed functions without human intervention to maintain a preset condition. The operator set the desired temperature, and the controls maintained this temperature by turning the gas on and off. The controls measure the temperature and determine if the heat requirement is met. If this temperature is met, the control turns off the flow of gas; if heat is required, it turns on and ignites the gas. In the event of system failure, the system shuts down.

Controls can be used to regulate the flow of heat, cold, volume of gas flow, oil, water, steam or electricity. The primary function of a control is to regulate the flow of energy and its conversion from one form to another. It is in this process of transportation, distribution or conversion of energy that one or more components of the system fail, resulting in an explosion and/or fire.

Each control system, regardless of its complexity, consists of a number of components which, working together, form the system. The fire investigator is interested primarily in the components, since one or more of them can be the possible cause of a fire. If all components in the system are functional after a fire, it is a strong indication that they were functioning before the fire. The opposite is not as apparent. If the controls do not function *after* a fire, it does not necessarily follow that they were not functioning *before* the fire; they may have been damaged by the fire.

BASIC CONTROLS

Controls vary from a pipe, which exercises control by its size, shape and length, to complex electronic, electrical and mechanical controls. They all have one thing in common of interest to the fire investigator. This is the manner in which they fail. The mode of failure is dependent upon the type of control, the material from

which it is manufactured, the environment in which it operates and
its age.

Mechanical Controls

Mechanical controls are the easiest to recognize and are generally unaltered by a fire. Pipes exert control over the fluid (either
liquid or gas) by a change in shape, length or internal friction.
All of these factors affect the flow of materials through the pipe.
When gas pressure is reduced, such as occurs when gas is brought
from an exterior line into a building, the size of the pipe carrying
the lower pressure gas is increased to reduce the friction of the gas
as it flows and rubs against the walls of the pipe.

The ultimate goal of any pipe is to deliver its contents to a location where they can be used. In the case of flammable liquids or
gases, the purpose is to ignite the fuel to obtain heat generated by
this process. At the end of the pipe, the quantity of the gas released is controlled by a barrier with a small hole. This barrier is
called an *orifice*. The flow rate of gas emanating from an orifice
regulates the height of the flame and the quantity of BTUs released per unit of time.

Orifices

Orifices are used to provide a constant flow for a gas- or oil-
burning appliance. On oil lines, they regulate the atomized spray
which is ignited to provide heat. Orifices are also used to regulate
the quantity of flow in certain portions of a line to slow down the
velocity of the fluid flowing past a certain point.

Valves

A valve is a mechanical device installed in a line to regulate the
flow of fluid. The valve is an adjustable barrier which can be
placed in a line to give a variable flow similar to a water faucet.
Valves can be very simple mechanisms such as an "A" cock found
upstream from a residential furnace. Its purpose is to enable a service man to cut off the flow of gas temporarily while furnace work
is done.

These are simple devices with a round, tapered shank with a
hole through the center. When the hole is in line with the pipe,

gas can flow; when the valve stem is rotated 90°, the sides of the stem cut off flow through the valve.

VARIABLE CONTROL VALVES. Valves can be used to gradually increase the flow, such as is required when lighting a large furnace. If the gas valve were fully opened, the inrush of a full flow of gas would cause an explosion. It is therefore desirable to control the flow of gas into large combustion chambers. This is accomplished by using slow-opening valves. The delay mechanism consists of a geared-down motor so that the process of opening the valve takes a minute or longer. This time delay permits the smaller volume of gas to achieve ignition without the risk of a major explosion within the combustion chamber.

Since the slow-opening valve is geared to a motor, it follows that the slow-opening valve would also be slow-closing. The slow closing of the valve would permit too much gas to enter the combustion chamber, increasing the risk of an explosion. A fast-closing valve is used in series with a slow-opening valve to immediately cut off the flow of gas to the combustion chamber, thus eliminating the possibility of an explosion. Both of these valves are controlled by sensors which react to the presence or absence of flame within the combustion chamber.

Pressure Regulators

Pressure regulators are mechanical devices inserted into a line to reduce or control pressure. These are commonly used when flammable gases are conveyed into a building. Building codes require that gas be reduced to 7 or 11 inches W.C. (water column), which is approximately ½ pound per square inch, inside a building. The pressure regulators used for this purpose are generally located on the exterior of a building and consist of several types. The difference involves the amount of pressure reduction achieved by each regulator.

On gas lines (LP or natural gas), a primary regulator is used to reduce tank pressure to line pressure (100 psi to 10 psi). A secondary regulator is used to reduce the lower line pressure to 7 inches W.C. or 11 inches W.C., the normal operating pressures used to convey gas to appliances within a building. An additional appliance regulator may be used to reduce the pressure to 4 or 9 inches W.C., the operating pressure of the appliance. The reason for using an appliance regulator is that the pressure drop from the second stage regulator to the appliance may be too small, and the resultant pressure would change the flame characteristics. The pressure is again reduced at the appliance to the design pressure of the appliance.

On some installations, a single regulator lowers the pressure from tank pressure (100 psi) directly to the 7 inches or 11 inches W.C. in one operation. These are termed *single stage regulators*. It should be noted that most secondary regulators are designed to function in the event of failure of the primary or first stage regulator. In other words, each secondary regulator can function as a single stage regulator; if the primary regulator fails, the secondary regulator would still function.

Thermostats

A thermostat is a mechanical device used to regulate electrical circuits. The most common type consists of a strip of dissimilar metals (iron, brass or copper) welded together. The expansion of the copper, due to its high coefficient of expansion, will warp the metal strip away from its fixed end, opening or closing an electrical circuit in the process.

As the bimetal strip cools, it will return to its original position, again opening or closing the circuit, depending upon the desire of the designer. The metal strips are usually a flat bar but quite often take unusual shapes, such as the spiral shape of a clock spring or the helical coil found on stack switches. The only difference between the two is the motion of the end of the bimetal strip. Movement is in a linear direction in the flat bar. In the spring shape, it follows the arc of a circle; in the helix, the end of the bimetal strip describes a helix. Once the circuit is reversed, the bimetal strip cools, reversing the process.

Limit Controls

A limit control is a device used to control the range of a variable. A typical example is a limit switch on a furnace. A furnace has a low limit and high limit representing the highest and lowest temperatures that the appliance will maintain. Limit controls may also be used as a safety device to prevent overheating of the appliance in the event of failure of the thermostat.

For example, if the thermostat contacts became fused together, without additional limit controls the appliance would continue to heat and would reach a balance between the heat dispersed by the appliance and the energy input. If the output of the appliance raised the temperature of adjacent combustible materials above their ignition point, a fire would result.

To prevent this, a second limit switch may be provided. This limit switch will be set below the ignition point of materials normally found in the vicinity of the appliance. In order for an appliance to cause a fire, the thermostat control and the limit control would have to fail.

A common type of limit switch is a metal snap switch which is normally closed in series with the thermostat. When the limit control is overheated, a bimetal disk in the control will snap from concave to convex, causing movement of a plunger which opens the contacts of the limit switch. The limit switch is typical of most mechanical components used to control an electrical circuit.

Electrical Controls

The flow of electricity is analogous to the flow of fluids in a pipe. As in water systems, smaller pipes carry lower flows of water, so the size of the wire required will decrease as the voltage decreases. Electrical controls vary from variations in the wire size (analogous to changing the size of pipes in a fluid system) to highly complex electronic circuitry for controlling the energy released in an atomic pile.

The fire investigator should be familiar with the basic types of electrical controls so that he can identify them and understand their purpose. Armed with this knowledge and simple tools, he can check the function and condition of the circuits before calling in an electrical engineer.

Switches

The simplest switches are mechanical devices that open or close an electrical circuit. Current flowing through an electrical wire has the capacity to leap across small breaks in the conductor. This small space is called an *air gap*. The air gap that the current is capable of leaping is a function of the voltage and current. For example, the high voltage circuit on an automobile ignition system can pull an arc approximately ¾ inch in length before the air resistance breaks the arc.

If an arc between two components of an electric circuit is permitted to continue for any length of time, the heat generated by the arc will heat the ends of the electrical conductors and raise them above their melting point. Should the switch contacts come back together in the molten state, the molten metal would fuse together, effectively welding the switch in a closed position. The strength of the weld may exceed the force normally used to open the switch so that it can no longer function.

Switches are designed to overcome this problem in several ways:

Type 1. Positive and rapid separation of the contact points.

Type 2. Enclosing the contacts and area involved in oil.

Type 3. Flooding the area of arcing with a blast of an inert gas for the duration of the arc. A blast of air can be used to accomplish the same ends.

Low voltage and normal residential or industrial switches handling voltages up to 120 volts are usually of *Type 1*. *Type 2* is normally used in voltages up to 440 volts. *Type 3* switches are normally used for switching and distribution systems.

Relays

A *relay* is a form of switch not requiring manual control. A relay consists of a coil of fine wire wrapped around a core. This forms an electromagnet to activate a lever to open or close a circuit. When the coil is energized, the magnetic force pulls on the switch lever (acting against a spring) and opens or closes the contact. When the coil is de-energized, the spring moves the lever away from the electromagnet and opens or closes the switch. The power of the electromagnet is a function of the voltage or the current of the circuit feeding it. The power must be strong enough to resist the force of the spring. This makes the electromagnetic relay a useful device for protection against overcurrents or undercurrents.

For overcurrent protection, the relay can be in the *normally open* position. The resistance of the spring which holds the contacting lever away from the electromagnet will govern the amount of force required by the coil to close the switch. When an overcurrent occurs, the magnet would overcome the resistance of the spring, closing the relay. This in turn can either make or break the contact controlled by the relay.

Conversely, the relay can be used to regulate undercurrents. In this situation, the relay would normally be closed. As the current drops, the force of the spring is sufficient to open the contacts. Relays are sometimes used in series, as on the starting circuits of electric motors. As the current requirement drops, the relays cut out the amount of current supplied to the motor.

Solenoids

A *solenoid* is a specialized form of relay. Instead of having a solid core, the core is hollow with coils wound in a cylindrical path. A shaft slides within the hollow core and is acted upon by the magnetic field when the coil of the solenoid is energized. This causes the solid shaft to move in reaction to the electromagnetic force exerted upon it by the energized coils of the solenoid. This movement is used to activate a control or open or close a switch.

As in a relay, the current which energizes the solenoid coils can be either on or off, and the action of the plunger can either turn a switch on or off, depending upon the desires of the designer.

For example, in the *gas valve train* (a series of valves) of a large furnace, a solenoid valve would be used to hold a spring-activated fast-closing valve in the open position. The electrical energy supplied to the solenoid would be holding the plunger against the force of the compressed spring. In turn, the electrical circuit which supplied power to the solenoid coils would be controlled by a photoelectric cell sensitive to the flame in the combustion chamber. In the event of a flameout, the effect of the absence of light on a photoelectric cell would open the circuit to the solenoid coils. The electromagnetic force on the plunger would decrease to the point where the spring would close the solenoid, cutting off flow of gas to the furnace.

Resistors

A *resistor* is an electrical device placed in a circuit to lower the voltage in a circuit. It accomplishes this by offering resistance to the flow of electricity. It is analogous to a pressure-reducing valve in a hydraulic line or a pressure regulator in a gas line. Failure of a resistor would permit excess voltage to flow past the resistor, upsetting the electrical balance of the circuit.

Timers

In control systems, the timing of certain events is regulated by devices of ingenious combinations of mechanisms. One common type of *timer* consists of a heater, a bimetal strip and a relay. The heater is activated by a high current and raises the temperature of the bimetal strip, which expands and closes a contact to activate a relay and control the desired appliance. This type of timer can regulate operations for long periods of time.

A common application of this type of device is the household toaster. When the operator inserts a piece of bread and depresses the operating bar, it closes the circuit and heats the element which toasts the bread. In turn, the heat from the heating elements warms the surface of a bimetal strip. When the bimetal strip warms enough, it expands and trips a spring, permitting the toast to pop up, and shuts off the flow of electricity to the heating elements. The bimetal strip cools, and the cycle may be repeated by depression of the operating bar, which resets the spring.

For longer sequences, electrical clock timers are utilized. Instead of the normal hands of the clock, wheels with indentations or knobs are used. The protruding knob will press against a switch lever, closing a contact, or the notches in the wheel will permit a contacter to fall into the notch, opening a circuit. The length of the notch or the length of the bump on the circumference of the wheel will regulate how long the circuit will stay in its secondary position.

The constant frequency of the electrical current is used to regulate the timing of the wheel. The gearing of the wheel determines the time between repeated cycles of the circuits controlled by the indentations or bumps on the wheel. A number of wheels can be mounted on one shaft driven by a master time clock. Combinations of circuits can be open and closed simultaneously and timed precisely by the spacing of the control wheels.

Transformers

A *transformer* is a mechanical device linking two or more electrical circuits. A magnetic circuit is created by the *primary* current. This induces a current in the *secondary* windings inversely proportional to the number of windings on the primary and secondary coils. If the number of windings on the secondary circuit exceed the number of windings on the primary, it is a *step-up* transformer; the secondary voltage is higher than the primary voltage. If the number of windings on the secondary are fewer than on the primary, it is a *step-down* transformer; the secondary voltage is lower than the primary voltage.

Transformers are commonly used in electrical transmission systems, since it is more economical to transport electrical energy via high voltage lines than low voltage. Therefore, step-up transformers are commonly used at a power plant to transmit electrical energy to the area of distribution. At the distribution point, step-down transformers are used to distribute the power. A block of residences may characteristically have a step-down transformer to distribute power from the pole to the houses.

In the control industry, lower voltages are favored due to the insignificant amount of arcing and smaller wires required. To facilitate this, step-down transformers are used to reduce voltage to a more manageable voltage of 25 volts. Buses and trucks use 25 volts or less; automobiles use 12 volts; motorcycles use 6 to 12 volts.

Thermistors

It has been pointed out that under normal conditions, very little change in the electrical resistance of materials occurs under normal heating of the materials commonly used in electrical conductors. However, one class of solids (semiconductors) has physical characteristics that make them temperature sensitive. In fact, the term *thermistor* comes from the description of the device which has been called a *thermally sensitive resistor*.

A *thermistor* is an electrical device in which resistance changes due to a change in temperature. This fact is used in control systems as a means of regulating temperature-sensitive installations. A thermistor is much more sensitive to changes in temperature than other types of thermometers. With the proper circuitry, thermistors can perform a control function much more accurately and efficiently than any other type of installation.

Thermocouples

If two dissimilar conductors are fused together, heat applied to the junction of the two dissimilar metals generates a current in the two wires. The device is called a *thermocouple*. The induced current is proportional to the amount of heat applied to the junction and can be used to control relays, solenoids or other electromechanical devices which are dependent upon the heat source to activate the current. The most common application of this device is in gas furnaces. The junction of the thermocouple is inserted into the flame of a pilot light. The current generated is used to hold a solenoid valve open, which permits gas to continue flowing. Should the pilot flame be extinguished, the current would cease to flow. The spring on the solenoid would close the gas valve, cutting off the flow of gas to the pilot line.

Other applications of the thermocouple principle are found in pyrometry. *Pyrometry* is the measurement of temperatures through the use of electric current generated by the heat being measured. Since the voltage is proportional to heat, the voltage is converted into temperature on a visual scale so that the temperature of the heat source can be measured.

Fluidic Controls

Recent developments in the control industry have seen the phenomenal rise of *fluidic device controls*. These devices have no moving parts and possess a high controlling ability. The control medium can be either gas, liquid or liquid-carrying solid particles (coal slurry). They are not normally found on residential or industrial heating systems but find wide use in large industrial plants.

The principle is simple and effective. The flow of the moving stream can be used to control itself. Since there are no moving parts and no exterior power is required, the only failure of these devices is the failure of the system itself. For example, the control lines may break, resulting in an undesired reversal of the controlled valve function, causing flammable material to be pumped into an undesirable atmosphere, which may result in an explosion.

Should the fire investigator become involved in a fire where a fluidic device control system is used, an expert in this field should be consulted. Although the device and its applications are very simple, the process system involved can be extremely complex.

CONTROL SYSTEMS
Combination Controls

By combining the various fluidic, electrical and mechanical controls, complete systems to perform a function or control a system can be devised. Very seldom is one isolated control found by itself. They are usually combined to serve more than one purpose with one control unit. An example is the 100 percent safety shutoff control unit found on modern residential furnaces (Fig. 63). These combine principles found in three or four separate units, including the following:

1. Main line shutoff valve
2. 100 percent shutoff on safety pilot line
3. Pressure regulation at the appliance
4. Solenoid activated by a thermocouple to provide shutoff for the entire valve in event of flame failure of the pilot light

Control Circuits

A control circuit can be comprised of mechanical, electrical, pneumatic, liquid or other media. Quite often they are an integrated combination of all of these, exercising control over an appliance or several appliances.

In industrial applications, circuits are used to control the manufacturing process as well as the environment within the buildings. In many of these processes, highly flammable substances are used.

Control System Elements

A control system consists of six basic elements:

1. A sensor which measures some physical aspect of the system (temperature, pressure, humidity, length, acceleration, de-

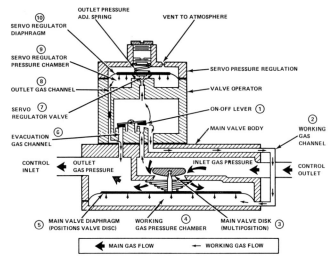

Figure 63. In this diagram of a typical gas valve, the on-off Lever (1) is in the position it assumes when the valve operator is energized by the thermostat on a call for heat. The left-hand on-off port is closed, and the right-hand port is open. This permits the flow of working gas into the Main Valve Pressure Chamber (4). Admission of working gas into this chamber causes upward movement of the Main Valve Diaphragm (5). This lifts the Main Valve Disk (3) to allow gas flow to the main burner. From *Honeywell Residential Controls Training, Gas Controls Reference Manual,* No. 71-97091. Courtesy of Honeywell, Inc., Minneapolis, Minnesota.

celeration, movement, color, light, radiation or some other measurable physical phenomena).

2. The element which receives the signal from the sensor and converts it into a readable signal by converting it into electrical, pneumatic, hydraulic, electromagnetic or other usable form.

3. An error detector element used to compare the signal received from the sensor with a desired value. If the difference between these two signals is zero, then the system has reached its *set point*. This is the desirable value required of the system. Once this has been attained, the system is functioning properly.

4. An amplifier used to increase the sensitivity of the interpret-

ing mechanism. Normally, the error signal must be kept at a very low power so that the system will not be weakened. After receiving the signal, the error is amplified and interpreted to the next stage.

5. A motor operator to operate the final control element in the system.
6. The control element which receives the amplified signal and is operated by the motor operator. After the final control element has been affected, the process is continued by checking the output of the controlled variable against the set point or desired result of the control element. This is illustrated in Figure 64.

This forms a closed loop for the system, and the system will continue to operate as long as the set point is not changed. Any change in the set point will cause corresponding changes in the controllers.

This simplified concept of a control circuit can be compared to the ordinary home heating. The desired result is the temperature at which the thermostat is set. This is the *set point*. When the temperature is within one or two degrees of the set point, the furnace control system will not activate. However, if the temperature drops below the lower range of the thermostat setting, the thermostat will call for heat. The furnace will generate heat until the room temperature raises the control thermometer of the thermo-

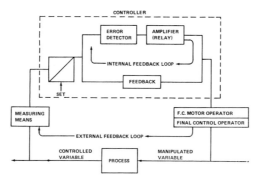

Figure 64. Typical control loop, diagram and identification of components. From *Measurement and Control Handbook and Buyer's Guide,* 1974-1975. Courtesy of Measurement and Data Society, Pittsburgh, Pennsylvania.

stat above the upper limit set on the thermostat. At that time, it will turn off the heat and enter another inactive phase.

The variable involved is the room temperature. The set point is the desired temperature. The primary control is the thermostat. The sensor is the thermometer which continually registers the temperature. When the temperature drops below the minimum temperature permissible, the sensor calls for heat by closing a circuit to the gas controller. The gas controller opens the valve, permitting the flow of fuel into the combustion chamber where it is ignited. Ignition of the fuel raises the temperature in the combustion chamber until an additional signal (activated by a low limit switch) causes the fan to activate and distribute heat throughout the building.

Following these general principles, highly complex processes can be regulated. The fire investigator should become familiar with circuits and their role in the process. He should also be familiar with the function and use of the components in control systems so that he can identify and recognize irregularities in the system.

Flame Control Systems

In any heat-consuming appliance, it is imperative that the flammable fuel introduced into the combustion chamber is ignited before the flammable mixture fills the combustion chamber completely. Should this occur and ignition is introduced at that time, an explosion would result which would damage the combustion chamber and possibly cause rupture of the fuel supply line and possibly fire. To prevent this, ingenious combinations of controls are used.

Gas Flame Control Systems

The normal flame control system on small residential-type furnaces is the omnipresent pilot light. Should the pilot light be extinguished, the system will automatically shut down, cutting off flow of gas to the pilot light and main burner.

However, in larger installations, it is imperative that the flame is *proven* before gas is turned into the combustion chamber. Therefore, the presence of the pilot light is monitored by a light-sensitive device which corresponds to the thermocouple of a pilot

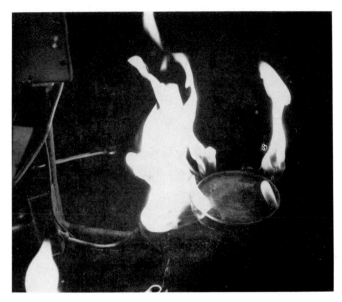

Figure 65. When the jet and ports of a burner become clogged, the resultant gas-air mixture burns improperly, causing impingement upon the surface of the hot water tank. This causes the formation of soot and compounds the problem. A slight downdraft can cause the yellow flames to impinge upon the aluminum pilot tube, causing it to fail. The yellow flames can also be blown back through the combustion air-intake passageway and impinge upon any combustible material in the vicinity.

light. The *cadmium sulfide cell* is exposed to light from the pilot light. When the pilot light is lit, the resistance of the cadmium sulfide cell is lowered so that current flows through the system, signaling to the control system that the gas mixture may enter the combustion chamber. If the pilot light is out, no light is sensed by the cadmium cell, the resistance is too high and the system will shut down.

Oil Flame Control Systems

The same type of system is used for an oil-burning system. However, in an oil-burning system, ignition is provided by a set of high-voltage igniters. The igniters are electrodes set approximately ¼ inch apart so that an electric arc can be formed between the two electrodes. As soon as this arc is detected by the cadmium sulfide cell, the circuit is closed, starting the oil pump to deliver atomized fuel to the combustion chamber. If no arc is observed, the electric circuit on the pump does not close and no oil is delivered to the combustion chamber.

Coal Control Systems

The same type of application is used for coal-burning furnaces. The function and application is similar to that of oil. No coal is delivered to the combustion chamber until a fire is *proven* by the cadmium sulfide cell or a similar heat-sensitive device.

Electric Motor Control Centers

Large electric motors require high currents to overcome their own inertia and the inertia of the load until the motor and driven portion reach normal operating speed.

By a combination of relays, the current is started at its maximum and is reduced as the inertia of the system is gradually overcome. This is accomplished by a series of electric relays and heaters.

Air Conditioning Control Systems

An air conditioning system consists of a forced-air circulating system driven by an electric motor. The cooling is accomplished by a compressor compressing gas and removing heat that is liberated as the gas expands. The cooled gas is carried through cooling coils, where heat is picked up from the moving airstream. Cold air is then circulated through the building.

The fluid used in air conditioning systems is nonflammable and presents no particular problem to the fire investigator. His main concern is with the *locked-rotor current* draw or starting current which is controlled by the motor control center and the circuits supplying them.

Industrial Processes

Industrial processes may involve hazardous materials. These constitute potential explosion and fire hazards. In highly industrialized areas, fire departments anticipate fires in complex industrial plants. A member of the fire service will contact plant managers and determine the flammable substances used in the manufacturing plant. Scale drawings of the plant layout will be obtained and saved, to be referred to in the event of a fire. On this drawing will be indicated the storage and movement of highly flammable substances throughout the plant. Since a fire usually destroys all drawings of the buildings and piping systems, these drawings, in the possession of the fire service, are invaluable to the fire investigator.

More detailed drawings of the piping system may be obtained from the engineering firm responsible for the design of the plant. They will usually retain the original drawings. The fire investigator should be cautioned that drawings obtained from the engineering designer may not contain changes made by the plant maintenance and operation personnel. If drawings are not available with the recent changes, the extent and detail of changes will have to be obtained by questioning employees.

The fire investigator should not attempt to check a complicated system without a detailed drawing as a guide. With a detailed schematic of the system, the investigator can save many hours of sketching. By following the process lines on the drawing, any apparent changes can be noted and verified by questioning personnel in charge.

APPLIANCE MALFUNCTIONS

Modern appliances have a number of automatic control systems. Although most of them are similar, each appliance has typical malfunctions which occur often enough to be noted. The control systems on industrial processes are so varied and complex that they cannot be covered except in a general manner. However, some of the principles involved in investigating the following appliances could also be applied to the investigation of complex industrial processes (Fig. 66).

Furnaces

The modern furnace is equipped with an automatic shutoff valve, regardless of the fuel used. The furnace consists of a combustion chamber and a fuel supply valve which controls the flow of fuel to the combustion chamber when commanded to do so by a thermostat. The thermostat is centrally located in a residence.

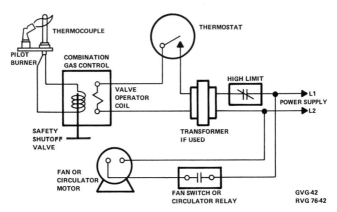

Figure 66. Typical residential gas control system. From *Honeywell Residential Controls Training, Gas Controls Reference Manual,* No. 71-97091. Courtesy of Honeywell, Inc., Minneapolis, Minnesota.

The thermostat is operated on a low voltage current and is seldom a cause of failure by itself. A short in a thermostat could cause it to continue to call for heat. This in itself would not cause a fire unless heating in the building exceeded the heat loss through the walls and roof. Eventually, the combustible materials could reach ignition temperatures.

If the thermostat has not been destroyed by fire, it should be checked immediately. A close-up photograph should be taken of the thermostat setting. A note should also be made of the set point. The thermostat cover should then be removed and examined for its condition. A multimeter can be used to test whether the circuit is open or closed. A check of the ambient temperature should be made to see if the thermostat should call for heat. The upper and lower limits of the thermostat can be determined either from examination of the limit switch settings on the thermostat or by rotating the thermostat and checking temperature settings at which the switch goes on or off.

The system supplying fuel should be tested for leaks up to the furnace control, as outlined in Chapter 7. The possibility of fuel coming from an external source should not be overlooked, particularly if an explosion accompanied the fire. The gas valve should be thoroughly examined. It is imperative that the fire investigator understand how control valves work.

It is sometimes necessary to obtain a similar valve or controller to familiarize yourself with its operation. Disassembling the duplicate valve should give a thorough understanding of the function of its component parts. Once the function is understood, the inspection of the suspect valve can be undertaken.

The suspect valve should be photographed before any tests or checks are run. All testing should be conducted, when possible, with the valve in place at the fire scene. The position of the valve controller, or any knobs or buttons on the valve, should be carefully photographed and noted. The electrical wiring from the valve to any components on the furnace should be recorded and photographed. An electrical check of continuity of all wires, noting the condition of all circuits, should be made.

Then, following the principles outlined in Chapter 7, the appropriate gas, at the operating pressure of the furnace at the time of the fire, should be applied to the system and the system tested for leaks. If none are found and the system functions, as indicated by the valve control knob location, the orifices on the appliances should be examined. The condition of the combustion chamber should be observed and documented. The condition of connecting flues to the exterior of the building should be checked for products of combustion (soot, oil, condensation, rust), leaks or improper connections.

The backdraft diverter should be examined to see whether it is free and operational. If the furnace is unscathed by the fire, the area around the draft diverter should be checked for signs of periodic backdrafts. This would be characterized by soot deposits on the exterior of the diverter. Signs of the products of combustion should be identified. These include rust stains, condensate stains or deposits of heavy soot on the interior of the diverter and chimney flue. If no excessive deposits are found, it can be assumed that the flame pattern was properly adjusted. If possible, the furnace should be lit for short periods of time by setting the thermostat at its highest position (calling for heat). The furnace should then be lit according to the manufacturer's instructions, and the pilot flame or arc (in the case of oil furnaces) should be examined.

After these are checked, ignition can be made and the burner left on for a brief period. During the ignition sequence of the furnace the flame should be carefully observed to see if a minor explosion occurs. Quite often a puffback will occur when lighting an oil furnace, which can blow accumulated oil into the room in front of the furnace.

If a fire investigator lacks confidence in his ability to perform these tests, an experienced furnace man should be contacted to assist and demonstrate the operation of the furnace.

With the furnace lit, an attempt should be made to extinguish the pilot light by simulated downdrafts. If it cannot be extinguished, the thermostat should be turned down to shut down the furnace. The pilot light should be extinguished and the thermostat turned up again, calling for heat. The action of the gas valve should be carefully observed to avoid an explosion. Under normal conditions with the pilot light out, the furnace valve should cut off the flow of gas to the pilot light and the main burner. The same principle applies to the cadmium sulfide detector cell on an oil system. If the system does not shut off and, upon call for heat by the thermostat, fuel flows into the combustion chamber, a defective valve is the cause of the gas flow. Should this occur, the valve manufacturer or an independent testing laboratory should test the valve (under the fire investigator's direction) to determine the cause of the malfunction.

Typical Control Valve Malfunctions

The most common malfunction of a control valve is from accumulation of dirt on the seating surface of the safety valve cutoff. The seat of the valve is exposed to the flow of gas through the valve. Dirt, scale or other debris carried by the gas stream can accumulate on either face of the sealing elements. When this occurs and the valve attempts to seat, it is held open by these minute particles, permitting gas to continue flowing through the valve.

Since the valve is normally open, gas is continually being drawn through the valve, exposing it to buildup of debris on the valve faces. The only time that the valve shuts off is when the owner manually closes the valve, or in an emergency, such as the flameout of a pilot light. When this occurs, the current from the thermionic generator decays, permitting the solenoid spring to close the valve. However, the presence of debris between the two valve faces holds the valve open enough for gas to pass through the crack created by the debris.

Another common gas valve malfunction is caused by clogging of the control orifices on the controlling valves for the main burner. This occurs in the same manner as in the pilot light valve, i.e. the dirt accumulates on the orifice, preventing the valve from sealing tightly and permitting enough gas to enter the system to hold the valve open. Should this occur, the main gas flow can flood the combustion chamber and pour into the room, giving rise to the possibility of an explosion.

One such case involved a control valve which had left the manufacturer with an unseated valve seal. The valve seal was normally attached to a spring-loaded lever attached to a solenoid spring. In this case, however, the valve seat was never inserted into its holder. The defect was not discovered at the factory and evaded sampling and testing procedures. The appliance installer evident-

Figure 67. Appliance controls, such as the water heater control illustrated in this photograph, are often totally destroyed by a fire so that no probative value can be obtained from their examination.

ly did not make a pilot light check and the furnace operated properly for almost ten years. However, in the tenth year, the pilot light went out. This was the first time that the pilot light had been extinguished in the history of the valve. Since the valve could not seat properly, when the thermionic generator shut down, the solenoid closed properly but did not impinge the valve against the valve seat. Therefore, gas continued to flow to the pilot light. Enough gas accumulated in the combustion chamber to spill out into the room, forming an explosive mixture. A source of ignition was found, and a violent explosion occurred.

Since no fire ensued, the furnace was still available for testing. Testing soon indicated a leak in the pilot system. Subsequent dismantling of the valve revealed the cause of the malfunction.

Water Heaters

Water heaters function similarly to furnaces except that the thermostatic control is located on the tank instead of in a separate room. Most water heaters, whether they are gas, electric or oil, are controlled by a thermostat, which is controlled by the internal temperature of the water in the heater. On top of the water heater is located a safety valve which reacts to overheating of water. Should the water overheat, the temperature may rise above the boiling point and form steam. Since the water system is a closed system, pressure can increase from steam pressure. The normal outlet setting on a water heater is about 125 pounds per square inch. The safety valve will normally release at this pressure. Should it fail, the pressure within the tank can relieve itself by pushing backwards against the incoming pressure of the water system. However, if a back pressure valve is located in the line, pressures will build up, rupturing the tank. Rupture of the water heater tank is usually caused by overheating of water and failure of the safety valve.

Therefore, as pointed out previously, two separate events must occur to have failure. The safety valve must fail, and the fuel controls must also fail, providing a continuous supply of heat.

Should the tank explode from internal pressure, the ruptured tank will be self-incriminating. The next step is to determine the cause of failure. The safety valve should be carefully observed concerning its orientation and position. Careful note should be made regarding whether the safety valve was mechanically prevented from opening (any weights set upon the activating lever).

If the tank is still intact, examine the combustion chamber, baffle and flue carefully. Any signs of heavy sooting should be noted and photographed. If heavy sooting occurred, the orifices should be inspected to see if the incorrect orifice size was used. If a natural gas orifice is inserted into an LP-consuming heater, the resultant oversupply of gas will form considerable soot, which may ignite and cause a fire. The accumulation of soot will result in a soft, lazy, yellow flame which can reverse flow through the flue, causing a downdraft. The flame can impinge upon the aluminum line leading from the gas controller to the pilot light in the combustion chamber, melting the line and igniting the resultant gas flow from the pilot tube. This can result in an external fire. A downdraft can also cause external fire by reversing flow of the flame out of the front of the water heater. The same problems experienced with a malfunctioning furnace gas valve also apply to a water heater control valve.

One additional caution should be stated. In the wintertime in subfreezing weather, a completely filled water tank will freeze solid. The result is that the hot water control valve will be held firmly by the ice. The normal water heater control valve is equipped with an emergency cutoff which forms an irregular humped projection into the ice. Any attempt to remove the control valve by unscrewing it will result in the destruction of the emergency cutoff. Therefore, the fire investigator is cautioned against removal of the valve until the water has thawed.

It is far wiser to test the valve in place for malfunctions. If possible, a flame test should be conducted to establish the condition of the flame in its setting with its own flue. Quite often the burn pattern will indicate that the fire started as a result of a malfunction of the ventilation system for the flue, due to a downdraft. This is covered later under *Downdrafts.* Water heaters are particularly subject to this type of malfunction, and the fire investigator should pay careful attention to the problems encountered with downdrafts.

Electric water heaters consist of immersion-type heaters con-

pletely contained within the water heater. Problems encountered
with these heaters cannot cause a fire as long as they are contained within the heater and covered with water to dissipate the
heat generated. The problems commonly associated with electric
water heaters usually encompass the controls which are mounted
externally on the heater.

These controls can be easily checked using a multimeter. Even
after being involved in a fire, the controls can still be found functioning properly, providing that the insulation has not been destroyed. The method used to check an electric water heater fire is
to locate the short or overheated wire which was the source of the
fire.

Dryers

A gas dryer consists of a rotating drum and a source of heat to
transmit, by conduction, convection and radiation, heat from the
combustion chamber of the gas burner to the articles to be dried.

The most common malfunction is overheating of the flue and
associated lint trap. These are easily checked by determining the
low point of the fire. If it involves the dryer, the flue system
should be carefully inspected to determine whether overheating
or ignition of lint in the lint trap was the cause of the fire.

Another common cause of dryer fires is overheating of the system by failure of a blower fan to carry away heat as fast as it is
generated. The shaft on the motor that drives the drum and exhaust fan should be checked to see if it functions after the fire.

Consideration should be given to the extent of heat involvement
of the motor in relation to the fire. It is difficult, if not impossible,
to tell whether the motor froze as a result of the fire or was the
cause of the fire. One simple test would be to examine the interior
of the motor housing. If molten lead has been thrown around the
interior, it is evidence that the motor was turning during the fire
and before the bearings froze. The lead, which has a lower melting
point than any of the other materials in the motor, would be
thrown out before the bearings seized. Another check is the examination of the remains of the V-belt which drives the fan. If the
fan was still rotating during the fire and the belt was burned
through, the belt would fall to the floor after separating due to

flame weakening. However, if the electric motor and fan were stationary, quite often a small remnant of the belt will be left at the top or uppermost point on the sheave.

A check of the gas controls, including the upper and lower limit switches, should be performed as a routine matter following the same procedures as outlined under furnaces.

Electric dryers can be checked in the same manner as gas controls. The most probable cause of an electric dryer fire relates to the lint and exhaust system. If the motor ceases to operate the drum, the exhaust fan also stops, as it is usually driven by the same motor. This permits the heat generated by the electric coils to build up and ignite the contents, lint or combustible materials near the dryer.

Another common cause of electric dryer fires is overheating of the distribution wiring at the back of the machine. The back of the machine is usually covered with a sheet metal back which, if overheated, can ignite adjacent combustible materials. Since most electric dryers operate on 220 volts, sufficient current is usually available so that the I^2R factor can start a fire when the circuit becomes overloaded due to stoppage of the motor. Machine overloading can cause overheating of the motor and result in an overload on the circuitry. The complete circuitry of an electric dryer should be carefully checked with a multimeter to determine whether the controls are functional after the fire.

Ranges

Gas ranges usually malfunction through operator error. This includes leaving boiling pans of grease on a range. The grease becomes ignited and can start a fire. These fires are very readily traced to the range by talking with the owner or person who last used the range. The position of the control knobs on the open burners should be noted and is usually the best indicator that an unattended range caused the fire. The low spot or burn pattern should be at the top of the range itself, with heavy destruction directly above the range. Quite often the pan will be found with remaining grease, and an identification of the "culprit" pot can be made.

Oven fires can be traced in the same manner. For example, a ten-pound ham can yield enough grease to completely destroy a home. The ovens as well as the ranges are sometimes controlled by automatic temperature regulators. Malfunctions of these devices are readily checked even after a fire. When the controls are under the oven or burners, the heat rises above the control and it is preserved unless total destruction of the building occurs.

With electric stoves or ovens, the "culprit" pans can be examined as well as the burn pattern to determine whether the open burner or oven was the cause of the fire. Once this determination has been made, the electrical circuitry should be examined and tested with a multimeter to discover if a defective control was responsible.

Electric Motors

Electric motors of one horsepower or less are commonplace in the modern residence or commercial building. The motor, regardless of which appliance it operates, should be inspected internally and externally. Its condition will usually reveal its involvement in the fire.

If the motor was the cause of the fire, it is usually due to the seizing of bearings and jamming of the appliance which the motor drove (pump, fan belt, compressor or other mechanical device). If the motor stopped for any period of time while it was operating by jamming, and the motor could not slip, the large current occasioned by the locked rotor will immediately overheat the internal windings of the motor. These can heat combustible materials by convection, radiation and conduction, causing a fire even before the circuit breaker trips.

Evidence of the motor involvement will consist of melted solder, melted or baked insulation on the windings within the motor, and a generally burned-out condition of the motor housing as well as the bearings and commutator. Comparison of the motor involved in the fire with an identical undamaged motor will provide a good basis for comparing components.

The second most common cause of fires by electric motors is overheating of the circuitry supplying power to the motor. As the electric motor labors, it draws more current. This increase in current heats up the wires to the motor, and quite often the wires start a fire before the motor windings do. It depends upon the size and design of the electric motor in relation to the wiring supplying the current. Therefore, the wiring leading to the motor should be carefully inspected to see if the wiring started the fire, with failure of the motor as a secondary cause.

Chimneys and Vents

Although chimneys and vents have already been discussed, it is necessary to explore their role in the function of a heat-generating appliance. The most common failure of a chimney or vent is overheating. This is caused by the accumulation of soot, creosote, oils or other byproducts of combustion on the walls of the chimney, flue or vent.

When a source of heat is introduced into a chimney or flue laden with unburned byproducts of combustion, the temperature of the flue lining can be raised to the point where these byproducts will ignite. The resultant fire can overheat the flue lining, which can ignite combustible materials near the flue.

Leaks or broken brick or tile in the flue can permit burning embers from the chimney to ignite combustibles adjacent to the flue. In addition, when a fire is ignited in a chimney, sparks and burning embers may be expelled from the chimney and ignite the roof of the building or adjacent buildings. These fires are generally easily identified by a low point near the chimney and the absence of any other possible cause. The only requisite for identification of this type of fire is that heat had to be present in the system to ignite the chimney.

DOWNDRAFTS

A *downdraft* is caused when the column of air in the chimney reverses its direction, comes out of the chimney into the fireplace and into the room or into the appliance and out the draft hood. In order to understand this phenomenon, it is necessary to understand the function of the draft hood, the chimney and associated vents.

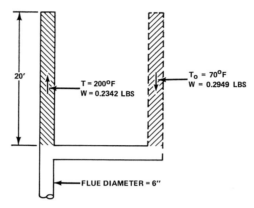

Figure 68. *Furnace draft.* The draft provided by a furnace is a function of the density of the flue gas and the density of the ambient gas or air within the building. From Earl A. Clifford, *Practical Guide to LP Gas Utilization,* 1969. Courtesy of Harbrace Publications, Inc., Duluth, Minnesota.

Stack Effect

Fire in a fireplace, water heater, oil furnace, gas furnace, electric dryer or similar appliances produces heat and products of combustion. To prevent contamination of the atmosphere in a building, these products of combustion, some of which are toxic, must be exhausted into the atmosphere. This is done by connecting the combustion chamber of the appliance, or the fireplace, to a vent which leads to a chimney. A chimney is a vertical structure which conducts the heated gases upward and is vented to the atmosphere above the roof line. This operation of the chimney and the surrounding atmosphere is termed the *stack effect.* Stack effect is illustrated in Figure 68.

For example, a chimney measuring 20 feet high and 6 inches in diameter will be used to illustrate the stack effect. The difference in the weight of the two air columns provides the force necessary to create a draft. In Figure 68, the available force is 0.06 pounds.

Draft Hoods

The above condition holds only after the flow of gases has stabilized. When the fire first starts, however, a column of cold air exists in the chimney which must be heated before it will form enough pressure differential to put the stack effect into motion and form a draft. In the meantime, however, the products of combustion are being formed instantaneously from the time the furnace goes on. In order to accommodate these products of combustion during the start-up, a draft hood is provided, as shown in Figure 69.

The products of combustion, which consist primarily of carbon monoxide and water vapor, are driven up the walls of the combustion chamber of the appliance and out the vent until they meet the heavier air in the chimney. At this point, their passage is blocked, and a means of egress is found out of the draft hood, as shown in Figure 69.

These products of combustion continue to circulate into the room until enough air in the chimney is heated by convection and radiation to establish a draft. This may take as long as five minutes in wintertime. Without the draft hood, the products of combustion would have no way of escaping and would accumulate in the combustion chamber. The carbon monoxide would block the flow of

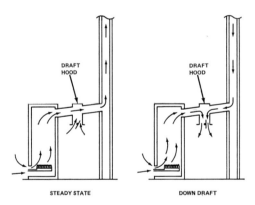

Figure 69. *Steady State:* A heating appliance properly vented with a draft hood. Varying draft conditions in the vent cannot interfere with the satisfactory performance of the appliance.
Downdraft: Under downdraft conditions, the products of combustion leave the combustion chamber and meet the downdraft at the draft hood, where they are both expelled into the room. From Earl A. Clifford, *Practical Guide to LP Gas Utilization,* 1969. Courtesy of Harbrace Publications, Inc., Duluth, Minnesota.

secondary air and would effectively extinguish the fire in the combustion chamber.

Turning again to Figure 68, the pressure differential necessary to establish a draft can be calculated. As can be seen, the pressure differential is very small, taking only 0.3 pounds per square foot pressure differential to establish an updraft or stack effect in the chimney.

Causes of Downdrafts

A downdraft can be caused by any change in pressure which would overcome the pressure differential formed by the stack effect so that the direction of gas flow in the chimney is reversed. The usual cause of a downdraft is a strong wind which blows the products of combustion down the chimney, effectively stopping gases coming out of the chimney. The downdraft reverses the flow of gases, which then flow down the chimney until they meet the gases from the combustion chamber, as these gases will continue to flow regardless of what occurs in the chimney. These two columns of gas meet at the draft hood deflector and are diverted into the room.

A downdraft will continue as long as atmospheric conditions prevent the reestablishment of the stack effect, i.e. the air in the chimney has now cooled due to the intrusion of exterior air, which must be heavier than air originally exhausted through the same chimney.

Should a downdraft occur during a period when the furnace is out, or has been out for a considerable length of time, the downdraft can establish itself not only through the deflector hood but also through the combustion chamber of the furnace. When this occurs, and the downdraft is well established, the furnace may reignite but the products of combustion will be carried out through the combustion air intake. Since the combustion chamber was not designed for this mode of circulation, the flame characteristics will be changed; yellow flame will be drawn out through the combustion air intake. If combustibles are located near the intake, a fire will result. This condition is virtually impossible when the combustion chamber is equipped with a combustion air intake fan preceding the combustion chamber. The pressure provided by the fan will reverse the downdraft and permit the products of combustion to leave the furnace via the draft hood.

Another cause of downdrafts is the presence of air-exhausting fans in a building. For example, if a large fan located over a range in a restaurant is used to exhaust fumes and heat from the range, it also exhausts air from the building. The exhausted air would also have to be supplied from within the building, so the pressure in the building will start to decrease. Normally, the pressure in the building is maintained by air leakage into the building through doors, windows and cracks in the siding and elsewhere.

However, in cold climates where buildings are tightly sealed by storm windows, the amount of air exhausted can soon start to draw air from the exterior and cause a lower pressure in the building than on the outside, causing a pressure differential which could be sufficient to negate the stack effect, drawing air down the chimney, creating a downdraft.

Other fuel-burning appliances may also cause downdrafts. These appliances not only exhaust air through a stack but also consume large amounts of combustion air in the process. This air must come from within the building. After lowering the building pressure, the appliance will start to pull air down chimneys which are not in use.

Significance of Downdrafts to the Fire Investigator

The draft hood is evidence of the heating industry's recognition of the existence of downdrafts. Downdrafts can start fires if the draft hood permits heated gases to come in contact with combustible materials. Therefore, the construction and heating industry specifies clearances of appliances to prevent these heated gases from starting fires.

In furnace fires where the furnace is suspected of having started the fire, the possibility of downdrafts should be considered as a possible cause. Check the clearance of vital openings for the presence of combustible materials. The owner should be questioned to verify that the vent hood of the draft hood was not blocked by materials stacked around it or on the combustion air intake. These actions would most certainly result in a fire.

The presence of abnormal or unusual wind or high pressure

areas should be checked with the local weather service to determine whether a downdraft could have been possible. The status of other heat appliances, or air-moving appliances, should be verified for their possible effect on downdrafts.

While downdrafts rarely cause fires due to the ease of their prevention, the possibility should not be overlooked, particularly since the elimination of this possibility is easily accomplished.

CHAPTER 10

Motor Vehicle Fires

A CCORDING TO NFPA statistics, motor vehicle fires comprised 19.8 percent of the total number of fires reported through the years 1970-1974; 16.6 percent of the reported motor vehicle fires were of incendiary or suspicious origin. The classification *Other, Unknown* consisted of 72.5 percent of the total number of motor vehicle fires. By using the same principle of adding incendiary and suspicious fires plus one half of the unknown fires, the total percentage of incendiary fires is 42.85 percent. These statistics emphasize the need for more thorough motor vehicle fire investigations (see Tables I, XXIII and XXIV).

America is rapidly approaching a time where there will be one motor vehicle for every two people. With a population of more than 200 million, this means that 100 million motor vehicles (automobiles, buses, trucks and motorcycles) are on the road. It is not surprising, therefore, that 20 percent of all fires are motor vehicle fires.

Few statistics are available to analyze the causes of these fires. In 1974, a study was made by the Insurance Institute for Highway Safety. A total of 2,637 motor vehicle fires were surveyed. Of these, 2,325 were passenger car fires. The distribution of the causes of the fires is shown in Table XXIV. It indicates that the majority of fires originate in the engine compartment as a result of collisions.

STATISTICS

Statistics are lean, but those available indicate the problems encountered. They also reflect the lack of in-depth investigation into automobile fires and the need for more intensive and productive investigation and analysis of these fires. Table XXV is a tabulation of the Minnesota State Fire Marshal's record for the years 1973 and 1974.

It is obvious from these figures that the number of fires result-

TABLE XXIII

ANNUAL INCIDENCE OF MOTOR VEHICLE FIRES*

	1970	1971	1972	1973	1974
Motor vehicle fires	497,700	501,600	550,300	574,000	640,000
% of total fires	20	18	20	21	21
Total fires	2,549,550	2,728,200	2,757,600	2,694,100	2,982,000

* Courtesy of National Fire Protection Association.

TABLE XXIV

FREQUENCY DISTRIBUTIONS OF PASSENGER CAR FIRES
BY TYPE OF INCIDENT AND AREA OF FIRE ORIGIN*

| | Type of Incident | | | |
| | Noncollision | | Collision-Related | |
Area of Fire Origin	Number of Fires†	Percent	Number of Fires†	Percent
Engine	1,085	59	39	54
Passenger compartment	647	35	3	4
Fuel tank	59	3	24	33
Trunk	31	2	3	4
Tire/brake	29	2	3	4
Total	1,851	100‡	72	100‡

* Courtesy of the Insurance Institute for Highway Safety, Washington, D.C.
† Figures do not include fires of unidentified origin. The 1,923 passenger car fires in which the area of origin was identified comprised 83 percent of the 2,325 passenger car fires surveyed and 73 percent of the 2,637 total motor vehicle fires surveyed.
‡ Column does not add to total due to rounding.

ing from undetermined causes averages 50 percent. The next highest cause of fires is mechanical. Following that is electrical, carburetion and incendiary, which is less than 2 percent. Suspicious fires, which may have been incendiary, total between 2 and 5 percent for this two-year period. These statistics, compared with NFPA statistics and the study made by the Insurance Institute for Highway Safety, indicate the sample may be too small and localized to extrapolate for the entire United States. It is clear that more work must be done on the investigation and determination of the causes of automobile fires.

Table XXIV indicates that approximately 17 percent of the

TABLE XXV

CAUSE OF AUTOMOBILE FIRES IN MINNESOTA, 1973, 1974

| | *1973* | | *1974* | |
Cause	*No.*	*Percent*	*No.*	*Percent*
Carburetion	56	4	74	8
Electrical	108	9	92	10
Gasoline	27	2	28	3
Heaters	4	1	7	1
Incendiary	14	1	22	2
Mechanical	130	10	124	13
Miscellaneous	69	5	40	4
Overheating	10	1	5	1
Smoking	44	4	41	4
Suspicious	31	2	62	5
Undetermined	762	61	471	49
Total	1,255	100	966	100

fires were of unidentified origin. Neither Tables XXIII nor XXIV reflect the number of arson fires. Compounding the problem are the questions of whether (1) the vehicle was moving or parked, (2) the engine was running or idle and (3) the vehicle was attended or unattended. These factors make the determination of the origin and cause of a motor vehicle fire a difficult problem to solve.

TYPES OF MOTOR VEHICLES

Motor vehicles include motorcycles, automobiles, buses and trucks. These vehicles have one common characteristic: they are propelled by a flammable substance. The vehicle consists of a chassis with upholstered seats for the operator, an engine for propulsion, a gear system to transmit power to the road, tires, wheels, brakes, a fuel system and an electrical generation, storage and distribution system.

Motorcycles

Motorcycles are commonly two- or three-wheeled vehicles with an open-air seat above the vehicle. The engine is between the legs of the driver. The fuel system consists of a tank located immediately in front of the driver which feeds the carburetor by gravity. An electrical system consisting of a battery, a generator or

Figure 70. Examination of a motorcycle fire includes determining the position of the vehicle at various stages of the fire. This can often be determined by the burn pattern, fuel stains, pavement gouges and statements from eyewitnesses.

alternator and a distribution system to the spark plugs and lights is located beneath the driver, on the engine.

Accidental motorcycle fires, either collision-related or non-collision-related, are the result of a gas system failure with fuel spillage ignited by the heat of the engine, or an electrical or friction spark. A fuel leak can be ignited by backfiring of the engine or a spark from the hot exhaust pipe. The result is total destruction (Fig. 70).

Flexible fuel lines leading from the gas tank to the carburetor can fail by connection failure or weakening of the nylon or rubber hoses by overheating. Once these lines fail, a source of ignition is all that is required to start a fire fed by the gravity line from the fuel tank. Once the vehicle is at rest, a small leak can accumulate a quantity of gasoline from constant gravity flow (Fig. 71).

The fuel tank is vented at the cap and, unless a large fire is

Figure 71. The fuel lines leading from the fuel tank to the carburetor are usually flexible, combustible lines. However, they are held in place by metal clips which, if not properly attached, may loosen and permit gas leaks which can be ignited by the heat of the engine.

started, will vent itself with no danger of an explosion. However, if a fire starts from a broken gas line, other gas lines may be destroyed by the fire, resulting in an increase in the quantity of fuel available. This additional fuel may create fire and heat which exceeds the venting capacity of the gas tank. This could result in a pressure relief explosion. Motorcycles are universally propelled by gasoline-fueled engines.

The heat from the gasoline fire associated with a motorcycle will result in melting the aluminum and pot metal parts of the engine. The puddled liquid may contain broken parts from the engine which may have been the cause of the fire. These puddles of molten metal solidify, as shown in Figure 72. X-rays can be taken to determine whether any other parts are contained within the solidified mass.

If the fire breaks out while the cyclist is riding, or if the cyclist

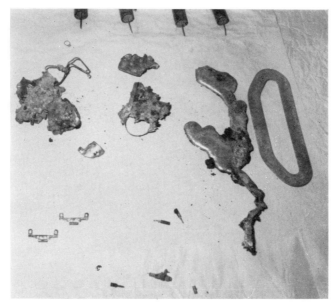

Figure 72. In any fire, but particularly in motor vehicle fires, lead or pot metal is found. The missing clamps or parts which cannot be accounted for may have been covered with the molten material. If this occurs, an X-ray should be taken to determine if the missing parts are contained in the solidified metal.

is sprayed by gasoline, the burn pattern on his clothing will often give a clue regarding the source of fuel and the volume sprayed on the driver. As can be seen in Figure 73, the cyclist was sprayed on the upper right portion of his body in this accident.

Automobiles

Since no breakdown of statistics is available for the number of automobiles relative to motorcycles, buses or trucks, this chapter is written with the assumption that the majority of investigations would involve automobiles.

Automobiles are propelled by gasoline-, diesel- or propane-fueled engines. The engines can be located either in the front or rear of the vehicle.

Buses

Buses are specifically designed for mass transportation of people. They are generally classified as two types: front end and rear

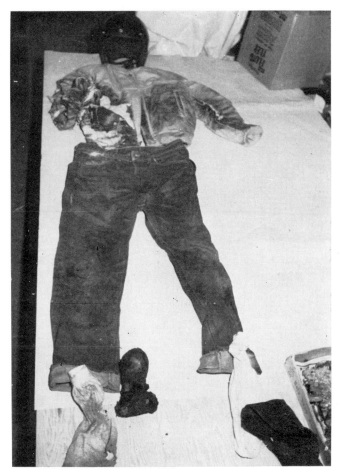

Figure 73. Quite often, a victim's clothing will reveal the burn pattern and cyclist's relationship to the origin of the fire.

end driven buses. A typical school bus is a common application of a front end driven bus, while a Greyhound or Transcontinental bus is typical of the rear end driven vehicle. Fuel systems in buses are either gasoline or diesel. The fuel supply tanks are located beneath the floorboards on the bus at various locations. Due to the large physical separation of the various possible causes of fires on buses, it is easier to isolate the *low point* and cause of the fire.

Trucks

Trucks are designed to haul freight, cargo or commodities. They vary in size from the small pickup to multiengined dump trucks designed for mining applications in off-the-road use. The most dominant feature of a truck is its size. The larger the truck, the easier it is to isolate the origin of the fire, due to the physical separation of the possible causes of the fire. Trucks may be gasoline, diesel or propane fueled.

The larger semitrailer consists of a tractor (containing the engine and propelling mechanism) with a trailer connected to the tractor by a fifth wheel. They are characteristically eighteen-wheeled vehicles.

The fuel system consists of saddle tanks located on each side of the tractor, forward of the rear axles of the tractor. The trailer, if equipped with a refrigerator unit, may contain an engine with a separate fuel system to maintain temperatures in the trailer.

The same tractor which is used to haul semitrailers can also be used as a truck by equipping it with a truck body. The configuration of the fuel system, brakes and transmission will remain the same.

Large trucks and buses utilize an air brake system as opposed to a hydraulic or mechanical system found on passenger cars or small trucks. The absence of flammable brake fluid eliminates one possible cause of a fire.

MOTOR VEHICLE CONSTRUCTION

Motor vehicles are similar in construction in several respects:
1. The vehicles are self-propelled. The only portion of a motor vehicle which is not self-propelled is an attached trailer.
2. Each motor vehicle has an engine compartment, separated from the passenger or operator's compartment by a firewall to block fumes, heat and fire.

The automobile is no exception. The passenger or driver compartment is a separated area designed and constructed solely for the occupant's comfort and safety. It consists of a comfortable seat for the operator and additional seats for passengers (Fig. 74).

The dashboard, or instrument panel, is often covered with a

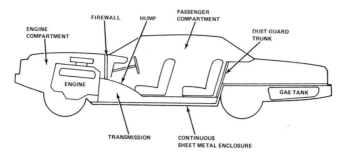

Figure 74. Typical automobile compartments. The gas tank is separated from the passenger compartment by a continuous sheet metal enclosure. Note also that the passenger compartment is separated from the trunk only by the rear passenger seat and a dust guard, usually made of cardboard.

soft, shock-absorbing combustible material. The floor, doors and roof are lined with a sound-absorbing combustible material. The chassis, or remainder of the vehicle, consists of the engine, transmission, steering mechanism, fuel supply, electrical and brake systems.

Engine Compartment

The engine compartment is separated from the passenger compartment by the firewall. The engine compartment contains the engine, fuel supply to the engine, electrical system, a source of electrical energy in the form of a wet cell storage battery and a lubricating oil system. The lubricating oil system is usually contained completely within the engine except for an external replaceable oil filter.

Exceptions are those lubricating oil systems on larger vehicles or specialized equipment, where the lubricating oil must be cooled by running oil through a radiator. This radiator is mounted at a location where it has access to fresh, cooling air which has not previously passed through the air cooling system of the engine.

Fuel Supply

The fuel supply consists of a fuel storage tank and a pump to move fuel from the storage tank to the engine. A carburetor or fuel injection system delivers vaporized fuel to the cylinders of the engine.

The fuel tank is seldom located in the engine compartment but is located beneath the vehicle so that the sheet metal floor acts as a physical barrier between the fuel tank and passenger compartment. In American cars, it is located in the rear of the vehicle to isolate it from the engine and passenger compartments. Fuel is delivered to the engine through a piping system by means of a fuel pump. The fuel pump may be located in the fuel tank or mounted on the engine.

Mechanical Fuel Pump

Since the fuel tank is lower than the engine or its carburetor, fuel must be pumped to the carburetor. The most common pump is a diaphragm pump, located on the side of the engine and operated by a cam-driven lever. In order for this fuel pump to function, the engine has to be running. If the engine stops, the fuel pump stops.

Another common type of fuel pump is the electric motor driven pump. This pump is submerged in the gas tank and operates by electrical current when the ignition switch is turned on. If the engine stops, this type of fuel pump can continue running and deliver fuel through a broken gas line or defective carburetor as long as current continues to flow through the motor. If a fire has started due to a broken fuel line, the engine may starve from lack of fuel. However, the fuel will continue to be pumped onto the fire as long as electrical current continues to operate the pump.

A fire can start from any break in the fuel system supply line or failure of the carburetor which spills the fuel, if a source of ignition is provided. In gasoline fires, an electrical spark is the usual source when the engine is running. The exhaust manifold can provide enough heat to ignite gasoline. A *backfire* of the engine (ignition of the gasoline within an engine cylinder while the intake valve is still open) can cause ignition.

In the event of failure of a diesel fuel supply, the source of ignition can be the same as in a gasoline fire, but the diesel fuel is more difficult to ignite. Sparks from the friction of metal rubbing on metal can provide ignition for diesel fuel.

In the event of propane gas failure, the vapors can be readily ignited by any sparks or heat.

Electrical System

The electrical supply in motor vehicles is a system of wet cell electrical storage batteries located in the engine compartment. In larger trucks or buses, they are found beneath the passenger compartment and in a separate compartment near the fuel tanks. In some foreign cars, the battery is located in the trunk. Although electrical fires do occur, the usual role of electrical energy is to provide ignition for the fuel.

The electrical storage battery should be tested to determine the state of charge and condition of the battery after the fire. A direct short, sufficient to cause a fire, would completely drain the charge from an electric storage battery. If the battery was not destroyed by the fire, the charge can be tested with a hygrometer to check the specific gravity of the battery fluid.

Power Steering

The power steering unit consists of a belt-driven pump to deliver power steering fluid to the power steering mechanism at a

Figure 75. The burn pattern on the automobile hood is of vital importance in identifying the main heat source of the fire.

Figure 76. This photograph of the engine of the vehicle shown in Figure 75 indicates that the cause of the fire was a leak in the high pressure system of the power steering unit. Examination of the burn pattern in Figure 75 and Figure 76 reveals no other possible cause of the fire.

pressure of up to 1200 psi. Should a pinhole leak develop in the high pressure side of the system, it would emit a vaporized, heated oil into the engine compartment. This can provide temperatures high enough for auto-ignition of the vapor, or a spark could provide the necessary ignition. Once ignition occurs, the heat may weaken the hose surrounding the pinhole to the extent that internal pressure would break the line, flooding the compartment with heated oil, thus adding more fuel to the fire (Figs. 75 and 76).

Another common cause of power steering unit fires is the loss of the cap from the power steering reservoir. Overworking of the power steering unit may result in overheating of the power steering fluid. This would foam or boil out of the power steering reservoir and flood the engine compartment with overheated oils and vapors. With the cap missing, the turbulence caused by the pump could splash oil onto the hot exhaust, where it can be ignited. A missing cap does not always indicate the cause of the fire but is one possibility which should be checked by testing the power steering fluid to determine its fire point.

Lubricating Oil System

The majority of American vehicles have a self-contained oil system within the engine. The oil is stored in a sump located directly below the engine. On specialized and larger engines, an exterior oil-cooling system is added and a separate oil reservoir used to store the oil. This is located towards the rear of the vehicle and normally will not become involved in a fire except by puncture of the container in a collision.

Lubricating oil operates near its flash point but below its fire point. When excessive wear occurs in an engine, the lubricating oil temperatures rise, so that maximum pressures and maximum wear occur within the engine. The oil temperature rises, and any break in the oil system due to the increase in pressure and temperature may cause failure of the oil system line between the engine and oil coolant. Seldom will an oil fire break out between the oil cooler and the storage reservoir. Leaks may develop, but the oil will not ignite due to its high flash and fire points.

Passenger Compartment

The passenger compartment consists of a seat for the operator and may or may not include additional seats. Combustible material found in the passenger compartment consists of the upholstery, cushions, floor coverings, dashboard and lining of the compartment. Combustible materials located in the passenger compartment are readily ignited. The only available source of energy contained in the passenger compartment is the electrical system. No fuel lines are located within the passenger compartment. They are carried beneath the floorboard of the passenger compartment from the fuel tank to the engine compartment.

The electrical supply is located beneath the floor covering in special channels provided in the sheet metal of the floor, or in grooves cut in the sound-deadening material, and under the dashboard.

Accidental causes of fire within the passenger compartment are either electrical or caused by foreign materials brought in by the operator or passengers. These accidental causes include electrical shorting, arcing, overheating, smoker's fires, spontaneous combustion and arson.

Electrical Supply

With the engine running, power delivered by the generating system of modern vehicles is shown in Table XXVI.

The electrical output of the modern vehicle has increased substantially over the last two decades, due primarily to the added convenience items. The increased use of electricity includes the addition of power windows, power vents, power seats, signal lights, heating and ventilating controls, etc. Increasingly complex ventilating systems have tripled the number of fans used in ventilating, heating and air conditioning systems. This increased use has resulted in the need for higher voltage circuits; a larger alternator or generator is required to meet the demand in ampere-hour size of the electrical storage batteries.

This means that when the generator is running at capacity, more than enough energy to start a fire is generated. It should be noted that the output of an automobile generator or alternator, as shown in Table XXVI, equals the heat output of ten pilot lights. This is sufficient energy on a sustained basis to melt the insulation on the electrical wires and ignite nearby combustible materials. The electrical storage battery of increased ampere hours will store enough energy to accomplish the same ends, should a massive short occur.

TABLE XXVI

ELECTRICAL ENERGY PRODUCED BY GENERATORS
OR ALTERNATORS

Vehicle	Voltage	Maximum Amperes	Watts	BTU/Hr.
Motorcycles	12	40	480	1,984
Autos	12	100	1,200	4,961
Buses	24	600	14,400	59,530
Recreational vehicles	12	150	1,800	7,441
Trucks	24	120	2,880	11,906

Chassis

The balance of the vehicle, excluding the engine and passenger compartments, is considered the *chassis*. The chassis consists of the frame of the vehicle, the brake system, electrical distribution system, transmission, drive shaft, differential or rear end, suspension system, wheels and tires, and the fuel tank with its accompanying fuel and vapor return lines.

Brake System

Three general types of brake systems are found in modern motor vehicles. These are mechanical, hydraulic and air-operated brake systems.

The compressed air and mechanical systems seldom cause trouble in the transmission of the braking force. Fires resulting from defective systems of this type are usually the result of dragging brake shoes on brake drums or failure of the mechanical portion of the system.

The braking system of a motor vehicle is a mechanical system wherein the energy of motion is transformed into heat energy of friction, which results in the slowing of the vehicle. The heat is generated by dragging a shoe that holds the brake lining against the brake drum or disc, which creates the necessary heat. If the heat is not dissipated by the forward motion of the vehicle, or by adequate ventilation of the brake drum, the heat can accumulate, heating the hydraulic brake fluid to the point of ignition should a leak develop.

The brake system can cause overheating of the drum, which in turn will transfer heat to nearby combustible materials (usually the tires) or the brake fluid. Increasing the temperature of the brake fluid causes an increase in pressure that may rupture the brake line, spraying brake fluid on the overheated drum, resulting in ignition and further fire spread. Once a leak has developed, the operator of the vehicle may pump the brakes to overcome the fading that results from loss of pressure. This results in pumping more fluid onto the fire, causing the fire to spread rapidly.

Electrical Distribution System

The electrical distribution system under the chassis will normally not cause a fire because of the low wattage in the system. Lines found under the chassis are usually distribution lines to the emergency lights, taillights, stop lights or other low-wattage appurtenances. These lines originate at a distribution panel located in the passenger compartment and are protected by fuses or circuit breakers. Seldom will a short circuit develop of sufficient magnitude to cause a fire because of the low wattage involved and the fact that few combustible materials are located under the chassis (Figs. 77 and 78).

The importance of the electrical distribution system in the underchassis portion of the vehicle is the possibility of an electrical short as a result of an accident which might supply ignition in the presence of flammable vapors. This is common in rear end collisions, when a fire results from a ruptured fuel tank in combination with electrical shorting.

Figure 77. This burned-out portion of a trailer had three possible causes: vandalism, smoker's carelessness, or electrical. (See Fig. 78.)

Figure 78. This photograph is the underside of the trailer bed shown in Figure 77. The fire started in this area and spread upward, as indicated by the significant difference in the depth of char between the two areas. Since the only available energy in this area is the electrical wiring, it had to be the cause of the fire.

TABLE XXVII

CHARACTERISTICS OF LUBRICANTS

Component	Lubricant	Operating Temperature Degrees F	Flash Point Degrees F
Power steering fluid	Oil	200	350
Front wheel bearings (disc)	Grease	300	450
Four-wheel drive, front transfer case	90 weight gear oil	200	400
Engine lubricating oil			
10W	Oil	220	400
20	Oil	220	440
30	Oil	220	460
40	Oil	220	480
50	Oil	220	500
Universal joints	Grease	220	520
Rear axle	Grease	220	520
Transmission oil			
Automatic	Oil	220	370
Manual	90 weight oil	220	400
Rear end differential	Gear oil	200	400

Overheated Mechanical Components

The modern automobile is a mechanical marvel. It can propel itself along highways at speeds up to 100 mph with complete comfort and safety for its passengers. This marvelous machine accomplishes this feat through the miracles of modern manufacturing methods and modern lubricating theory and practice. Without a thin film of oil separating the faces of gears which propel the vehicle, it could not travel more than a mile before seizing up due to overheating of the mechanical components.

The engine's transfer case, gear box, wheels, axles and bearings are designed to operate at a temperature below the fire point of the lubricant. The components are designed so that when the design temperature is reached, the heat is removed at the same rate it is generated. This results in a constant temperature of the components within their designed operating range. The operating ranges of the various lubricants are shown in Table XXVII.

Lubricating oils and greases have a wide range of flash and fire points. If the bearings, pistons or cylinder walls, due to wear or a factory defect, cause excessive friction, the friction will cause excessive heat, which may result in a fire.

The friction causes heat, and the oil will remove heat from the two interfaces. The oil in turn, carrying away the heat, will experience a temperature rise. Should this temperature rise exceed the fire point, a fire may result within the engine compartment. This fire will be an oxygen-starved fire due to the lack of air in the engine compartment. The heat will cause a pressure rise, forcing oil vapors from the vent where they will mix with air and ignite, being above their auto-ignition point.

This flame front will be fed by additional vapors from the vent and may ignite any available combustible material. If the fire is detected and extinguished, oil will still be present in the crank case unless the fire destroyed the gaskets which sealed the oil in the crank case. Floor insulation located directly above the source of heat will be subjected to the heat of the fire by conduction, convection and radiation. Directly on either side of this point, a distinct line of demarcation will be found.

Since the temperature of the oil is less than that of the ignition point of paint, quite often the paint on the material will not be blistered below the top level of the oil in its compartment. This phenomenon of overheated mechanical components can occur in any of the gear boxes, such as the transmission, transfer case, differential, or wheel bearings.

Tires

Tires can become overheated in several ways. Overheating can occur from underinflation, overloading or a combination of both. These conditions result in flexure of the sidewalls. This flexion will create heat faster than the forward motion of the vehicle can remove it. As a result, the temperature of the sidewalls rises, breaking the bond between the sidewall fabric and rubber material. This separation creates additional friction between the loose strands and the rubber casing of the sidewall. This results in heat generation. Should the vehicle stop, the accumulation of heat will quickly raise the temperature of the sidewall above its auto-ignition point, resulting in a spontaneous fire.

This can occur on any type of vehicle but is particularly hazardous on dual wheels of a truck or trailer. This may occur when one of the dual tires loses air. Since the adjacent tire now carries twice the load, it may be overloaded and the same phenomenon will occur.

However, the usual case involves the flat tire which rotates and flexes in a nonrigid position. Without the advantage of the internal pressure to hold its shape, flexing of the tire increases temperatures rapidly and results in overheating and ignition of the tire.

The danger in this case is that the heat is generated within the tire and the tire will become ignited as it is being driven along the highway. The forward motion of the vehicle will prolong the time required to ignite the tire. However, once it is ignited, it is difficult to extinguish. Should the vehicle stop so that the driver could extinguish the fire, the forward motion of the vehicle no longer carries heat away and the temperature of the tire increases rapidly. Unless copious amounts of water are available to cool the tire, it will spread to the trailer and any combustible cargo.

Fuel Tank

The fuel tank on modern American vehicles is usually located beneath the floor of the trunk and is a separate entity from the vehicle. It is a solid tank attached by metal straps independent of the floor of the trunk. The tank is normally isolated and safe from accidental fires, except in the event of a collision (see Fig. 74).

Another type of tank found in smaller vehicles and some foreign vehicles is termed a *drop-in* tank. Characteristically, the drop-in tank is mounted in a hole cut in the floor of the trunk of the vehicle. As such, the tank itself becomes part of the structural body of the trunk and is subject to any stresses imposed on the rear portion of the vehicle.

In a rear end collision, the structural aspects of the drop-in gas tank make it more susceptible to collision damage than the under-chassis tank. The under-chassis tank will normally break away because of failure of the supporting straps, and the tank will not become involved in a fire.

The drop-in tank, however, is highly susceptible to rear end collisions where compression and deformation of the rear end of the vehicle results in deformation, rupture and penetration of the drop-in tank. This will result in gasoline being splashed into the passenger compartment via the shattered rear window as a result of trauma of the collision.

INVESTIGATION OF A MOTOR VEHICLE FIRE

Investigation of a motor vehicle fire involves more facets than a building fire. Some questions to be determined immediately are factors leading up to the fire.

All circumstances surrounding the origin, discovery and extinguishment of the fire should be determined before examination of the vehicle, with emphasis on the following questions:
  1. Who discovered the fire?
  2. What was the condition of the vehicle at the time of discovery?
     A. Was the vehicle at rest?
     B. Was the engine running?
  3. Was the vehicle involved in a collision?
     A. Was the vehicle at rest?

B. Was the car on fire before the collision?

C. Was the fire the cause of the collision?

Answers to these questions will aid the investigator in seeking clues regarding the cause. It should be noted that once the cause of the fire has been found, other causes must be eliminated.

The Fire Scene

After the above questions have been satisfactorily answered, using the Witness Matrix, the fire investigator should examine the fire scene. If the vehicle was moving, the scene may involve a considerable length of road on which the vehicle was traveling. This area should be carefully examined. Pieces may have fallen from the vehicle (brackets, gas cap, drain plug or vital engine parts). If nothing is found during the preliminary examination of the vehicle's route or at the scene, it should be remembered that all vehicle parts must be accounted for either at the scene or leading up to the scene unless deliberately removed or missing from the vehicle before the fire. Therefore, any parts found at or leading up to the scene should be impounded for later identification and possible connection to the fire.

Examination of the area around the fire scene should be carefully made. All footprints, tire prints and material should be identified and impounded, with casts made of all impressions on the road surface.

If gasoline, kerosene or other flammable liquids are found beneath the vehicle, samples of the earth or pavement should be obtained so that the material can be identified. Tire prints of other vehicles should be carefully preserved by photography or in-place casting.

Spacing between wheel treads and tire widths should be measured for comparison with other vehicles which may become involved. The front and rear of the wheel tread on cars are quite often different. If two distinct sets are found, the wheel tread on all wheels should be determined.

The area should be searched for containers which may have been used to transport an accelerant. It should never be forgotten that arson can be a possible cause of a motor vehicle fire.

If the engine was running upon discovery of the fire, the scene should be carefully examined for skid marks. The existence of skid marks would indicate that the brakes were functioning properly at the time the vehicle was brought to a stop. The absence of skid marks does not prove that the brakes were inoperable; the driver may have stopped the vehicle without locking the wheels.

If the vehicle is at the scene when the investigator arrives, the undercarriage of the vehicle should be examined on the spot for signs of a burn pattern resulting from a ground fire. The ground beneath the vehicle should be examined to determine whether the burn pattern on the underside of the vehicle corresponds to the burn pattern on the ground.

For example, if a fuel line broke at a low point in the fuel line, even with the engine stopped, fuel would gravitate to the break in the line and spill on the ground. If this material is ignited beneath the vehicle, a corresponding burn pattern would appear directly above the fire.

If the burn pattern on the vehicle does not match the burn pattern on the ground, it indicates the vehicle was moved or moving as fuel spilled from the break in the line. The dripping fuel will leave a trail which can be traced back to its origin.

Before removal of the vehicle, the entire scene should be carefully photographed, with close-ups taken of any burned areas on the ground as well as photographs of the underside of the vehicle. The best time to take these photographs is when the wrecker lifts the vehicle into the air to be towed. Before the vehicle is raised, however, the lubrication level of all major components of the vehicle should be checked. These include the lubricating engine oil, automatic transmission oil, brake fluid in the master cylinder compartment, power steering reservoir oil and the electrical charge of the battery.

Before the vehicle is moved, the various compartments should be tested for leaks and broken lines, particularly under the vehicle.

Once the vehicle is raised, the remaining treads of the tires can be examined. If the tires have been burned, the footprint of the tire, which will not have been consumed by the fire, will usually be visible directly beneath the wheel. These should be carefully examined. If the owner states that new tires were recently installed on the vehicle, the remaining tread can be checked regarding depth and type of tread, which should correspond to the make, model and type of tread described by the owner. If the owner had anticipated burning his vehicle and had replaced the new tires with an old worn set of tires, the remnants of the tires will reveal this information.

The location of the fire scene may give a clue as to the possibility of arson. If the scene is located in a remote area, the fire investigator may conclude that the owner was seeking an isolated area to destroy the vehicle. Time and location will go hand-in-hand. If the location is secretive, the time will usually be in the early hours of the morning to avoid detection. The owner should have a valid reason for the presence of his vehicle at that time and location.

Recall Campaigns

The National Traffic and Motor Vehicle Safety Act of 1966 established a system whereby any defect related to motor vehicle safety must be reported to the purchaser and the Secretary of Transportation.

The National Highway Traffic Safety Administration publishes quarterly summary reports of defect campaigns conducted by manufacturers, available from the Superintendent of Documents, U.S. Government Printing Office, Washington, D.C. 20402.

When a motor vehicle fire is investigated, the recall campaigns for that particular make and model should be studied. If the motor vehicle involved in the fire has a reported defect which may result in a fire, the fire investigator will have an immediate clue to one possible cause. Assuming that the investigation reveals that this defect was the cause of the fire, all other possible causes should then be eliminated.

Origin of the Fire

A determination of the origin of the fire needs to be made if the cause of the fire is to be found. The fire had to start somewhere within the burned area, not a long distance away.

A determination of the origin of the fire should be made before the vehicle is moved, if possible. If the fire investigator does not have the opportunity to examine the vehicle at the fire scene, a comparison should be made of the burn pattern on the underside of the vehicle with the burn pattern at the fire scene, as though the vehicle were still at the scene. Determining the origin of the fire follows the same general *modus operandi* as in any other fire. Starting from the areas of least destruction, the fire investigator should eliminate the areas in which the fire could not have possibly started. This is done by observing the degree of destruction of the vehicle. The condition of the fire is considered in relation to the combustible material available and the presence or absence of wind. A fire has a difficult time burning against the wind. This is particularly true with a moving vehicle. However, once the vehicle has stopped and the engine is no longer running, the fire will burn upward as in a structural fire. With the engine running, the equivalent of a 30 to 35 mph wind blows horizontally across the engine and carries at lesser velocities toward the rear of the vehicle. If the vehicle is moving along the highway, its velocity is equivalent to that of the wind blowing from the front towards the rear of the vehicle.

Once these factors are considered, the low point of the fire can usually be determined. Exceptions to this low point rule include the tires and dripping combustible materials. For example, the tires are readily combustible. With the exception of a small footprint found on the ground, the tires will usually burn downward from the top until all of the tire has been consumed. If all four tires burn completely, four low points will be found. By comparing the relative combustion of the four tires and allowing for the presence or absence of wind, the fire investigator can determine whether the tires are low points or merely the result of the entire consumption of the vehicle by the fire.

The other consideration of a false low point is the presence of a burn pattern underneath the vehicle caused by dripping tar, nylon, styrofoam, melted insulation or other materials which, when heated, melt, flow and finally ignite. These can form low points which can be misinterpreted as accelerant puddles left by an arsonist. This error is usually easily detectable. The material which melts and runs will usually solidify after the fire has been extinguished, and remnants of the material will be found near the suspected spots.

Low Point of the Fire

As in any fire, an automobile fire usually has a *low point*. For example, if an overheated brake drum started a fire at the front of the vehicle and ignited the tire and wiring, the obvious low point would be the wheel and overheated brake drum. This would show cracking and burning of the brake lining, which would be discovered by comparing the brake drum, tire, wheel and linings with the other tires and wheels on the vehicle.

In the event of total destruction, the quantity of heat generated by the consumption of all combustibles in the vehicle may hide the "culprit" wheel. However, by pulling each wheel, if no other cause of the fire can be found, a comparison of the char on the brake drums will indicate the difference in the brake linings destroyed by friction and those destroyed by radiant and conductive heat from the fire.

The brake linings destroyed by the fire will have been heated by conductive and radiant heat from the adjacent heat sources. Since the brake linings are not normally in intimate contact with the brake drum, the conducted heat will not be as great as the heat generated by friction, which initiated the fire.

This same reasoning applies to a flat tire which started a fire. However, in the case of a flat tire, the destruction of the tire and subsequent destruction of all the other tires will obscure detection of the "culprit" tire. The "culprit" tire can only be identified by the fact that the tire would normally not heat up unless run for a long period of time in an underinflated or overloaded condition. The driver of the vehicle would become aware of the problem due to the difficulty in handling the vehicle and the presence of smoke and fire emanating from the tire as it ignited. Passing motorists would normally alert the driver, and the "culprit" tire would be identified by eyewitnesses.

Every rule has an exception, and the exception to this rule is a tire on a truck/trailer which may be unnoticed as the truck is speeding along the highway. However, if the driver stops for any period of time, the accumulation of heat rising to the top of the tire may just be enough to heat the tire to the kindling point as the truck comes to rest. The fire then starts unnoticed by the driver or passing motorists. This situation is rare, but it can and does happen, particularly if the tire is an inside tire on a dual axle. In this case, the tire has become overheated and ignited. The flame impinges upon the deck of the trailer, igniting the wooden deck and any cargo immediately above it (see Figs. 77 and 78).

Extinguishing a fire of this type is very difficult, due to the large amount of thermal energy contained in the tire. The tire can be extinguished but will immediately rekindle itself. The most effective method of extinguishing the fire is to remove the wheel from the trailer, but with an inside tire it is too dangerous and time consuming.

Again, eyewitnesses will be the best source of information by which to identify the "culprit" tire.

Another low spot associated with automobile fires is overheating of the lubricating agents in transmissions, gear cases or, at times, wheel bearings. These are usually easy to identify. The first step is to check the oil in the gear box or transmission case. If the oil has been overheated, it will have a characteristic "burned" odor which is unmistakable. The oil should be checked, using the dipstick, to determine the amount of oil remaining in the transfer case. The oil can also be smelled and rubbed between the fingers to feel whether it is full of grit or gound-up metal particles. If the oil has a gritty feeling, a dark color and a burned odor, a sample of the oil should be taken after determining the quantity of oil contained in the transfer case. Analysis should determine engine wear.

A fire in the engine compartment will usually destroy all hoses, and the power steering fluid reservoir will be dry. Check the interior of the reservoir to make sure that it contains the coiled spring and the dipstick. These are held in place by the plastic cap of the dipstick. As the fire destroys the plastic cap, the spring and dipstick will fall into the reservoir.

If the dipstick and spring are not found in the bottom of the reservoir, then the cap was not on the reservoir. With the cap missing, the power steering fluid, heated to its normal operating temperature, can be blown out of the reservoir under full stall conditions of the pump. This occurs when the steering wheel is turned all the way to one side or the other. If the oil is thrown on the exhaust manifold (usually right to the rear of the reservoir), the oil may ignite from the heat of the exhaust manifold, causing a fire fed by the contents of the power steering fluid reservoir.

This type of fire normally occurs when the power steering unit has been overworked, such as in parking a vehicle. The driver, unaware of the problem, turns off the ignition shortly after the fire has started, but before it has had a chance to create noticeable smoke and flames. The only requirement for this type of fire is that the cap is missing, the engine running and the power steering unit overworked. The positive indicator for this type of fire is lack of the cap being in place. Again, this is identifiable by the fact that the spring and the dipstick will not be found in the bottom of the power steering fluid reservoir.

Motor Vehicle Examination

It is seldom that a motor vehicle will be completely destroyed by fire. The compartmentalization of the vehicle will usually confine the fire to the compartment of origin. This simplifies the search for the cause of the fire because the fire has been restricted to one compartment.

Engine Compartment

Fires confined to the engine compartment should be investigated the same as any other fire, with emphasis on whether or not the engine was running at the time the fire was first discovered. The investigator should determine the circumstances surrounding the time of discovery of the fire, particularly at what time the engine was turned off, what electrical appliances were operating and the general condition of the engine.

The engine oil, transmission, power steering and master brake cylinder reservoir should be examined to determine the condition and level of fluids in these compartments. It is prudent to obtain samples from each oil reservoir and have the oil analyzed to see whether excessive wear, moisture or contaminants are found in the compartments. The condition of the oil on the dipsticks of each of the major compartments should be checked for a black color and burnt odor.

The top and bottom of the engine compartment should be carefully examined. It will be obvious if the fire started at the bottom or top of the engine compartment. If it started at the top, no burning will be evident under the car unless dripping accelerants are found as a result of an overzealous arsonist or a broken fuel line.

The majority of engine compartment fires are started through failure of some portion of the fuel line to confine the fuel to the carburetor. A connection in the fuel line may leak, spraying gasoline around the engine compartment where it can be ignited by a casual spark, a backfire from the carburetor or by a hot exhaust manifold. The second most likely cause is the failure of the diaphragm on the fuel pump. With each revolution of the cam driving the activating lever, a spray of gasoline emits from the fuel pump to ignite at the top of the engine compartment as well as the side where the fuel pump is located.

The engine compartment contains a considerable amount of combustible material. This includes insulation (including electric insulation) and neoprene hoses for the heater, air conditioner, power steering unit and cooling system. It takes a hot fire to ignite most of these materials. The fuel commonly causing ignition of these materials is gasoline.

The carburetor has been blamed for many fires but is probably no more guilty of starting fires than any other part of the engine. A carburetor's problem is usually a construction feature. When carburetors are injection molded, an outlet must be provided to let air escape from the mold. After the carburetor is assembled, the hole is plugged with a "soft plug." During the use of the vehicle the plug may fall out, permitting raw gasoline to escape into the engine compartment.

The other source of energy contained within the engine compartment is the battery. Since all of the current-controlling devices (fuses and circuit breakers) are located in the passenger compartment, the lines leading from the battery to the main power harness can carry the maximum amount of power available from the battery. This can result in a massive short and create enough heat to ignite the insulation or any combustibles located near the wires. These causes are readily checked if the battery has not been destroyed by the fire. A massive short would result in heavy arcing and burning, with the accompanying beading of wires. While these symptoms can also be caused by a fire, the distinction must be made by eliminating other probable causes of the fire. In the absence of any other cause, the electrical shorting must be the "culprit."

Passenger Compartment

Fires originating in the passenger compartment will most probably be man-made fires. No source of energy is sufficient to cause a fire in the passenger compartment other than electricity. Motor vehicle fires may be incendiary. This is a problem for the arson investigator. Characteristics of accidental and incendiary motor vehicle fires will be covered at the end of this chapter, listing important clues to look for when arson is suspected.

Accidental fires in the passenger compartment can be caused by the careless use of matches, cigarette lighters, cigarettes, pipes, tobacco, cigars, or various types of smoking materials. Tests conducted on various motor vehicle interiors have proven that the initial fire had been smoldering fire deep in the upholstery. This had gone undetected for hours and broke out into a flame fire long after the car had been vacated.

Since a smoking fire will normally start in the upholstery, it is important to carefully examine the burn pattern to determine whether the origin of the fire started at seat level or floor level. It is highly improbable that the fire would start at the roof level in the roof lining. The presence of a smoldering fire over a sustained period will result in a substantial formation of soot. This will normally be deposited on the windows in the interior of the vehicle, blackening the roof liner and interior. This type of heavy sooting is also found in an electrical fire but is seldom, if ever, observed in an arson or incendiary fire where accelerants are used.

The reason for this is that the amount of heat caused by the use of an accelerant will raise the temperature of the windows to the breaking or melting point long before accumulation of soot can occur. This rapid increase in temperature will usually cause the windows to shatter from unequal thermal expansion. This provides ventilation for the fire in the event that the arsonist neglected this aspect.

Smoking materials are commonly brought into the passenger

compartment by the driver and passengers. These can become inadvertently lodged in the upholstery of the seats or embedded in the long-nap carpeting of the floor. A smoldering fire will develop and in time can develop into a flame fire. Should this be a suspected cause of the fire, duplicate seating material should be obtained and the circumstances reconstructed to determine whether smoking materials could have caused the fire. Carefully examine the interior for the remains of the cigarette (or other smoking material). The burning cigarette will turn to ashes, but the ash will retain the original shape of the cigarette. Since the fire will not move these items around, carefully examine all ash patterns for the "culprit" smoking item.

ELECTRICAL CAUSES. All motor vehicles use electricity. One of the most common causes of an electrical fire within the passenger compartment is massive shorting through the wire harness where it penetrates the firewall into the dashboard. The dashboard contains a distribution panel for the wiring elements. All of the control wiring terminates at the dashboard to provide electrical controls for the driver.

The fans which drive the heater and air conditioning system are of such small capacity that, even with fully locked rotors, the current draw is too small to cause a fire. However, the locked rotor of a fan can increase the current draw enough to melt insulation and cause shorts of other higher current carrying wires, resulting in a fire. Any fire originating in the dashboard can usually be traced to either an electrical short or a locked rotor on a fan motor.

When a fire is confined to a passenger compartment, the search for the origin must first answer the question as to whether the fire started at floor level, seat level or at the instrument panel. If the fire started in the dashboard, it is undoubtedly an electrical fire. If it started on the seats, it is most probably a smoking fire. If the fire started on the floor, there can be any number of causes, including arson.

Not to be overlooked is the possibility of spontaneous combustion. A check with the owner will determine whether the vehicle was recently cleaned with a flammable liquid. If this were the case, the contents of the flammable liquid should be analyzed chemical-ly. If the material is of the drying-oil type, tests will reveal whether the material contained a flammable liquid and thus whether spontaneous combustion could have been a causative factor.

The owner of the vehicle should be asked whether any flammable liquids were transported in the vehicle immediately preceding the fire. If the owner denies the presence of any such flammable liquids and containers are found, this should be pursued. Check the trunk for any containers which might be suspect. It should be remembered that hard liquors (gin, vodka, scotch, etc.) are flammable liquids and can cause an explosion with a resulting fire.

Chassis

If the fire's origin is found on the chassis, the low point should be located to pinpoint the origin of the fire. The investigator should be wary of confusing the burning of tires with a low point, as cautioned earlier. The vehicle should be carefully examined from one end to the other to determine the low point. Examination of the interior side of the backing plates on the drums will usually give evidence of an overheated brake drum. This can be verified by comparing the interior of all four backing plates. The "culprit" brake drum backing plate will be burned, with all traces of road film turned to char and all incidental oil burned off. From there, the fire will follow a course upward into the engine compartment (in the case of front wheels) or into the trunk (in the case of rear wheels). The burn pattern on the "culprit" wheel side will be more intense than that on the opposite wheel.

This same principle applies to any low spots in a vehicle, as covered under *Overheated Mechanical Components.* Once the low spot or origin of the fire has been identified, the next step is to determine the cause.

Cause of the Fire

The cause of the fire is the heart of the matter to a fire investigator. It is the sole purpose of his investigation. Several factors are involved in determining the cause, but the cause is usually accidental or incendiary—seldom natural.

Fires Caused as a Result of a Prior Accident

When two vehicles collide, the energy is dissipated by deformation of metal. This deformation can break fuel lines, rupture containers of flammable liquids and cause sparks which are sources of ignition, either by friction or massive shorting of electrical systems.

In order to determine whether the fire caused the accident or the accident caused the fire, it may be necessary to retain an automotive accident reconstruction expert. He can reconstruct the accident, develop the paths of the vehicles leading up to the collision and determine whether the fire was burning before the collision.

The most common collision resulting in a fire is a rear end collision. The impacting vehicle penetrates the gas tank of the other vehicle, resulting in rupture and immediate ignition of gasoline, leading to a fire. The impact will spray gasoline from the ruptured tank throughout the interior of the impacted vehicle. As a result of deformation of the impacted vehicle, either the rear window is broken, permitting an avenue for the gasoline to penetrate the passenger compartment, or the rear firewall (usually cardboard) is ruptured, so that gasoline enters below the rear window through the trunk area.

Experiments conducted by the Insurance Institute for Highway Safety indicate that rupture of the gas tank occurs when the speed differential between the two impacting vehicles is around 35 mph.

Some foreign cars and older model lightweight vehicles manufactured in the United States are built with drop-in gas tanks. These are made a structural part of the trunk floor so that protection, such as is afforded by the sheet metal trunk floors on larger American cars, is nonexistent. In addition to the lack of a barrier between the gas tank and the vehicle interior (the trunk), it leaves nothing to prevent the intrusion of gasoline into the passenger compartment after an impact ruptures the gas tank. The fact that the gas tank is part of the structure of the trunk floor increases the probability that the tank will rupture when struck from the rear.

Mechanical Causes

Once the origin of the fire has been determined, a mechanical cause is usually self-evident. The actual cause can sometimes be determined by inference, once the origin has been located. For example, when the brake drum is the origin, the cause can be found by dissecting the brake system until the "culprit" part which failed and caused overheating can be found. At this time, a specialized mechanic should be contacted for a professional opinion. His expertise should encompass not only the mechanics of the vehicle but also experience in the determination of causes of vehicle fires. College professors, mechanical engineers or automotive technicians are often capable of providing this dual expertise.

Electrical Causes

The presence of an electrical cause is usually determined by several factors. First, evidence of massive shorting and an abundant supply of electricity must be present. If the fire investigator determines that it was an electrical fire and then discovers that the engine was not running and the battery had been dead for three weeks, then obviously he has erred. Confirmation of these factors is necessary to make a determination of an electrical causation. Electrical fires are usually confirmed by the elimination of all other possible causes rather than by direct evidence. A fire can cause electrical shorts, resulting in the same evidence which could appear if the fire had been caused electrically. If the battery is still intact, it should be checked for state of charge. After causing a fire, the battery will be fully discharged, as the energy is expended in raising combustibles to the ignition point.

Explosions

It is quite common for witnesses to state that they heard explosions at the scene of the fire. What they usually hear is the sound of compressed air in the tires escaping in a pressure release explosion as the carcass of the tire disintegrates from burning, permitting the escape of enclosed air. In a fast-burning fire, these explosions can occur with a sharp report. Under normal burning, they may sound like a big "whoosh." Since a vehicle is equipped with four tires and a spare, it is not uncommon to hear all five tires explode.

Vehicles equipped with air conditioning may also emit sounds similar to an explosion when the freon escapes. Other explosions occur in hollow drive shafts due to the internal pressure within the sealed drive shaft; this explosion is also a pressure release explosion. Other containers such as sealed bottles of beverages, food and aerosol-type containers may explode periodically as they become overheated.

The most common source of an explosion is usually considered the gas tank. However, since it is vented, it rarely explodes. If the vent is insufficient to relieve the buildup of pressure within the tank, the source of heat causing the increase will destroy the flexible rubber or neoprene connection between the tank and filler cap. Once this rubber connection is destroyed, the tank will be vented and no explosion will occur. The fumes released from the tank may add fuel to the fire and be visible to fire fighters. However, the tank will not explode, if vented.

Gasoline must form a flammable mixture in order to explode. Since the fumes in the tank are normally above the upper limit, the fumes cannot explode. The explosions associated with gasoline tanks occur by rupture of the tank, exposing the contents to the atmosphere so that a flammable mixture may be formed, resulting in a low order explosion. However, rupture of the tank must precede this explosion. The rupture can be caused by penetration, causing a pressure release explosion, which results in a low order flammable liquid explosion after the tank is ruptured by pressure release.

Arson

Every automobile fire should be investigated for the possibility of arson. The circumstances regarding the fire should be obtained before the vehicle is moved from the scene of the fire. Circumstances which point toward arson would include the following:

1. The fire occurred in the early morning hours, when no one would be around to witness the fire and which would ensure

a later arrival of the fire department, guaranteeing total destruction of the vehicle.

2. The fire scene is in a remote area where surrounding houses would not provide eyewitnesses to the fire, nor passing motorists who could observe the fire.
3. The vehicle is totally destroyed, with the fire involving the engine and passenger compartments.
4. Abnormal heat pattern, indicating the use of an accelerant. The presence of a fine ash from the plastic and upholstery of the vehicle, plus loss of temper in seat springs, plus a sagging roof due to excessive heat, are all good indicators that an accelerant may have been used.

If any of these circumstances are present, it is likely that arson has been committed. All other possible causes should be eliminated.

Other indications that could be found at the scene are the remains of burned matches in the engine or passenger compartment where the arsonist stood back and lit matches, throwing them into the vehicle compartments.

If gasoline was used as an accelerant, the resulting explosion may singe the hairs on the hands and face of the arsonist. If the arsonist is the owner, he will most likely explain that the fire, rather than the explosion, was the cause of the singeing.

Should the owner of the vehicle claim that the vehicle had been stolen, a diligent search for the remains of the ignition key is in order. The ignition key will usually be found in the lock, which characteristically falls from its position on the dashboard or steering column and can be found in the debris located on the floor. The owner may trap himself by claiming that the vehicle was stolen and that the keys were not left in the ignition, as this would be self-incriminating.

If the owner of a vehicle decides to burn it to "sell it to the insurance company," he may choose to claim that the car was stolen and will proceed to make arrangements to drive the car to a remote location, burn the car and have a friend or relative pick him up and drive him home. For this reason, the tire tracks and footprints around the vehicle should be carefully examined to see whether another vehicle could be identified as having been at the scene. All tire tracks and footprints should be photographed and identified and compared with vehicles and shoes owned by the owner, relatives and friends.

The whereabouts of the owner at the time of the fire should be verified carefully. The stories of friends and relatives should be entered into a Witness Matrix for verification and cross-checking of their stories. His friends or relatives may have been accomplices who returned him from the scene of the fire.

The ground where the vehicle rests should be tested for flammable liquids present which may have spilled on the ground when the vehicle was drenched with gasoline.

The time of year should be considered regarding the disposition of the vehicle windows. In order to ventilate the fire, the windows may be left open. This would be most unusual if the temperature was below freezing. It may also be unusual if the vehicle is equipped with air conditioning and temperatures were in the nineties or higher.

A fire in the passenger compartment, unless ventilated, will usually smother due to lack of oxygen. The inside surfaces of the vehicle and windows will be stained jet black by heavy soot formation. Presence of an oxygen-starved fire does not exclude arson as a cause. The arsonist may not have realized the necessity for ventilation in the passenger compartment. If the fire has smothered itself, the vehicle should be tested for the presence of an accelerant. The presence of an accelerant could confirm arson.

Totally Destroyed Vehicles

The majority of American cars have a firewall separating the engine compartment from the passenger compartment and a floorboard made of sheet metal which isolates the gas tank from the trunk and/or passenger compartment.

A fire originating in the engine compartment is generally caused by a failure in the gas supply system, causing gasoline under pressure to spray on the engine, igniting combustibles within the compartment. Another cause is a massive electrical short, which will ignite combustibles, leading to a much slower burning fire than one fed by gasoline.

A fire originating in the engine compartment can spread to the passenger compartment if a massive short occurs with the engine running. In this way, a constant supply of electricity is channeled through the wires which penetrate the firewall and enter the intricate electrical harness beneath the dashboard. This will ignite combustibles under the dashboard and quickly involve the passenger compartment.

Under ordinary circumstances, providing the fire is discovered early, the fire originating in the engine compartment will be confined to that area unless it is permitted to continue burning. Then, radiant and conductive heat will ignite insulation and plastics on the passenger side of the firewall.

By the same token, a fire starting in a passenger compartment will usually be contained by the firewall and may even smother from lack of oxygen if all doors and windows are closed. However, if a window or door is left open, convective currents will feed the fire so that it can be transmitted through the firewall by conduction and radiation.

Therefore, should an accidental fire occur, it is seldom that it spreads past the firewall from either the engine or passenger compartment. Only when the fire has burned for a considerable length of time does penetration of the firewall occur and total destruction of the vehicle result. The most common occurrence of a totally destroyed vehicle is through an incendiary fire. Therefore, the totally destroyed vehicle should always be carefully examined for the possibility of arson.

CHAPTER 11

Spontaneous Ignition

SPONTANEOUS IGNITION is the cause of approximately 1 percent of the total number of fires as reported by the National Fire Protection Association (NFPA). *Spontaneous ignition* is the classification given by the NFPA to those fires resulting solely from the uninhibited chemical or biological heating of materials to the point of burning.

This classification is unique in that the material ignited and the source of ignition are the same. Under the right circumstances and combination of chemical and biological action, some combustible material will heat and ignite itself.

Spontaneous ignition is caused by chemical reactions, biochemical actions or microbiological organisms that alter the physical and chemical characteristics of the combustibles. Materials subject to spontaneous ignition cover a wide range of organic materials, chemicals and agricultural products. In this classification are agricultural products such as hay, silage and grain; organic materials such as oil, fats, metals, charcoal and coal; and combinations of chemicals such as sodium, potassium, ammonia, chlorine, oxygen and hydrogen. Another classification, though not combustibles themselves, that may cause ignition of combustibles which they may contact, is the oxides (elements combined with oxygen).

Another unique characteristic of spontaneous ignition fires is the varying length of time they take to develop. The formation of pyrophoric carbon from wood (which is a form of spontaneous ignition) can take as long as fifteen years to heat and ignite the wood. Fires of spontaneous origin, starting in silos or coal storage yards, can take months to develop, while rags soaked with volatile fluids can cause ignition within a matter of hours and highly volatile fluids can start a fire within minutes.

An arsonist who understands the principle of spontaneous ignition can use this to start a fire at a desired time. The variations in the rate of chemical reactions can delay ignition from a few minutes to as many hours as needed to establish an alibi.

The unique characteristics of spontaneous ignition are variations in the time and chemical processes required to achieve ignition. This phenomenon is termed *spontaneous heating*.

SPONTANEOUS HEATING

Spontaneous heating is defined as the chemical reaction wherein the *oxidant* (usually air) combines with the *reactant* (the combustible material) in an *exothermic* (giving off heat) reaction. Exothermic reactions give off heat as a result of chemical reaction. A sufficient amount of air must be available for the reaction to occur and be self-sustaining, but not enough so that the generated heat is carried off faster than it is generated. The reaction results in a temperature rise in the material in which it is occurring. The rate of reaction will vary according to the substances involved and will accelerate as the temperature rises. The rate of increase varies from doubling to quadrupling or greater for each $10°C$ rise in temperature until the auto-ignition temperature of the combustible material is reached.

Spontaneous heating may attain an equilibrium temperature at which no rise in temperature will occur. This occurs when the heat generated is equal to the heat lost to the environment. This condition will prevail until a change in the environment occurs. If this equilibrium temperature is lower than the auto-ignition temperature, no fire will ensue. Not all spontaneous heating results in fire. However, the potential for reaching ignition temperature does exist when a sufficient amount of oxygen is present with spontaneous heating.

SPONTANEOUS IGNITION

Spontaneous ignition is defined as the attainment of auto-ignition temperature of materials subjected to spontaneous heating. Due to the rise in temperature, the rate of heat generation accelerates until the ignition temperature of the substance is reached.

Quite often, heating will result in chemical alterations of material being heated so that its normal ignition temperature will be lowered dramatically. An example would be the heating of wood over an extended period of time. The normal ignition temperature of wood is in the range of 450° to 460°F. As heat transforms wood to pyrophoric carbon, the auto-ignition temperature of wood may be reduced to as low as 200°F.

Once the ignition temperature is reached, spontaneous ignition occurs and either a glowing or flame fire can occur as a result. The sustenance of a fire is a function of the factors explained in Chapter 4, i.e. there must be a continuing supply of fuel and air with heat and continuous feedback in order to sustain a fire.

SPONTANEOUS COMBUSTION

Spontaneous combustion is the end product of a spontaneous heating and ignition process. Once the combustible has been heated above the ignition point, flaming or glowing combustion occurs and a full-fledged fire is in progress. The spontaneity which created the fire may continue until all of the material has been consumed and no evidence of the cause of the fire remains. All of the elements which contributed to the spontaneous heating, ignition and combustion will have been destroyed.

Substances Causing Spontaneous Combustion

In order to determine whether spontaneous combustion was the cause of a fire, the investigator should know those substances which can cause spontaneous combustion. A study by Underwriters Laboratories, Inc., defines four general groups of substances that are known to cause spontaneous combustion.

Group I—Substances Not Themselves Combustible but May Cause Ignition

The most common compound in this category is calcium oxide (unslaked lime), which has been found to cause fires when exposed to combustibles. One pound of lime mixed with water will emit 493 BTUs, which is approximately the amount of heat emitted by a pilot light burning for one hour. Similar reactions are experienced from barium oxide, sodium peroxide, phosphorus, chlorine and sodium. These chemicals have been used by arsonists to create the impression of a spontaneous fire. These chemicals serve a dual purpose in that they can also be controlled and serve as a timing mechanism, providing an arsonist time to leave the scene of the fire.

Group II—Compounds with a Low Ignition Temperature

These compounds are those which have an ignition point below ordinary or ambient temperatures. Examples would be phosphorous hydrides, silicon in the presence of air, sodium and potassium in the presence of water, turpentine and ammonia in the presence of chlorine and hydrogen, and chlorine in the presence of light.

This classification includes substances which are not acted upon by oxygen alone but which may simultaneously combine with another compound and act as a catalyst. Sodium and potassium react with water with the evolution of hydrogen. Hydrogen, for example, is spontaneously ignited at ambient temperatures.

Group III—Combustible Substances which Undergo Oxidation at Normal Temperatures

This classification includes vegetable oils, fats and certain finely divided metals. It includes the metal sulfides, pyrophoric carbon, charcoal and coal. Linseed oil, soybean oil and olive oil at ordinary room temperatures oxidize and produce heat. The key factor is a thinly exposed oil film so that the ratio of exposed oil surface to air is a maximum.

The same principle applies to finely divided metal. The decreasing size of metal particles accelerates the generation of heat and thus the rise in temperature. Pyrophoric carbon and charcoal are formed by the destructive distillation of wood or its byproducts. Charcoal has the property of adsorbing large amounts of gases exothermically. This in turn raises the temperature of the charcoal, resulting in the spontaneous ignition of the mass. Coal is in the same classification. The factors in the spontaneous ignition of coal are the size of the coal and its moisture content.

Group IV—Agricultural Products

Spontaneous heating can occur in practically all organic materials. The rise of temperature to the ignition point is accomplished through respiration, growth of microorganisms and the chemical combination of oxygen with the plant material.

Almost every harvested plant material can provide these three essential ingredients. Again, as with any other form of spontaneous ignition, the physical factors involved are the temperature of the material, the atmosphere, ventilation, microorganisms and water. Water acts as a catalyst for the chemical reactions involved.

The process is accomplished in three separate stages. The source of heat is that of respiration. This is similar to the process carried out in all living organisms where carbohydrates are combined with oxygen resulting in an end product of carbon dioxide, water and energy or heat. In other words, respiration is an exothermic process that can raise the temperature of the plant material to the point where the second stage of heating can occur.

The second stage of heating is accomplished through the respiration of microorganisms, called *mesophiles,* whose metabolic processes are exothermic. This is similar to the body heat of a group of persons raising the temperature in a room. Mesophiles start to reproduce and cause increases in the temperature up to 155°F.

At about 150°F, the rate of reproduction of mesophiles diminishes and the reproduction of thermophiles starts to increase. *Thermophiles* are microorganisms that thrive at a higher temperature than mesophiles. The growth of the thermophiles increases until the temperature of the substances (heated by the metabolic heat generation of the thermophiles) increases the temperature to approximately 165°F. At this point, the production of thermophiles reaches a peak and the third stage of heating occurs.

The third stage in the spontaneous ignition process is a chemical reaction with the plant material. Chemical reaction with the plant material starts at approximately 160°F and continues upward until the ignition point of the substance is reached or the process establishes a heat balance.

However, if conditions are conducive to spontaneous heating, the process will continue to raise the temperature until the ignition point is reached. Water generated by the mesophiles and thermophiles acts as a catalyst to increase the rate of heat production. If a decrease in the rate of heat dissipation occurs, the system will be arrested and ignition will not occur.

The key to spontaneous ignition of agricultural products is the moisture content. Water provides the catalyst necessary for spontaneous ignition, which can occur without the presence of micro-organisms. The heat of respiration alone can raise the temperature sufficiently to sustain the chemical reaction.

Materials Subject to Spontaneous Ignition

An excellent source for materials subject to spontaneous ignition or heating is found in the *Fire Protection Handbook,* 14th Edition, by the National Fire Protection Association. It lists the names of seventy-seven materials, classified by their tendency toward spontaneous heating. The list includes the usual shipping containers, storage methods, precautions against spontaneous heating and comments on each item. The materials listed in Table XXVIII have a high tendency towards spontaneous heating.

The absence of a material from this list does not preclude its tendency to spontaneous ignition. The exact condition and circumstances surrounding material stored, its container and the ambient temperature and drafts before the fire should be thoroughly investigated. If the containers had been used previously to store other materials, the nature of the materials should be determined for the possibility of a reaction with the newly stored materials. Quite often, the previously stored materials will act as a catalyst when combined with the new material.

The conditions preceding the fire should be reconstructed as

TABLE XXVIII

MATERIALS WITH A HIGH TENDENCY TO
SPONTANEOUS HEATING*

Alfalfa meal	Fish oil	Oiled rags
Charcoal	Fish scrap	Oiled silk
Cod-liver oil	Linseed oil	Peanuts, "redskins"
Colors in oil	Menhaden oil	Perilla oil
Corn-meal feeds	Oiled clothing	Tung nut meals
Fish meal	Oiled fabrics	Varnished fabrics

* From Gordon P. McKinnon and Keith Tower (Eds.), *Fire Protection Handbook,* 14th ed., 1976. Courtesy of National Fire Protection Association, Boston, Massachusetts.

carefully as possible to see whether spontaneous combustion could develop from the combination of materials involved.

It should be remembered that the heat of the fire itself may have caused the material to ignite spontaneously. This possibility can best be resolved by determining whether the origin of the fire contained the material suspected of spontaneous heating or whether the origin of the fire was adjacent to the suspect material.

ELEMENTS OF SPONTANEOUS HEATING

The environment in which spontaneous heating can occur is very sensitive. The reactant must be exposed to a maximum amount of air so that as high a rate of reaction as possible can occur. The air supplied to the reactant must be sufficient to sustain the maximum rate of reaction and not dissipate the heat generated by the reaction. This results in increasing temperatures which ultimately reach the ignition point. The critical parameter in this reaction is the ventilation of the substances.

Temperature

The ambient temperature of the constituents, or the reactants and air, has a bearing on how rapidly the spontaneous heating occurs. Quite often, it will determine whether the ignition temperature will be reached.

For example, assume that some oily rags are lying in the corner of a home in the dead of winter. The furnace has been set at 70°F, and ventilation in the corner of the room is sufficient to carry away the heat generated in the oily rags. The temperature of the oily rags will not rise. This situation could continue until all of the vapors in the corner have diffused into the atmosphere and danger of spontaneous combustion no longer exists.

Further assume that the owner of the house had increased the temperature to 75°F. This could be sufficient to raise the temperature of the oily rags to the point where self-generation of the chain reaction would occur. In a short time, the fire would evolve and could destroy the building.

Moisture

The moisture content of the reactant plays an important part in the generation of spontaneous heating in organic materials. This applies not only to agricultural products but also to charcoal, wood, coal or any organic substance. The presence of moisture acts as a catalyst to accelerate the rate of heat generation.

CHEMICAL CHARACTERISTICS

The reactant usually found in spontaneous ignition will have a great affinity for oxygen. Oxygen will usually be taken from the air, but in some cases, a chemically released oxygen can be provided. This means that a continuing supply of air is not necessary.

Under these circumstances, a much tighter enclosure can be tolerated and still allow the process to continue. This is an advantage to an arsonist who does not need to rely on stable ventilation conditions for the successful completion of a *set* that he plans to disguise as a spontaneous ignition fire. Instead, he can rely upon the chemical reaction of the substances to supply the necessary oxygen, ignition temperature and combustible material to touch off his fire.

In conditions just mentioned, the addition of catalysts can speed up the process and fine tune the timing of the reaction.

Organic material is normally the reactant in spontaneous heating fires. Usually the material ignited will be wood, cloth, fabric, plastic or any of the hydrocarbon liquid solvents belonging to the drying oil group.

PYROPHORIC CARBON

Spontaneous combustion has been assumed to be the cause of many fires where a steady source of low temperature heat has caused the decomposition of surrounding wood, resulting in the formation of pyrophoric carbon. This results in lowering the ignition temperature of the wood and will cause a fire at much lower temperatures than would normally be associated with wood.

For example, wood normally ignites spontaneously at temperatures in the range of 500°F. However, when pyrophoric carbon has been formed, the ignition temperature is lowered to 200°F. This is accomplished through degradation of the wood to pyrophoric carbon, which results in spontaneous combustion fires at a much lower temperature than would normally be thought possible.

INVESTIGATION OF SPONTANEOUS IGNITION FIRES

Once a spontaneous ignition fire has started, it will take on the characteristics of any other fire. From the low spot, it will burn upward and spread, depending upon the availability of fuel and air to determine its course. Therefore, the origin of a spontaneous ignition fire will still be found at a low point, although the low point may be a much larger area than normally associated with the origin of a fire. The origin of a spontaneous ignition fire will be found by using the same procedures as with any other fire. At the origin, however, which may be quite large, no other possible cause may be apparent. If the material and reactant are stored in a bulk storage warehouse, for example, the entire warehouse could spontaneously ignite in various areas simultaneously, giving the appearance of a "set" fire. This occurs because of minor variations in temperature and moisture content of the stored material.

Multiple Origins

A spontaneous ignition fire can be of multiple origins, one of the few cases where it can occur naturally. In a recent case, a woman used a product containing beeswax and turpentine to clean upholstery in three different rooms in her home. Five hours later, the cleaning fluid had spontaneously ignited the three pieces of furniture. Fortunately, eyewitnesses were available to observe the three separate fires and verify that they were of accidental origin.

Had the three separate fires started after the family retired, it would undoubtedly have resulted in the death of the family. This would indeed have been tragic, as the family members were the witnesses of the origin and cause of the three fires. Had witnesses not been available, it is doubtful that the spontaneous nature of the fire could have been determined, as the evidence (an accelerant and multiple origins) would have strongly indicated arson.

Determining the Cause

Assuming that the origin of the fire has led to a corner where the only evidence is the charred residue of cloth, a possibility exists for spontaneous heating. The cloth, soaked in volatile oils, would provide a surface for the oil to react with air. The only missing element after the fire would be the oil itself.

Therefore, after photographing the scene, a Sniffer should be used to determine whether hydrocarbons were present. If none are detected, then the material should be carefully photographed and removed in an airtight can to be tested for the presence of hydrocarbons.

The occupants of the room should be questioned to determine whether any flammable substances may have been contained in the fabric. Empty containers may yield information regarding the presence of any flammable liquids.

The same reasoning applies to bins or storage compartments where a fire started and spontaneous combustion is suspected. If any portion of the unburned material is available, it should be examined microbiologically for the presence of mesophiles and thermophiles. The possibility that the appearance of spontaneous ignition could have been provided by an arsonist should always be considered.

Heavily smoked windows in and around the area should be scraped, and the carbon and soot deposits on the window panes should be tested for hydrocarbons as well as residue from leaded gasoline.

Developing and Testing Your Theory

To eliminate the possibility that spontaneous combustion has occurred, a search of the premises should be conducted to locate all cleaning fluids or flammable liquids. If the fluid containers are not destroyed by the fire, the contents can be readily identified. Tests should be conducted to determine whether the contents will cause spontaneous combustion, particularly when applied to the combustible material found at the origin of the fire.

Tests should also be made on the residue of the cloth at the origin to determine whether any of the by-products from the fluids found on the premises are in the residue.

Interviews with neighbors will reveal whether the householder made a practice of using cleaning fluids on upholstered furniture, rugs or drapes in the home. Once these inquiries are made and a link between the cleaning fluid and the origin of the fire has been established, your work is still not complete. It is possible that an arsonist has used spontaneous ignition as his timing device.

CLUES TO A SPONTANEOUS IGNITION FIRE

The area of origin of a spontaneous combustion fire may be devoid of any other source of ignition or fire cause. The material first ignited and the source of ignition will be the same. The area of origin will be a function of the amount of material involved in the spontaneous ignition and combustion at the origin of the fire. If other sources of ignition are possible, they must be eliminated by examination, testing and interviewing witnesses.

Origin

The origin will be defined by a burned-out area consisting of the material first ignited, which can usually be identified by the owner or by chemical tests of the residue. If accelerants are found at the origin, the occupants should be able to explain their presence. If they are unable to do so, then the possibility of arson must be thoroughly investigated.

The debris at the origin should be thoroughly sifted and examined for the presence of substances which would not normally be found in conjunction with the materials present. The chemical relationship between any foreign substance and material normally found at the origin should be investigated and analyzed. Should the residue contain indications of accelerants, some of the original material should be obtained. This should be burned to determine whether a normal residue of hydrocarbons is a part of the material involved. This simple test can save a fire investigator many hours of work in pursuing blind alleys and can avoid prosecution of an innocent person.

The time involved between the deposition of the materials at the origin and the outbreak of the fire should be checked carefully. For example, if oily rags had been placed at the origin an hour before the fire started and they contained paraffin-based oils, it is highly improbable that spontaneous combustion caused the fire. Other causes should be investigated.

The time element is of vital importance and can be verified by duplicating the origin of the fire by reconstructing the composition of the materials found at the origin and duplicating the time, temperatures and ventilation leading up to the fire.

Cause

The cause of any spontaneous combustion fire is the result of the chemical reaction between two or more materials. The cause of the fire can best be determined by chemical tests of the materials involved to prove that the chemicals will react, given the circumstances involved in the origin of the fire. The fire should be reconstructed by duplicating the materials, oxidant and reactants concerned in the fire. Close attention should be paid to the pressure, temperature, humidity, drafts and ventilation surrounding the origin of the fire. Emphasis should be placed on the ventilation of the fire. This should be duplicated as precisely as possible, since the ignition of spontaneous combustion fires is a function of the ventilation characteristics of the fire.

CHAPTER 12

Miscellaneous Fire Causes

OPEN FLAMES AND SPARKS

O PEN FLAMES AND SPARKS, according to the NFPA, account for 6 percent of the total fires in the United States. Subclassifications of this category, along with the respective numbers of fires caused by each, are listed in Table XXIX.

Under this overall classification, the subclassification *Other Open Flames* includes fires caused by such ignition sources as candles and sparks from locomotives, incinerators, and chimneys (except when fires originate on roofs). The NFPA goes on to explain that many fires in this classification could probably be included in another group if more detailed information on the cause was reported.

The fires of interest in this category, however, are those which are readily classified and have identifying characteristics. For example, the second subclassification, *Sparks, Embers,* includes fires caused by ignition of roof coverings by sparks from chimneys, incinerators, rubbish fires, locomotives, etc. These fires are readily identified by the fact that the origin is on the roof or exterior of the building. This can usually be verified by eyewitness reports, and further verification would be needed to confirm that a fire existed in the chimney or incinerator, that a rubbish fire was burning or that a locomotive passed shortly before the fire started. These are rather easily proven by positive evidence.

Furthermore, a fire starting on the exterior of a building could indicate that it is of natural or accidental origin and is subject to rapid and easy discovery by passersby. An arsonist would normally work inside a building to hide the fire as long as possible and to ensure involvement of the fire before discovery.

Friction, Sparks from Machinery is a subclassification which includes fires caused by friction heat or sparks resulting from impact between two hard surfaces, at least one of which is capable

340

TABLE XXIX

OPEN FLAMES, SPARKS*

Subclassification	Number of Fires, 1974	Percentage of Total, This Classification
Other open flames	34,900	45%
Sparks, embers	13,300	17%
Friction, sparks from machinery	11,900	15%
Welding, cutting	11,600	15%
Thawing pipes	5,800	8%
Total, this classification	77,500	100%

* From Gordon P. McKinnon and Keith Tower (Eds.), *Fire Protection Handbook,* 14th ed., 1976. Courtesy of National Fire Protection Association, Boston, Massachusetts.

of striking a spark. Friction heat can result from the interaction between two hard surfaces. Heated bearings on rotating machinery and belts which become overheated due to slipping can start fires.

Sparks from dropping steel tools on a concrete or steel floor, or from pieces of metal caught in grinding machinery, may ignite vapor fires. The presence of a flammable vapor is all that is necessary; the vapor does not need to completely fill the room. It can be a trail of vapor that can lead back to a room full of vapor or a more concentrated supply of a flammable mixture that can flash back to the larger source as a result of a spark igniting a flammable mixture at floor level.

The most interesting subclassification for a fire investigator in this category is *Welding and Cutting.* The act of welding and cutting provides a source of ignition and the necessary heat to raise the temperature of combustible material; thus, the chances of a fire occurring are greatly increased in cutting and welding operations.

The temperature of an acetylene-oxygen flame is approximately 6000°F. This is a sufficient temperature and volume of heat to cause melting and oxidation of steel or iron. The very act of cutting with an acetylene torch necessitates heating the metal to above its melting temperature, enabling completion of the cutting process. Usually, a round piece of metal (plug) is burned free as the torch cuts a circular hole in a larger piece of metal. The *plug* falls to the floor. If it drops or rolls into a combustible material which affords insulation, the metal at 2000°F may start a smoldering fire which may not break into a flaming fire for hours.

Gas cutting is particularly susceptible to this type of fire-starting mechanism due to the fact that a piece of slag is usually involved in the cutting process. Unless this piece of slag is allowed to cool in a noncombustible atmosphere, it has the potential of starting a fire.

Gas welding is not quite as hazardous as the cutting operation, since less slag is involved. The greatest danger in gas welding is the heating of materials being welded. This creates the possibility that their conductive capabilities could store enough heat to raise adjacent combustible materials to the kindling temperature.

These fires are readily checked regarding the origin and cause. The proximity of the welding and cutting operation to the origin of the fire can be readily determined and verified by eyewitnesses. If destruction is not complete, following the "V" pattern back to the low spot may reveal the plug, or slag that fell from the burner's torch, and the material which ignited should be apparent by the location of the slag.

Welding by electric arcs falls in the same category as gas welding. The greatest danger of fire exists with the shower of sparks resulting from the welding process. These sparks can and have fallen on combustible material and smoldered for hours before finally breaking out into flame combustion. The technique for discovering their presence is the same as delineated in gas burning and welding.

Soldering

Two types of soldering are found in general practice in the United States and are classified as *silver soldering* and *lead soldering*. Silver solder is composed primarily of a 1.5 to 3.5 percent silver-lead combination, with a melting temperature of approximately 590°F.

The common plumber's solder, or lead solder, is usually comprised of 50 percent lead and 50 percent tin. These solders are commonly used in the installation of copper piping in residences, factories and commercial buildings. The melting point of 50/50 solder is in the 420°F range. The melting temperature increases as the amount of lead increases; the highest percentage of 98 percent lead and 2 percent tin has a melting point of 611°F. This would be the highest lead content normally used for residential or industrial soldering. Melted solders of these temperatures could start a smoldering fire. However, the normal 50/50 solder would be incapable of starting a fire due to the small mass of metal involved and the temperatures attained.

If solder is suspected in the start of a fire, the exact duplicate of the solder and soldering conditions leading up to the fire should be tested. These tests are quickly and easily run. The most important variable is the chemical constituents of the solder, as pointed out previously. The solder should be analyzed for contents.

Soldering is a borderline hazardous process. It is questionable as to whether it can start a fire. Another heat process technique similar to soldering is called *brazing*. A fusible alloy, comprised of varying percentages of silver, copper, zinc and cadmium, is used to bind two metal surfaces together through coalescence. These materials usually have a melting temperature above 800°F. This is below the melting point of the metals being joined. Brazing temperatures range from 800° to 1600°F, depending upon the percentage composition. Any of these metals, when heated, can drip on combustible material and are capable of starting a fire, based on the temperature and mass of the molten metal and given combustible materials of the correct consistency (excelsior, loose paper, scraps, oily rags or other highly combustible substances).

The technique for determining this as the cause again lies in reconstructing activities of the brazer or plumber soldering in the vicinity. The presence of the spattered material will be evident in the combusted material, as the material itself (molten solder) will not have been destroyed by the fire. A word of caution: the molten solder, having started the fire, would then be reheated and remelted by the resulting fire.

Soldering Tests

Tests were conducted to determine the maximum temperatures encountered in the normal soldering process with ½ inch diameter copper tubing, as experienced in residential construction. Preparatory to the tests, a study was run on the time required by experienced plumbers to make a soldered joint. It was found that the maximum time required, even for the most difficult solder joint using 50/50 solder, would be a heat application of less than thirty seconds.

In the test, to allow for differences in mechanical aptitude of various plumbers, a heating time of sixty seconds was used. A

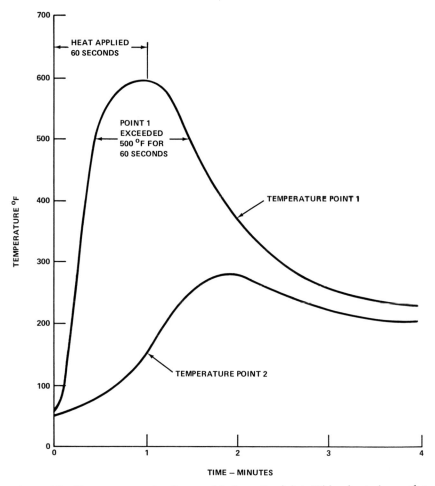

Figure 79. Temperature rise from soldering of a joint. This chart shows the temperature rise in respect to time of the application of heat, as shown in Figure 80.

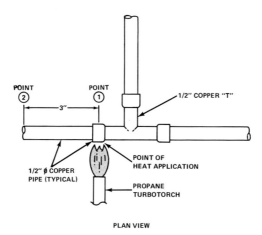

PLAN VIEW

Figure 80. Heat was applied at Point 1 for 60 seconds. Temperatures were measured at Points 1 and 2 of the ½ inch copper lines. Temperatures rose to 600°F at Point 1 with a maximum of 290°F at Point 2.

standard hand-held propane torch was used to apply heat in accordance with the manufacturer's instructions. Heat was applied at the joint for a total of sixty seconds, and the solder was allowed to run into the joint until it was well filled. Temperatures were recorded at ten-second intervals at the soldered joint (Point 1) and another point 3 inches horizontally from the joint (Point 2). At the soldered joint, the maximum temperature reached was 600°F. The temperature remained above 500°F for sixty-five seconds at Point 1. This is illustrated in Figures 79 and 80.

The temperature at Point 2 reached a maximum temperature of 280°F. This temperature was well below the ignition temperature of combustibles such as wood shavings, paper and oily rags, items similar to those found on construction sites.

The temperature at Point 2, located three inches from the heat source, does not come close to approaching the ignition points of wood, insulation, tar paper or any common materials likely to contact the pipe being soldered.

Experiments with molten solder also prove conclusively that it cannot start a fire. Molten solder poured on excelsior, wood shavings, oil rags and newspaper, even in large quantities, would not raise the temperature high enough nor provide a large enough heat sink to cause ignition of the material.

Should solder, or the conductivity of a copper, aluminum or other type of piping, be considered as a possible transfer medium with which to start a fire, it can easily be checked by reconstructing circumstances claimed to have started the fire and thus proving the point conclusively one way or the other.

Fires Started by Heat Transfer of Welded Materials

Extensive tests on the conductivity of copper, steel, brass, aluminum and other types of piping have proven conclusively that it is virtually impossible to start a fire by heating a pipe adjacent to combustible materials and have combustible material ignite by direct conduction of heat to the material, particularly if the material is wood studding.

Examination of Figure 80 will clearly indicate the logic behind this statement. As heat is applied to the copper pipe at the point of soldering, the conductivity of the pipe carries heat away from the pipe at a rate dependent upon the coefficient of conductivity of the material. This rate is shown in Table XXX.

As can be seen, even a good conductor such as aluminum or copper will dissipate the heat more rapidly than it can be conducted and transferred to the adjacent wood. One reason is that air and wood are both poor conductors while aluminum is a good conductor of heat. Heat is not transferred to the wood if any airspace is available. Assuming intimate contact with the wood, wood is such a poor conductor that the rate of heat transferred to it is insufficient to raise the temperature of the wood to the kindling point. A fire cannot occur from this source.

It is fairly well established that wood is a poor conductor. Any Boy Scout can tell you that it is easier to start a log fire with two logs than with one. The reason is that the second log stops the loss of heat through radiation, conduction and convection. The heat from the first log is reflected to the second and vice versa, so that a recycling of the heat generated by one log is absorbed by the adjacent log, thus raising the temperature of both logs.

This is due to the low thermal conductivity of wood. By the same token, a blowtorch can be turned on a two-by-four wood

TABLE XXX

THERMAL CONDUCTIVITY OF VARIOUS MATERIALS*

Alphabetical Listing Material	*Thermal Conductivity* $BTU/(hr)(sq\ ft)(°\ F/ft)$	*Order of Decreasing Conductivity* Material	*Thermal Conductivity* $BTU/(hr)(sq\ ft)(°\ F/ft)$
Aluminum	128	Silver	245
Asbestos	0.092	Copper	227
Bakelite®	9.7	Gold	172
Brick, building	0.4	Aluminum	128
Brass		Tungsten	116
Red	87	Brass	
Yellow	69	Red	87
Bronze	17	Yellow	69
Cellulose	0.033	Platinum	39.9
Charcoal	0.03	Tin	37.5
Copper	227	Iron, wrought	34.9
Concrete	0.54	Nickel	34.4
Earth (dry and		Iron, cast	27.6
packed)	0.037	Steel	26.2
Felt	0.03	Lead	20.1
Fire clay brick	0.58	Bronze	17
Glass	0.59	Bakelite®	9.7
Gold	172	Marble	1.5
Gypsum	0.25	Porcelain	1.3
Iron		Glass	0.59
Cast	27.6	Fire clay brick	0.58
Wrought	34.9	Concrete	0.54
Lead	20.1	Limestone	0.54
Limestone	0.54	Plaster	0.43
Marble	1.5	Brick, building	0.4
Nickel	34.4	Gypsum	0.25
Paper	0.075	Asbestos	0.092
Plaster	0.43	Hardwood	0.09
Platinum	39.9	Paper	0.075
Porcelain	1.3	Softwood	0.07
Silver	245	Earth, dry and packed	0.037
Steel	26.2	Cellulose	0.033
Tin	37.5	Charcoal	0.03
Tungsten	116	Felt	0.03
Wood			
Hardwood	0.09		
Softwood	0.07		

* Reprinted from *ASHRAE Handbook of Fundamentals,* 1972

piece and immediately start a glowing fire. However, when the blowtorch is removed as the source of heat, the fire will extinguish itself. Only after sufficient heat has been absorbed by the two-by-four to raise the temperature of the entire piece of wood to the kindling temperature will a fire be self-sustaining. This accounts for the difficulty in starting a fire by heat conduction or heat transfer from a piece of metal which has been welded or soldered through the activities of a burner, brazer, plumber, pipefitter, welder or other mechanic.

Thawing Pipe

Fires started while attempting to thaw pipes comprise 8 percent of this category and are usually easily detected because of their location. The classification includes fires involving torches, oil-soaked rags and other open-flame devices used in thawing pipes with an open flame. The practice itself is very dangerous and since a person is involved, the cause of the fire would be apparent to eyewitnesses, particularly the person who started the fire. Therefore, no problems should be encountered in detecting this type of fire.

SMOKING-RELATED FIRES

Smoking-related fires account for approximately 10 percent of fires according to the National Fire Protection Association figures. Children playing with matches account for approximately 5 percent of these fires. These two classifications are grouped under one heading primarily because of the ease with which these fires are identified.

Smoker's Fires

The majority of smoker's fires will be the result of persons smoking and falling asleep with a lit cigarette. In these cases, the origin of the fire will be very obvious and can be pinpointed to bedding or upholstery on chairs or sofas in the area where the smoker fell asleep. Death from smoke inhalation will often be the result of the smoker's carelessness. The location and position of the body will indicate whether the victim died in his sleep or awoke in time to take measures to escape the fire.

The clue to a smoker's fire is the fact that it usually occurs in a bedroom, with the bed as the origin of the fire. Smoking is the major cause of fires in bedrooms, but lamps, electric blankets and faulty wiring in appliances also cause fires. Other possibilities may present themselves, but the above would be the main causative factors to investigate.

The wiring and electric light bulbs can be checked very rapidly by determining whether combustibles were nearby. If so, and if the lamp bulb is intact, the fire could have been caused by contact of the hot bulb with a combustible material. Small particles of combustible material will be attached or fused to the light bulb. If the material is unburned, it can be tested and determined whether it is combustible under the heat of a duplicate lamp.

Even after the fire, tests can be run on the electrical cord to see if it has shorted. Quite often a fire does not destroy the insulative qualities of a cord.

The electric blanket can be tested for shorts in the wiring. If a short occurred, consult with an electrical engineer to more thoroughly determine whether the electric blanket caused the fire. This is covered in Chapter 6.

The most important aspect of any smoker's fire is to determine whether the victim or occupant was a habitual smoker and, if so, his habits concerning smoking.

Technical Aspects of Burning Cigarettes

Cigarettes are approximately ¼ inch in diameter and are made of rolled paper tempered with clay so that the burning rate of the paper corresponds with the burning rate of the packed tobacco. A burning cigarette is actually a case of smoldering fire where the combustion is occurring only at the surface of the ground tobacco and the air interface. As soon as the cigarette is lit, the glowing fire progresses toward the unlit end of the cigarette at approximately 0.21 inches per minute.

Cigarettes range in size from 2.75 to 3.94 inches in length and can burn for fifteen to twenty minutes. Cigarette tobacco in still air burns at a temperature of 500°F. Due to the small volume of the cigarette, the actual output in BTUs is rather small. However, under certain conditions, a cigarette is capable of starting a smoldering fire which can break into a flame fire with an increase in ventilation.

The Consumer Products Safety Commission is taking steps toward increasing the flammability resistance of mattresses, bedding and bedclothes to decrease the chances of a smoldering fire from a cigarette. However, until all motels, hotels and dwellings are equipped with fire- and flame-resistant materials, an occasional fire will be caused by careless smoking. Do not expect to find the cigarette. If firemen have turned the fog nozzle into the bedding, the cigarette ashes will probably be destroyed in the process.

While it is possible that a smoker's fire can occur anywhere, a fire found in a bedroom will most likely be a smoker's fire. The clues mentioned previously regarding the bed, chair and sofa apply equally to any room in a house. They also apply to any receptacle used by the smoker to discard matches. Fires can smolder in an ashtray or wastebasket for hours and break into flames after the volume of the glow fire has reached large enough proportions to support a full flame combustion fire. Again, it is the origin of the fire which pinpoints the fire as a smoker's fire.

Smokers not only have problems with cigarettes but also with the source of ignition (matches or lighters). Matches have been dropped in wastebaskets and ashtrays full of cigarette butts and other combustible materials, resulting in smoldering ashtrays and containers. If these are dumped into ash cans or wastebaskets while still smoldering, a fire will ensue when sufficient heat is built up to cause flaming combustion, igniting available material. Although a pipe smoker may not be associated with starting a fire through ashes or laying his pipe down, the fire could have started through his use of matches. The same applies to cigar smokers. About the only tobacco user who can escape accusation is the user of chewing tobacco.

Should it appear that the fire is not a smoker's fire and that the material first ignited would be unlikely to ignite, then by all means run tests. Use identical cigarettes and material under the same ventilation circumstances which supported the fire. This will prove whether a fire could start in this manner.

Children and Fire

Studies conducted by the NFPA reveal that the majority of fires started by children playing with matches originate in closets. A closet is a very small, compact area where a child can find seclusion and feel secure. If a child wishes to perform a forbidden act, the closet offers the needed security. Children are, of course, told not to play with matches. Consequently, they will invariably hide in a closet with the forbidden matches.

If a fire started in a closet with no source of combustible materials, energy source or ignition, the next line of inquiry would be to determine whether children were in the building shortly before discovery of the fire.

COOKING FIRES

According to *America Burning,* cooking fires are the most prevalent household fire, but they are discovered and extinguished by the housewife. Those which are not extinguished are usually identified by her as the cause of the fire. She would know that the oven or stove was on and the contents that could result in ignition of combustibles on or near the appliance.

Therefore, any kitchen fire should be investigated by questioning the housewife regarding her activities in the kitchen. The cause of the fire may then be easily established.

Kitchen or cooking fires can also be caused by defective controls on automatic stoves. Strides have been made by the manufacturers which have put the stove in the category of a robot. The stoves turn themselves on, cook the dinner and keep it warm until the housewife returns. This is done by mechanical and electrical timing mechanisms which control the flow of heat energy to the stove. Since these are mechanical controls, they can and do malfunction. If the stove was set on an automatic control, a close examination should be made of the automatic controls as the first step in the investigation.

Vents located above stoves are another cause of fires in the kitchen. These vents become clogged with grease and condensed oil. Under the proper circumstances they can reach ignition temperatures, causing a fire between the kitchen ceiling and the roof of the house. A fire starting in this area becomes very difficult to contain before it gets out of control. Again, the housewife should be consulted to see whether the appliance was in use and the status of the heat output in terms of the burners in use and their settings. Examination of the controls will reveal their position.

Microwave ovens are a source of high energy which can also result in fires. Since they are a totally electrical appliance, questioning of the housewife is necessary to determine the status of the appliance.

All microwave ovens have a safety door which needs to be closed before current can flow. This is used to prevent radiation from being generated through an open door. Therefore, the latching mechanism of the door should be examined to determine whether it malfunctioned.

TRASH BURNING

This classification accounts for 13 percent of the total number of fires annually, according to the NFPA. This classification includes rubbish and waste material, although it is not a cause of fires but rather the fuel which was ignited. Since the cause of ignition is *unknown,* this could be classified under *Unknown Causes.*

Trash burning includes fires which originate in areas where rubbish and waste material is stored. The location of the origin of the fire plus the nature of the combustible material qualifies fires for this category.

Rubbish and waste materials are commonly located in specific areas, such as under stairways, in unused closets, corners of garages, offices and basements where little traffic is experienced. These areas are also excellent targets for the pathological firesetter, so it can be assumed that a reasonable percentage of these fires are of incendiary origin.

The source of ignition can be of various types: a careless smoker, discarded hot ashes, discarded hot smoking materials or even spontaneous ignition of discarded materials.

A clue to determining the true cause of these fires is the location of the source of ignition. Every possible source, such as sparks from adjacent chimneys or activities of anyone using open flames, should be carefully considered. Traffic in the immediate area of the fire origin should be carefully examined to determine whether any smokers have passed recently who may have discarded smoking material in rubbish. If these possibilities can be satisfactorily eliminated, the possibility of arson should be considered.

FLAMMABLE LIQUIDS

Refinery Fires

Refinery fires are included in this chapter because of their low percentage of occurrences. They are included in this book because of the seriousness in terms of property and dollar loss as well as injury and lives lost. The primary concern will be with loading rack fires caused by *switch loading* and *splash loading*.

Switch Loading and Splash Loading

Switch loading is a term given by the petroleum industry to the practice of loading a compartment in an oil tanker with a lower volatility hydrocarbon fuel (such as fuel oil) after having carried a fuel of a higher volatility in the same compartment. The result of switching from one fuel to the other results in driving off vapors from the walls of the compartment as the new material is introduced. It displaces the film of higher volatile fuel, forming a flammable mixture. It should be noted that the flammability limits of gasoline range from approximately 1 to 7 percent by volume.

The vapors are dense, so the entire volume of the empty tank could be considered to be 100 percent gasoline vapors, which will be forced out of the tank to make room for the new product. As the gasoline fumes are expelled from the container, they will mix with air, depending upon the air currents available and form flammable mixtures. The only requirement is a source of ignition.

Since this is a constant hazard in all refineries, extreme measures are taken to avoid possible ignition sources. Warning signs are posted. Drivers and operators are trained to stop the engines before any loading operation is commenced. No smoking is permitted in the area. All electrical pumps, switches, lighting fixtures and circuits are encased in explosion proof conduit and fixtures. This accounts for the low percentage of refinery fires.

One source of ignition is static electricity. This can be generated by friction between substances which creates static charges capable of discharging a spark with enough energy to ignite a flammable mixture. Electrically bonding the structural components of the loading platform and proper grounding to eliminate the buildup of static charges does much to eliminate and control the accumulation and discharge of static electricity.

Another way to generate static electrical charges is the flow of liquid through a pipe. The friction between the product and the walls of the pipe can produce an electrostatic charge so that at the point of contact with the air, an accumulation of charge discharges, creating a spark which can ignite a flammable mixture.

Grounding of the various components will have no effect upon the generation of these charges in the fluid stream. Grounding of the vehicle into which the product is discharged will likewise have no effect upon the static electricity generated by the flow of the product. Therefore, other means are used to control the accumulation of electrostatic charges through this phenomenon.

The method used is to place a device in the pipe as close to the point of discharge as possible. This device will negate the effect of the accumulation of electric charge and bleed off the static electricity before it can reach dangerous proportions. These devices are called *Electrostatic Charge Neutralizers* or *ECNs*. They are very effective but do nothing for the charge generated in the pipe from the ECN to the point of discharge at the tank truck. Friction within the fluid and between the fluid and walls can generate static charges downstream from the ECN to the point of discharge. At the point of discharge, friction between the product, the bottom of the tank truck and the air can add considerable electrostatic charges to the accumulation. This could ignite a flammable atmosphere created as vapors mix with air due to the splashing and agitation of the product falling into the truck. Thus the term *splash loading* originated.

Four requirements are necessary for an explosion and fire to occur as a result of *splash loading:*

1. A means of electrostatic charge generation
2. A means of accumulation of a charge capable of producing a spark with enough energy to ignite the flammable mixture
3. A means of discharging the accumulated electrostatic charge in the form of a spark
4. A flammable mixture within the vicinity of the spark so that ignition can occur from the discharge of the accumulated electrostatic charge

When these conditions exist, it is likely that an explosion and fire will occur. The vapors ignite and generate more vapor in a continuous fire. If the explosion is severe enough, it results in rupturing the tank and spilling the product on the ground, rapidly spreading the fire.

Splash loading (holding the downspout above the container floor so that the product falls eighteen inches or more to the surface of the liquid) and switch loading are operations which can lead to these conditions.

Investigation of Refinery Fires

A refinery fire should be approached on the same basis as any fire. The only added factor to be considered in a refinery fire is the high probability of static electricity as the cause of the fire.

If static electricity is the cause, the position of the loading arm in relation to the truck should be carefully reconstructed to determine whether the loading arm was capable of being inserted properly so that splash loading was not possible. A normal cause of splash loading is the failure of the truck operator to relocate his truck so that the loading arm can be properly inserted into the bottom of the tank to avoid splash loading.

In investigating a refinery fire, the presence of a flammable mixture should be determined first; secondly, the source of ignition should be determined. If no other source of ignition is found, a concentrated search for a source of static electricity should be made.

Aerosol Cans

Aerosol cans are the result of the consumer's demands for convenience in application of liquid products. In response to this demand, the packaging industry has provided pressurized cans of various sizes, usually in the one pint to one quart sizes, with an inert gas as the pressurizing medium, used to vaporize or atomize the contents under pressure through a small orifice.

While the pressurizing medium (the gas) is usually inert and usually a refrigerant (freon), the contents quite often are highly volatile and highly flammable. Indications of the degree of flammability are, by law, printed on the can. Various degrees of flam-

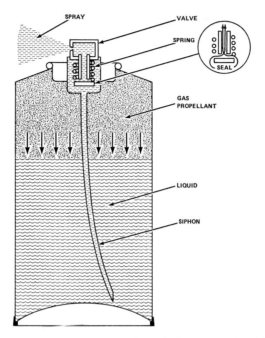

Figure 81. Aerosol System. The product, indicated as liquid, is placed in the container. A gas propellant under pressure is sealed above the liquid. Pressure of the gas forces the liquid up the siphon through the orifice when the valve is depressed, breaking the seal. When the valve is released, the seal is established, preserving gas pressure until needed. Should the spring break at any time during its life, the gas will propel all liquid out of the container. Should the spray find a source of ignition, the flammable aerosol spray can ignite, causing a fire. From *The Way Things Work: An Encyclopedia of Modern Technology,* Volume Two. Copyright ® 1971, by George Allen and Unwin, Ltd. Reprinted by permission of Simon & Schuster, a Division of Gulf & Western Corporation.

mability are listed under different descriptions according to the jurisdiction which regulates the packaging industry.

Any pressurized container which contains a flammable product can be the source of a fire. Should a leak develop in the valving mechanism, a flammable spray will emit from the orifice which, if ignition is found, can result in a fire.

Aerosol cans found near the origin of a fire should be carefully examined to determine whether they could have been the cause of the fire. They can be the cause if, for example, a leak develops in the valving mechanism. To demonstrate this, consider an aerosol can stored alongside a stove. If the leaking orifice was pointed in the general direction of a lit stove or pilot light, the cloud of flammable vapor could be ignited and in turn could ignite combustibles near the flame. The spray emitting from the containers can extend as far as 24 inches from the point of emission. This is illustrated in Figure 82. The resulting flame could ignite anything located in the general area above or adjacent to the flame.

Tests for flame projection and drum tests have been developed by the Interstate Commerce Commission. The procedures are listed below.

Flame Projection Test

EQUIPMENT: The test equipment consists of a base 4 inches wide and 2 feet long, marked in 6 inch intervals. A 30 inch rule (with inches marked) is supported horizontally on the side of the base and about 6 inches above it. A plumber's candle of such height that the top third of the flame is at the height of the horizontal rule is placed at the zero point in the base.

PROCEDURE: The test is conducted in a draft-free area that can be ventilated and the atmosphere cleared between each test. The self-pressurized container is placed at a distance of 6 inches from the ignition source and the spray jetted into the top third of the flame with valve opened fully for periods of fifteen to twenty seconds. The length of the flame projection from the source is read on the horizontal scale. Three or more readings are taken on each sample, and the average is taken as the result.

Drum Tests

EQUIPMENT: The equipment consists of a 55 gallon open-head steel drum or similar container which is placed on its side and fitted with a hinge cover over the open end that will open at a pressure of 5 psi. The closed or solid end is equipped with one shuttered opening at the top. This is for the introduction of the spray. The opening is approximately two inches from the edge of drum head and is two inches in diameter. There is a safety glass or plastic window 6 inches square in the center of the solid end.

A lighted plumber's candle is placed inside the drum on the lower side and midway between the ends.

PROCEDURE: The tests are conducted in the open and where temperature is between 60°and 80°F.

Open Drum Test

This test is conducted with hinged end in a completely open position and with the shutter closed. The spray from the dispenser, with the valve opened fully, is directed into the upper half of the open end and above the ignition source for one minute. Any significant propagation of flame through the vapor-air mixture away from the ignition source is considered a positive result, but any minor and unsustained burning in the immediate area of the ignition source is not considered a positive result.

Closed Drum Test

This test is conducted with the hinged cover dropped into position to rest freely against the end and to close the open end of the drum to make a reasonably secure but not necessarily a completely airtight seal. The shutter is opened, and the spray is jetted into the drum through this shutter with valve fully opened for one minute. After clearing the atmosphere in the drum, the jetting is repeated similarly three times. Any explosion or rapid burning of the vapor-air mixture sufficient to cause the hinged cover to move is considered a positive result.

Aerosol Can Testing

Examination of aerosol containers after a fire should consist of weighing the can to determine whether all of the contents have been consumed by fire. If some contents remain in the container, it is possible that the can did not start the fire. Once a leak has developed in the valving system, it is unlikely to seal itself. However, it is possible that use prior to the fire had decreased the proportion of gas to liquid content. It would then be possible for a leak to develop that would exhaust the gas pressure and still leave some liquid within the container. This would not be unusual if a gas leak had developed from a malfunction of the valve. It is quite possible that a smaller leak may have occurred which could have

Figure 82. *Flame Projection Test.* This test vividly demonstrates the potential hazard in aerosol cans containing flammable liquids. The test is performed by setting the nozzle of an aerosol can six inches from a plumber's candle. The valve is depressed, and the vapors are ignited by the candle. The distance of the flame projection from the candle is measured

permitted gas to escape with very little vaporization of the product, leaving a shortage of gas within the container.

The can should also be examined to see if any gas remains in the container. This can be done by carefully depressing the operating valve, providing it has not been destroyed by the fire. It is not impossible for the aerosol container to develop a leak, start a fire and remain undamaged by the fire. This is because the cone of the flame will usually start a fire some distance from the container. It is possible that the aerosol container will not be destroyed by fire, assuming the fire is extinguished at an early stage.

Usually, however, the fire does impinge upon the container, and the degree of impingement, temperature of the fire and length of time the container was exposed to the fire will determine the amount of destruction. Quite often, the container will exhaust itself of pressure and product. The fire may surround the container, burning the outside and heating the can to the point that interior pressure is increased faster than the leaking valve can relieve it. The can then explodes, relieving the pressure.

As a consequence of the fire, usually the plastic tube and rubber gasket are subjected to high temperatures and melt or burn, depending upon the access of oxygen to the material. Should this occur, there may be no evidence to prove or disprove the possibility that the aerosol can was the cause of the fire. In the absence of any other cause, however, this should be given serious consideration.

Figure 83.   Figure 84

Figure 83. The aerosol can control valve spring can be checked by X-ray. In this positive X-ray, the spring is located in the mouth of the can at the top of the photograph. The spring is in its normal position, indicating the valve is still functioning. (See Figure 84.)

Figure 84. In this positive print made from an X-ray negative, the valve spring has been dislocated, indicating failure of the nozzle mechanism, which may have permitted the contents to escape, causing a fire.

Should the container be intact, X-rays can be taken to show the relative position of the tube and the spring that acts as a retainer on the valve to keep the valve closed under normal inoperating conditions. These are illustrated in Figures 83 and 84.

By comparing the length of the spring with the length of a normal spring, taken from a similar container either by physical removal or by X-ray, it can be determined whether the spring is depressed, indicating the valve is closed and intact, or whether the spring is elongated, indicating that the valve mechanism has melted or been destroyed, permitting elongation of the spring. Elongation of the spring could also mean that the valve has failed and caused the valve to open, permitting gas to escape.

If a can is empty, a leak can be detected by submerging the can under water and watching for air bubbles, similar to the process for detecting leaks in a tire. Before using this method, make sure that submersion in water will not affect the can. Photograph the can prior to immersion and dry it carefully to ensure complete drying without disturbing the surface markings of the can from the fire.

OTHER MISCELLANEOUS CAUSES
Electric Irons

A common household appliance which has started fires is the electric iron. Irons reach temperatures of 700°F and are commonly left face down on the ironing board. The iron overheats the material on the board, starting a fire. This is easily checked, especially if the iron and ironing board have not been disturbed. If the iron was the cause of the fire, the iron can be lifted to reveal a highly scorched area in the shape of the iron.

This area will be scorched despite the fact that everything else on the ironing board may be burned to a cinder. The reason for this is that the iron that started the fire was covering this area and prevented complete oxidation of the material. Therefore, if the iron had been on and caused the fire, the pattern of the iron would be well defined by a scorched pad. It may also have burned if the iron was knocked from the ironing board during the course of the fire. Even in this case, the outline of the iron would still be apparent in the burned debris because the material had been scorched first and burned later.

If the iron was off but was found sitting face down, an unscorched and unburned area of the ironing board covering would be found because it would have been protected by the face of the iron. This would be positive proof that the iron was cold at the time of the fire. This is irrefutable evidence that the iron did not cause the fire, and thus it can be eliminated as a possible cause.

Light Bulbs

One type of slow combustion is that caused by unshielded electric light bulbs. The exterior temperature of an unshielded electric light bulb ranges from 250° to 600°F, depending upon the wattage. A light bulb can start a fire when combustible materials such as paper or cardboard cartons come in direct contact with the surface of the bulb.

Heat from a bulb can be transferred to combustible material by conduction, convection and radiation, but it is usually transferred by conduction. A considerable length of time is usually required for a smoldering fire to occur. This time is necessary to heat a combustible material such as paper sufficiently to raise it above the kindling temperature. Once this is accomplished, a smoldering fire will result. The fire will smolder until sufficient ventilation is provided and the temperature is raised enough to support and maintain a flame fire.

The following case illustrates the fire hazard created by exposed incandescent light bulbs. A steel frame warehouse was equipped with wooden shelving and illuminated by 500 watt incandescent bulbs with porcelain reflectors. These large bulbs protruded below the reflectors, exposing them to contact.

The lights were suspended from the ceiling over the aisles to a level below the top of cartons stored on the shelves. In some aisles, the lights were off center and within 6 inches of the edge of the shelving.

While an employee was loading boxes of lampshades on the upper shelf from the side opposite the lamp, he unwittingly pushed a carton in direct contact with a light bulb. Approximately two hours later, a full-fledged fire broke out which resulted in total destruction of the building.

The clue at the origin of the fire was the physical proximity of the hanging conduit, the porcelain reflector and the remains of the shattered light bulb and shelves. Verification that the employee had two hours previously placed cartons on the shelf from the blind side was sufficient evidence to conclude that the fire was caused by the light bulb contacting the lampshade carton.

Eyewitnesses verified that the fire started near the top of the shelf. The only other possibilities in this particular case would have been arson, cigarettes or spontaneous combustion. The circumstantial evidence pointed overwhelmingly to the conclusion reached.

Static Electricity

Under normal conditions, the free electrons of atoms of a substance are electrically neutral with the atom. However, when good conductors are brought in intimate contact with poor conductors, free electrons are given up to the poor conductor at their point of contact. This upsets the electrical balance of the atoms, and a *charge* accumulates on the two surfaces. This charge consists of the free electrons transferred from one to the other. The substance receiving the excess electrons is said to be *negatively charged;* the surface that is losing electrons is said to be *positively charged.*

This condition will prevail until the excess electrons are freed to equalize the condition and electrically neutralize the surface. This potential electrical discharge existing on the two surfaces is called *static electricity.*

Static electricity is a natural phenomenon. The most prevalent and dramatic form found in nature is lightning. This subject has been covered in Chapter 4. An example of static electricity experienced in human activities is the accumulation of static charges on the human body generated by walking across carpeting. A heavy charge will be accumulated on the body until a contact with ground or a neutrally charged body drains the excess electrons from the body. This discharge experience is often accompanied by an audible snap and brief jolt or shock.

The aspect of static electricity of concern to the fire investigator, however, is the accumulation of potential static charges which build up enough to cause a spark in the vicinity of a flammable vapor.

As a general rule, a static charge will be produced when two unlike materials are separated, causing electrons to leave a conductor for a nonconductor. This occurs in industrial and commercial plants on belt drives, rubber conveyor belts, grain chutes or other locations where conductors and nonconductors are intimately together (a pulley and rubber belt) and are separated. It occurs where the grain being transported by a metal chute leaves the chute and breaks contact with the metal sides of the chute.

This same phenomenon occurs when flammable liquids such as gasoline, kerosene or fuel oil are being discharged from a pipe. The flammable liquid itself can retain a static charge capable of causing ignition of the flammable vapors. This was outlined under *Refinery Fires* earlier in this chapter.

CHAPTER 13

Forensic Photography

THE PROCESS OF PHOTOGRAPHY began in 1820. After fifty-five years of development of a commercially viable product, photographs were introduced into evidence in 1875 in the case of Luke vs. Calhoun Co.; 52 Alabama, 115; 1875. These were monochrome prints, commonly called *black-and-white,* and it was not until the development of Kodachrome® by Eastman Kodak Company in 1935 that color film became a commercial reality. A short eleven years elapsed before the admission of color photographs in 1946 in Hanagan, Postmaster General vs. Esquire, Inc., February 4, 1946, U.S. 146. Since that time, few cases involving fires have been tried without color or black-and-white photographs introduced as evidence.

The reason is twofold. In no other way can the relationship of the details of the fire be shown to a jury as explicitly. The photograph can capture the scene as it was before it was altered by the fire investigator, whose work makes it impossible to duplicate the fire scene. It would be physically impossible to record in a notebook the many details which can be captured in one photograph.

THE IMPORTANCE OF PHOTOGRAPHS

When the fire investigator arrives at the scene, the building may be in a condition that would present a hazard to life and limb of passersby. Therefore, it is of the utmost importance that the condition of the building be recorded photographically so that demolition of the building can be completed to reduce this hazard.

In the case of supermarkets or processing plants where organic material may be contaminated (livestock or foodstuffs), the spoilage can create a health and nuisance hazard. It is imperative that the scene be recorded on film so that the menace to health is eliminated as quickly as possible.

Another reason for photographs is a matter of convenience. It is much simpler to present a photograph of a 10 by 20 foot wall which illustrates a burn pattern than it is to bring the wall into court or the jury to the scene.

In the process of examining debris, it is necessary to dig under the debris to locate evidence. The layers and disposition of the debris itself constitutes evidence. The routine of uncovering the debris must be photographed to show the relationship of the various layers.

It would be a great assistance to fire investigators if a professional photographer were available as he proceeded with his investigation. If such a photographer is available, the fire investigator should have complete control over his activities and supervise the photographing in an orderly manner.

Unfortunately, most fire investigators cannot afford this luxury and are obligated, by necessity, to become adept at photography. Fortunately, the present state of the art of photography is not too difficult to learn. With a small investment in time, a fire investigator can learn to use a camera effectively and develop the ability to photograph details as they are encountered, thus eliminating the expense of a professional photographer.

LEGAL REQUIREMENTS

The legal requirements of photographs are covered under the rules of evidence. The laws vary with the jurisdictions, but one prevailing ruling is that *the photograph must be relevant.* It must have some relationship or represent some feature pertaining to the case and illustrate a point which the fire investigator is presenting as evidence. The photograph may be an illustration of the burn pattern which led the fire investigator to the origin of the fire. While many words could be used to describe this point, the photograph would portray it visually.

Each photograph should represent some relevant feature of the fire. These features would include evidence of flashover, accelerants, deep char or a timer. The key word is *evidence.* Unless the photograph *tends to prove* what the fire investigator is attempting to illustrate, the photograph would be considered irrelevant and may not be admitted.

The second requirement for a photograph is that it should *accurately portray the scene at the time the photograph was taken.* The photographer (if he is also the fire investigator) can attest to this fact. The photograph need not reflect the scene as it was immediately after the fire. However, it should meet the criteria stated above, i.e. it should portray the scene as it appeared to the fire investigator during his investigation. This subtle distinction should not be overlooked. The scene may have been changed considerably between the time the fire was extinguished and the time of the photograph.

The third element of the photograph to be admitted as evidence is that *it must be verified as to its authenticity.* As indicated above, this can be done by the person who took the photograph. It can also be verified by others at the scene when the photograph was taken. It does not have to be verified by the photographer as long as somebody can attest that the photograph does accurately portray the scene as he viewed it.

SOURCES OF PHOTOGRAPHS

While it is a necessity for the fire investigator to take progressive shots of his activities during the investigation, it is just as important to have photographs of the building as it was before and during the fire.

Photographs of the building and its contents before the fire can be obtained from the owner, real estate agents involved in the sale of the property or from previous owners. These photographs will reveal the state of the building before the fire and show the location of doors, windows and other openings which may have played a role in the fire.

If the building is an industrial installation, the contractor or local press may have photographs. Photographs taken during the fire may be available from local television and radio stations or from the newspapers. Quite often, the local fire department will have a photographer-fireman whose duty is to take photographs of the progress of the fire. These are invaluable to demonstrate the fire spread. Their value is enhanced if the photographer-fireman recorded the time of each photograph. The press photographs of the fire can also be correlated regarding the time.

In addition to photographs taken during the fire, photographs will be taken by insurance adjusters, the owner, private parties or other experts involved in the fire investigation. These should be made available to the fire investigator. If there is one axiom that holds true for all fire investigators, it is that *you cannot take too many photographs.* Many a case has been won or lost by photographs.

Copying Preexisting Photographs

As mentioned before, photographs are quite often obtained from other sources. At times, these photographs are the only remaining photographs of a scene which cannot be reproduced. For example, a photograph of the building before it burned may exist in the form of a photograph for which no negative is available. The negative may have been lost through the years, or an instant process (Polaroid®) film may have been used, with no negative available.

These photographs may be too small to reveal details that may have some probative value. In this case, the photograph can be copied. The resultant negative obtained is then enlarged to show greater detail. The negative can also be used to make duplicate enlargements for court use.

In some jurisdictions, the *best evidence* rule is imposed. The jurisdiction may interpret the original photograph as being the best evidence. The original photograph and copies may need to be introduced together to show that no alteration has been made in the enlargements. Under most jurisdictions, the previously mentioned factors, i.e. relevancy, accuracy and verification of the enlargements, must also be met.

TYPES OF FORENSIC PHOTOGRAPHY

The basic photographic tool of the fire investigator is the camera. However, the camera can be used in many ways to produce the desired results. None of the methods should be overlooked as a means of preserving evidence.

Aerial Photographs

Aerial photographs may be available of a fire, particularly if the fire is of any significant size. Most press or television newsrooms will be on the scene in a helicopter or airplane and take excellent photographs showing the progress of the fire. The timing and sequence of the fire can often be traced by study of the films.

In large fires, it is often expedient to rent an airplane and take aerial photographs of the scene. These will often reveal details not readily apparent from a ground study of the scene (see Fig. 61).

Aerial photographs taken from a helicopter or light airplane are subjected to detrimental effects due to vibration or movement of the aircraft. To overcome this problem, a fast film and fast shutter speed is a necessity. If the fire investigator is not equipped with a camera capable of taking aerial photographs, the assistance of a professional aerial photographer should be obtained. The cost is reasonable and the results are well worth the additional expenditure. Aerial photographs should be taken as soon as the

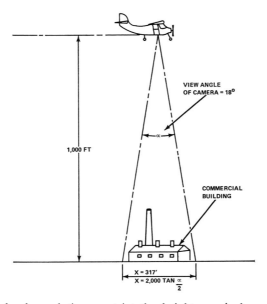

Figure 85. Federal regulations restrict the height an airplane can fly above a populated area. To determine the focal length needed to cover the scene, the photographer can determine the minimum elevation above the ground he can fly, the maximum area he wishes to cover on the ground and, by use of this formula, can compute the view angle of the required lens. He can then choose the appropriate lens with the correct view angle and be assured that full coverage of the area required will be obtained.

smoke clears and before overhaul has been started. Enlargements of these photographs are of great assistance in understanding the magnitude of the fire and its relationship to the building involved.

If the fire investigator has a camera capable of $\frac{1}{1000}$ second exposure time, successful aerial photographs can be taken. Aerial photographs should never be attempted on dull, cloudy or overcast days. The brighter the day, the better the photographs. Any color film having an ASA index of 100 can be used, although color films with a speed of 400 are recommended.

Architectural Photographs

The majority of photographs taken by a fire investigator will be architectural photographs. A basic requirement is that distortion is kept to a minimum. This is accomplished by holding the camera level. One method to eliminate distortion is to mark the building where the ground is level at a spot level with the photographer's eye. Then walk away from the building and position the viewfinder so that the center of the image is on this *eye level* spot. This will guarantee that the camera is reasonably level. The photographer must move away from the building until the entire building is in the viewfinder. It may be necessary to rotate the camera upward for tall buildings. A photograph should be taken with the camera in this position. If the entire building is not visible in the frame of the viewfinder, the camera should be tilted upward until the entire building is visible in the frame, and a second photograph should be taken. This will ensure an accurate representation of the building with minimum distortion. This technique is useful when wide angle lenses are used.

In the event that the photographer cannot stand back far enough to cover the entire building, he can, by photographing segments of the building and rotating the camera to ensure coverage of the entire building, form a mosaic of the building by overlapping these photographs. These can be matched and taped together to form a composite photograph. Care should be taken to ensure that the exposure is correct and that each photograph overlaps the preceding one. To facilitate this, a tripod can be used to provide a stable and level platform for each photograph.

Interior shots require the same techniques as outlined in the preceding paragraphs. Particular care should be taken to eliminate distortion. It should be remembered that most residential buildings have 8-foot ceilings and the average person's eye level is approximately 5 feet, 6 inches above the ground. Therefore, to equally show the floor and ceiling, it may be advisable to lower the camera to a 4-foot elevation. Good photographs show equally the top and bottom of the walls.

Another problem encountered with interior shots is in obtaining focus of the nearest object and the furthest object. This characteristic of a lens is called *depth of focus.* The maximum depth of focus is obtained by using the smallest aperture opening on the camera. This requires using a slower shutter speed, whereby movement of the camera may blur the image. A tripod can be used to eliminate movement, and slower exposure times will guarantee that the maximum depth of focus is obtained without blur.

Should the problem of depth of focus be encountered, the photographer should avoid using flash illumination. The intensity of illumination drops off proportionally to the square of the distance. If the distance from the camera is doubled, the light loss is quadrupled. This can be overcome with floodlight illumination. Floodlights can be placed to provide uniform illumination of the room. The uniform, intense, bright illumination afforded by floodlights also provides a convenient means of making a detailed examination of the room. After completing the examination, the lights can be used to photograph the scene.

Photomacrography

Every photographer has had the experience of desiring to photograph an object up close only to find that the focal design of the lens prevented him from achieving focus closer than 3 feet. The photographer may have then proceeded to take a photograph of a small object (broken piece of pipe) and was disappointed when the photograph, even when enlarged ten times, did not show the detail seen with the naked eye. Fortunately, the photographic industry has solved this problem with a completely new system of lenses and techniques called *photomacrography,* the art of small object photography.

By utilizing supplementary front lenses on a standard lens, or with a special lens which will extend the focusing range to within a few inches of the front of the camera, close-up photographs of small objects can be taken. No special techniques are required; the photographer must be extremely careful regarding focus, and a problem may be encountered with illumination and exposure. The normal hand-held exposure readings will not provide accurate exposures when used in conjunction with supplemental lenses or macrolenses on close-up work.

Since light has to travel further through the lens, an exposure increase is necessary depending upon two factors: the distance from the front element of the lens to the film plane and the distance from the film plane to the object.

If a through-the-lens metering system is used, the problem is nonexistent; the through-the-lens meter reads the exact exposure directly. However, with hand-held meters the problem of exposure must be calculated correctly or underexposure will result. For details on exposure and other problems involved in photomacrography, consult *Close-up Photography and Photomacrography* by H. Lou Gibson, Eastman Kodak Company, Kodak Technical Publication Number N-16.

Photomicrography

Photomicrography is the photography of objects through a microscope. The microscope is a valuable tool to the fire investigator, but it has one distinct disadvantage. It is impractical for a jury to examine evidence under a microscope. To overcome this, the fire investigator has the alternative of photomicrography. By using a slide projector or an 8 by 10 enlargement of the photomicrograph, the fire investigator can show the jury what was seen under the microscope.

Most crime labs are equipped with microscopes which have photographic capabilities. The fire investigator should be familiar with the types of microscopes available and their capabilities. Photomicrographs can be taken either in black and white or color and are an effective means of demonstration to the jury.

Two types of microscopes are commonly available. The first and most common is the optical microscope. This is a tool which provides magnifications varying from 20 to 400 power. The lower powered microscope can produce excellent photographs of opaque three-dimensional objects. The higher magnifications, from 100 power upwards, are more suitable for flat monoplane objects, such as cut sections or slides. These would be commonly used for metallurgic sections of materials, such as electrical wires, to show the growth of dendrites or other phenomena (see Figs. 38 and 39).

The other type of microscope is the Scanning Electron Microscope, commonly termed *SEM*. This instrument, although limited to black-and-white shadowgraphs, has a distinct advantage in the photography of opaque, three-dimensional objects. Since the SEM portrays a substance by electronic means, no depth of focus problem exists and the three-dimensional photographs are excellent renditions of the actual shape of the subject matter. The electron microscope, while limited to black-and-white photographs, cannot be excelled in power of magnification. Magnifications range from 100 to 65,000 power and can show minute details of any material encountered by the fire investigator.

X-Rays

The fire investigator should not overlook the X-ray photograph, commonly used by medical experts. The X-ray can be used to examine valve interiors or other mechanisms without destroying them. The interior of a pressurized can may be examined without destroying the container. The X-ray photograph is normally a transparency which shows a shadowgraph of the substance in shades of black and white on a film. This is the normal usage of an X-ray by doctors. The X-ray film is viewed by holding it in front of a light box; the film detail is observed by the transmitted light.

The above is not the only method in which an X-ray film can be used. Lay people are accustomed to black-and-white photographs which are viewed by incidental or reflected light shining on the photograph. An X-ray film can be converted into a black-and-white photograph by any good photographic laboratory. The X-ray should be submitted to the laboratory with a request for a positive black-and-white print from the film. The results will be in the opposite tonality of the original film. What appeared white on the film will appear black on the print and vice versa. However, as shown in Figures 83 and 84, a jury would have little trouble distinguishing the elements the fire investigator is attempting to show in this way. Should the court object to the submission of this black-and-white print, both the film and print can be submitted as evidence. The jury can then compare the two for any differences.

PHOTOGRAPHIC EQUIPMENT

The photographic equipment of a fire investigator is limited only by his budget. The quality of inexpensive cameras has improved over the last two decades so that satisfactory results can be obtained with them. Limitations imposed by an inexpensive camera can be overcome by hiring professional photographers for aerial photography, photomacrography and photomicrography.

Regardless of the initial investment (which should be as high as the budget can be stretched), the fire investigator should become thoroughly familiar with his photographic equipment. As a general rule, the acronymn *K.I.S.S.* (*K*eep *I*t *S*imple, *S*tupid) should be remembered when purchasing photographic equipment. The fire investigator should be so familiar with his equipment that it can be operated subconsciously. His mind should be on the immediate problem–the detection of the origin and cause of the fire–and not with problems of equipment, exposure, depth of focus, etc. Proper equipment and application of training techniques can evoke solutions to the problems of focus, exposure and the operation of the camera so that the fire investigator can concentrate on fire causation.

The selection of photographic equipment should be made with this objective firmly in mind. Contact with a good photographic store or laboratory will provide the necessary help in selection of equipment to meet individual needs. A brief course in photography at a local university or night school is helpful.

Camera

The choice of camera is entirely up to the fire investigator. Cameras available on the market today range from 35 mm to 4 by 5 press cameras or larger. Smaller cameras are available, but due to the problems involved in enlargements, they are not recommended. The minimum size suitable for fire investigative work would be a 35 mm camera. This is the most popular size and the most widely used. These cameras range in price from $20 to $2,000. The smaller and less expensive cameras are rather limited in their flexibility.

The minimum requirement for any camera would be that it have independent focus, exposure and aperture adjustments. It should also be adaptable to either strobe or flashbulb use. No particular advantage is gained in having a shutter speed faster than $\frac{1}{500}$ second unless the camera will be used for aerial photography or personal use (sports photography). Experience will teach that the slow shutter speeds, capable of giving depth of field by stopping down the aperture opening and decreasing the shutter speed, will be more advantageous than fast shutter speeds. The same logic applies to lenses; the use of fast lenses offers no advantages.

In the 35 mm cameras, a multitude of automatic exposure or automatic shutter controls are available. While these have distinct advantages, they are also a source of considerable repair and maintenance problems. Remember the acronym *K.I.S.S.* and select a camera which has a match/needle system of internal exposure control.

Match/needle systems are simple to operate, are immediately visible through the viewfinder and are a constant reminder that exposure must be checked. Their simplicity and convenience make the additional cost well worth the investment. Cameras boasting these accessories range from $200 to $500. However, with the simplest of 35 mm cameras, a hand-held exposure meter is necessary to ensure a high probability of success with each photograph. When the cost of the inexpensive camera plus the hand-held light meter are added together, the difference between a camera with a built-in exposure system does not present an overwhelming financial obstacle.

The next commercial size above 35 mm is the 2¼ by 2¼ single lens reflex camera. An advantage to this camera is the larger negative size which provides a larger contact print than that provided by the 35 mm. The same arguments regarding size, cost and accessories outlined in the 35 mm also apply to the single lens reflex in the 2¼ by 2¼ size.

A broad range of sizes larger than the 2¼ by 2¼ are available. This includes a 2¼ by 3¼ single lens reflex, which is a multiple of 8 by 10. When an enlargement is made from the original negative to an 8 by 10 size, no cropping must be done as in the case of the 35 mm negative or 2¼ by 2¼ negative.

Each time the negative size in increased, the size and weight of the camera are also increased. It also follows that the size, bulk and weight of the film are increased. All of these increases add up to increased sizes of negatives, contact sheets and storage space for the camera and accessories as well as storage space for the negative and contact sheets. Perhaps the largest feasible camera to be carried in the field is the 4 by 5 press camera. The advantage of this camera is obvious only to the dedicated professional photographer. The size makes it necessary to use a special carrying case to protect the camera and accompanying film holders. The greatest objection to the use of the 4 by 5 press camera is the inconvenience of loading and unloading film. Film pack adapters which hold sixteen negatives are available for the 4 by 5 press camera so that sixteen photographs can be taken before changing the film pack. However, the 35 mm camera will readily accommodate a thirty-six-exposure roll of film so that loading and unloading is less than half that of the 4 by 5 press camera. The cost per unit would double that of 35 mm prints.

As stated before, it is impossible to take too many pictures; the use of the 35 mm camera encourages taking more pictures, while the use of the 4 by 5 press camera discourages taking many photographs. The other camera sizes come in rolls of twelve to twenty exposures and are a compromise between the 35 mm and the 4 by 5 press cameras. With all of the paraphernalia carried by a fire investigator, it would seem logical that the lighter camera would be an advantage, both in convenience and cost per picture.

Lenses

In the less expensive cameras, lenses are not interchangeable. Cameras with interchangeable lenses make it possible to take more than one type of photograph. With a 35 mm interchangeable lens camera, a long focus 180 mm lens can be used for aerial photography so that the aircraft need not get within 1,000 feet of the ground to take photographs covering an entire fire scene. The same camera, using a 35 mm, or 28 mm wide angle lens, can be used in small rooms to photograph the entire side of a room in one photograph.

Normal Lenses

The *normal* lens found in any camera is a compromise in optical design. A rule of thumb used to define the normal lens is that the focal length of the lens is equal to the diagonal of the negative film size. A 35 mm camera uses a negative size of 24 by 36 mm. The diagonal of this negative size is 43.27 mm. A normal lens for a 35 mm camera is one with a focal length of 45 to 50 mm.

The normal lens is designed to give the most natural perspective to photographs. The photograph, when viewed by the eye, most closely resembles the perspective of the scene.

Wide Angle Lenses

A camera lens with a focal length shorter than the normal lens is considered a wide angle lens. Using the 35 mm camera as a basis for comparison, a 50 mm lens would be considered normal, and the first of a family of wide angle lenses would be the 35 mm lens. Even wider angle lenses are obtainable; 28 mm, 18 mm and 15 mm lens are manufactured for cameras with interchangeable lenses. Table XXXI gives the approximate data on a family of 35 mm lenses.

Table XXXI indicates the wide range of lenses available for the popular 35 mm camera. The same selection in comparable focal lengths is usually available for any good quality camera. Various manufacturers may vary the focal lengths (which also changes the angle of view and near-focusing distance).

Of these lenses, a normal lens comes with the camera. If the budget will not allow the purchase of other lenses when the initial purchase is made, this is the lens which will be obtained with the camera and will be of the most value, with one very important exception–the macrolens.

TABLE XXXI

CHARACTERISTICS OF 35 MM LENSES

Type of Lens	Focal Length (mm)	Angle of View (Degrees)	Depth of Focus (Feet)
Normal	50	46	2 – α*
Macro	50	46	0.25 – α
Wide angle	35	62	1 – α
	28	74	1 – α
	24	84	1 – α
	20	90	1 – α
	15	110	1 – α
	7.5	180	1 – α
Long focus	85	29	3.5 – α
	105	23	4 – α
	135	18	5 – α
	180	13	6 – α
Telephoto	200	12	10 – α
	300	8	16 – α
	500	2.5	25 – α

* α represents infinity.

Macrolenses

A *macrolens* is a normal lens designed to increase the near-focusing range so the lens can be used for close-up work. *Close-ups* are photographs taken of an object at or near life-size reproduction. A *photomacrograph* would produce a negative from one-half to life-size.

The macrolens is produced by providing a method of extending the front elements of the lens further away from the body of the camera, thus enabling it to focus on near objects. These lenses are invaluable to the fire investigator. If only one lens is being considered for a camera, rather than purchasing a regular normal lens, a macrolens should be considered. The additional cost is well worth the investment. The macrolens can be used for normal photographs, even though the focal length of the macrolens may be slightly longer than the so-called normal lens of the camera. As a rule, macrolenses will have a focal length of 50 to 60 mm. The slight increase in focal length does not significantly change the perspective.

Macrolenses are invaluable for taking close-ups of fuses, fuse boxes, electric switches, screws or bolts on control valves or objects where minute detail is required. As the macrolens is used close to an object, the depth of focus decreases proportionately. Therefore, in order to obtain the maximum depth, the lens should be stopped down and a slower shutter speed used. This will usually necessitate the use of a tripod to avoid camera movement.

SUPPLEMENTAL LENSES. For those investigators already equipped with a camera who do not feel that the purchase of a macrolens is justifiable, an alternative is available. Many manufacturers of optical devices manufacture a *supplemental* or *front* lens which, when attached to the standard lens of a camera, changes the design of a standard lens so that it may be used for macro or close-up work. These produce excellent results, particularly when used with through-the-lens metering and viewing. With these two facilities, no problems are encountered regarding parallax (discrepancy between what the camera and photographer see) or exposure. However, if the camera is a range-finder type of camera without through-the-lens viewing and metering, focal distance and exposure become critical and must be calculated. Exposure readings from a hand-held meter are not applicable and must be calculated. Suppliers of the front lenses will provide the necessary information to use their product.

Long Focus Lenses

Long focus lenses give a smaller field of view so that the photographer need not be close to the subject to fill the viewfinder. Long focus lenses are convenient for taking close-up photographs of inaccessible objects. However, their use is very limited in fire investigations. They are used primarily in aerial photography so that the pilot need not bring the aircraft too close to the ground and violate city or local ordinances in order to fill the viewfinder.

Telephoto Lenses

Telephoto lenses differ in design from long focus lenses and give larger magnification for a given focal length. Their use is the same as the long focus lens, and they find very little application in the field of fire investigation. Their use would be the same as for a long focus lens but would have less application.

Zoom Lenses

A zoom lens is a specially designed lens which has a variable focal length. The ranges vary from a normal lens to a long focus lens, such as 50 to 90 mm. Others are available from 90 to 300 mm. In other words, the investigator can focus on an object and then zoom in until the object fills the viewfinder, thus guaranteeing the maximum usable negative without trying several different lenses.

The zoom lens is a very useful adjunct to the photographer's kit and replaces three or four lenses. Its disadvantage is initial cost and the risk of damage. If one of the lenses which it replaced were damaged, other similar lenses would still be available to perform the task. If the zoom lens is damaged, the fire investigator will be "out of business" until it is repaired. In a way, it is "putting all your eggs in one zoom lens."

Exposure Equipment

Most good-quality cameras today have through-the-lens exposure meters. These have proven to be extremely reliable, simple to operate and inexpensive to maintain. Through-the-lens metering is advantageous to the fire investigator, particularly in macro or close-up work. Therefore, one of the considerations in purchasing a camera is that the camera is equipped with through-the-lens metering. The type or style is immaterial as long as the fire investigator understands the operation, keeps the batteries fresh and does not leave the meter turned on for long periods of time.

The alternative to built-in metering is a hand-held light meter. Many good brands are available which can be selected on the basis of price rather than function. The fire investigator will normally be using a reflected light meter rather than an incident light meter reading. The light meter selected should have sensitivity to all colors, and the fire investigator should not be concerned with low level light readings. When the light level becomes so low as to require time exposures, the fire investigator should turn to artificial illumination rather than depending upon natural illumination.

Lighting Equipment

The fire investigator will rely on all three types of illumination. These include natural, floodlight and flash lighting. The investigator should be aware of his limitations as far as hand-held photography is concerned. Under natural lighting conditions, if the lighting has become so dull and dark as to require slowing the shutter speeds enough to cause camera vibration and a blurred picture, a change should be made to floodlight or flash illumination. This limitation is easily determined.

Shoot a roll of black-and-white film of the same scene at increasingly longer exposures. Record the shutter speed for each shot. Examine the photographs for camera movement. This will be evidenced by uniform blurring of the entire photo. Never use a shutter speed slower than the speed which introduces blurring.

Whenever possible, the use of natural daylight illumination should be limited to hours between 10 AM and 4 PM so that colors are rendered more accurately. Any photographs not taken during these hours should be taken with fill-in flash to give better color-rendering of the scene.

Floodlights

The fire investigator should not overlook the possibility of incorporating equipment used for his visual examination with that used for photography. Burned-out buildings are at best a very poor reflective medium, and interior lighting is usually inadequate. Therefore, the fire investigator should equip himself with floodlights in the range of 1,000 watts to provide good uniform illumination of the scene to be examined. These lights can be mounted on portable tripods and electricity brought in from a neighboring building or from a portable generator. On serious fires, this investment will prove more than worthwhile. It will also provide excellent illumination for photography.

If color film is being used, floodlights which provide daylight illumination are available. These include 100 to 200 watt blue photoflood bulbs or professional lights such as General Electric FAY® 650 watt bulbs. Such lights make an excellent investment in terms of providing illumination for the detailed examination of burned-out areas as well as an effective light source for color photography.

Flash

Two options are available for flash illumination. The most common is the flashbulb, which has a one-time use. These are available in either daylight or tungsten illumination for color or black-and-white film. They provide uniform illumination but require computing the aperture opening for each change in distance from the camera to the object, due to the variation of light intensity with distance. To eliminate this problem, the use of automatic strobe lights is strongly recommended.

Strobe lights, or electronic flash tubes, are operated by portable or household electric current. The flash tubes are energized and discharged to provide daylight illumination suitable for either black-and-white or color photography. In addition to eliminating the time-consuming replacement of flashbulbs for each shot, the electronic flash tube has the capability of automatic illumination control. The strobe is preset so that once the shutter speed is selected, the aperture opening is set according to the film speed and the electronic control provides the proper amount of light when the shutter is released.

The latest models have a green light to indicate if enough illumination is present on the scene for proper exposure. If the photographer is too far from the object, or if the object does not reflect enough light for adequate exposure, the indicator will not respond. Thus the photographer knows he must open up the lens (and control) or move closer to the object.

Since many scenes taken with flash photography are a one-time opportunity, these added electronic devices are a prudent investment, despite the extra cost. Most of the new electronic flash gear comes equipped with rechargeable dry cells, an added advantage. An extra pack of fully charged cells can be carried at all times. This is particularly important in cold weather because batteries lose half their strength when exposed to low temperatures.

A strobe light or electronic flashgun which can be detached from the camera is an important feature to the fire investigator. With the flashgun removed from the camera, it can be placed to provide side lighting to the scene being photographed. This enhances the texture of a burn pattern or material being photographed. This technique is illustrated in Figures 86, 87 and 88.

Should softer light be desired on bright contrasting objects, a diffusion filter can be placed over the lens of the flash gun. This provides a much softer light with less shadow. It is not recommended for photographing burn patterns or dark, dull objects.

Filters

Filters have very limited use in fire investigative work. They are used primarily to emphasize contrast between materials when used with black-and-white film or to make slight color corrections on color film. However, most jurisdictions may look twice at a photograph that was taken with the use of a filter and demand that a duplicate photograph be submitted that was taken without a filter so any differences can be noted. Therefore, if filters are to be used, one photograph should be taken without the filter so that the judge and jury can examine the photographs and understand

Figure 86. The lighting of the material photographed will greatly influence the texture revealed. In this photograph, the strobe light was held adjacent to the camera, giving a flat lighting to the scene. Compare this with Figures 87 and 88.

Figure 87. The strobe light was removed from the camera and held at arm's length to the left of the camera, pointed directly toward the center of the photograph. The result is an overexposed scene on the left of the photograph and underexposed on the right, but the texture of the material shows up better than in Figure 86.

why a filter was used. Perhaps the only filter of any practical value is the haze filter. The haze filter can be used to reduce haze in aerial photography or long-distance photography. It will have very little effect on smoke, however.

Film

The film selected should be made by a reliable manufacturer who distributes the product everywhere within the range of the fire investigator's activities. Sooner or later, the investigator will run out of film in the field. If a popular brand is being used, no problem will be encountered in obtaining the same type of film, even in the smallest village. He should select one type of film and become familiar with the loading, unloading, care, handling, storage and correct exposure of the film.

In today's courts, color film has become so widely accepted that practically all photographs should be taken in color. While black-and-white film is admissible in court, color film does a much bet-

Figure 88. The strobe light was held at the extreme right of the photograph, resulting in overexposure on the right side of the scene and underexposure on the left. When in doubt, all three positions of the strobe light can be used to ensure that the best photograph will be obtained. By moving the strobe light to the right or left of the camera, the texture of the material is shown in greater detail (compare Figs. 86, 87 and 88).

ter job of rendering the details of a burn pattern. Black-and-white film can be used in cases of emergency.

One problem encountered with color film is *reciprocity failure.* Reciprocity failure occurs when the exposure limitation of the film is exceeded. Each film characteristically has a time limitation which, when exceeded, will cause dramatic changes in the color rendition of the film. The data sheet accompanying the film should be studied carefully for complete understanding of reciprocity failure.

Commonly, reciprocity failure occurs when exposures longer than 1/10 second are used with daylight color film. Therefore, time exposures should be limited to less than 1/10 second.

The use of black-and-white film is sometimes preferable. For example, black-and-white film will show the char pattern of a burn slightly better than color film. When in doubt, the fire scene should be photographed in both color and black and white.

Exposure problems may occur due to the difference in film speeds of color or black-and-white film. Color film is traditionally a slower film (less sensitive to light) than black-and-white film. By selecting a black-and-white film that has the same ASA rating as the color film, the photographer can eliminate the problem of resetting the light meter ASA index when changing films. Many color films are available with an ASA index of 400. This fast color film is well adapted for fire investigative work.

Other films available include infrared film, which is sensitive to the infrared portion of the spectrum. The use of this type of film applies primarily to photographs involving the detection of writing on burned or charred documents.

Store film in a cool place. If stored in a refrigerator, warm to room temperature before using to allow evaporation of moisture.

Development of Exposed Film

As soon as film has been exposed (particularly color film), it should be developed, including partial rolls. Film is, after all, inexpensive compared to the cost of reshooting scenes. In fact, many scenes cannot be reshot for the simple reason that they no longer exist. Therefore, it is good practice to sacrifice unexposed frames and have the roll developed immediately. The latent image on color film is subject to change and can be easily rendered useless by excessive heat, exposure to X-ray or other electromagnetic phenomena.

Any good commercial laboratory can develop color film and produce satisfactory results. Some photographic errors can even be corrected. The fire investigator should be aware of steps to follow in the event of incorrect exposure.

A common error made by semiprofessional photographers is to either overexpose or underexpose a roll of film by neglecting to change the ASA index on the exposure meter. As a result, a roll of 400 ASA rating film may be exposed at 100, which will result in four times the exposure necessary, or vice versa. In either event, the fire investigator should know what action to take.

If the film has just been started and only a few photographs have been taken that can be retaken, the ASA index should be reset to the proper setting and the scenes rephotographed.

However, if irreplaceable shots have been taken, they can be saved. The balance of the incorrectly exposed roll will be used by the photographic laboratory to determine the correct development of the irreplaceable portion of the roll. The following procedure will enable them to save the roll:

1. Note the frame number when the error was discovered.
2. Continue shooting the roll at the *erroneous* setting.
3. Shoot only scenes which can be reshot.
4. After completing the erroneous roll, send it to a film processor with the following instructions:
   (EXAMPLE) The enclosed roll of film was exposed at 400 ASA exposure meter setting. The film speed is 100 ASA. The first ten shots are irreplaceable. The following twenty-six shots were exposed at 400 ASA and are expendable. Please hand-develop the last twenty-six shots to determine proper development time to ensure proper development of the *first* ten shots.

The laboratory will then hand-develop the last twenty-six shots to determine the necessary development time for the first ten shots. This will guarantee successful development of these negatives. The same procedures apply to errors in selecting the correct aperture opening or shutter speed.

Background Material

Quite often the fire investigator will desire a close-up of a piece of evidence to show it in greater detail. The evidence should be photographed in its original position as found by the fire investigator. In order to show it more clearly, the distracting background can be blocked out by holding an appropriate piece of colored posterboard behind the object so that it is clearly outlined.

A piece of flat white artist's posterboard with a black matte reverse side can be used effectively for this purpose. It can be hinged by cutting the 3 by 4 foot material into quarters and then hinging them using transparent masking tape so that the posterboard can be folded to one-quarter size. This makes it convenient to store. For most subjects, a neutral gray or 18 percent reflectance card, such as printed and manufactured by Eastman Kodak Company, can be used. These cards are available in 8 by 10

Figure 89. To reveal details of the tank, gauges and hose, a simple background of white paper was used. The camera was moved away from the tank and gauges to reveal the principle. Had a photograph of the tank and hoses only been desired, the camera would have been moved forward to exclude distracting objects.

cardboard sizes and are also useful as a guide to exposure for the color laboratories. They give a reference point in developing and printing the negatives. When photographing evidence using background material, always take a duplicate shot with the background removed to point out that the background material was not used to hide something.

For larger background material, soft pliable rolls of material from 24 inches wide to 10 feet wide by 50 to 100 feet long is available for studio installation. These are useful when photographing appliances or automobiles. The background material comes in varying colors, but those of interest to the fire investigator are matte black, matte white, 18 percent reflectance flat gray and matte green. The matte green is useful when photographing in color. Since green represents the center of the visible light

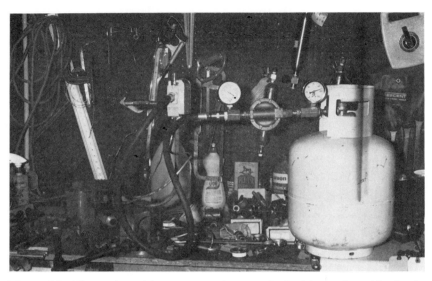

Figure 90. The tank and hoses were photographed without the white background used in Figure 89. The distraction of the other objects on the bench is obvious. These can be eliminated by using a plain white background paper, as shown in Figure 89.

spectrum, it is considered a neutral color in respect to other colors and is acceptable as background.

Evidence in the laboratory can be photographed without any distracting background, as illustrated in Figure 91. In Figure 92, the object to be photographed is placed on a glass surface and illuminated as desired. Two separate sources of illumination are used. The background illumination should be four times as intense as the object illumination. In this procedure, the photograph is taken with one exposure by setting the correct camera exposure for the objects, with the background exposed.

The same results can be obtained by using a double exposure. For the first exposure, the object should not be illuminated. With the camera on a tripod, the first shot is taken to achieve overexposure of the background material. The object is then illuminated and the second exposure made, based upon the proper illumina-

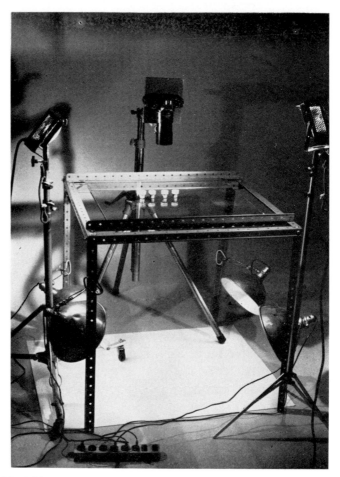

Figure 91. This procedure is used to photograph opaque objects with complete elimination of annoying background. A clean glass table is provided with high intensity lighting on a white background under the table. Objects are carefully lit so that intensity on the objects is less than the background lighting. Careful light meter readings of the objects are taken by turning off the illumination on the background material. Exposure is then made, and the results will be as shown in Figure 92.

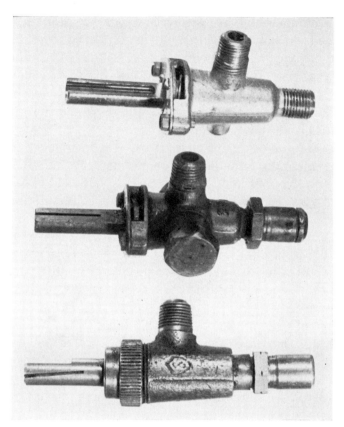

Figure 92. The objects have been photographed using the technique de-scribed in Figure 91. The background was exposed to high intensity light-ing, and camera exposure controls were set for the necessary lighting of the three valves. This method completely eliminates backgrounds. If a black background is desired, it can be attained by laying a piece of black matte material under the objects and lighting the objects so that no light falls on the black matte background. Exposure is made for the objects to be photo-graphed, disregarding the background. A spot meter is recommended for this type of photography.

tion of the object. Figure 91 was photographed using the first procedure.

WHAT TO PHOTOGRAPH

The importance of photographic evidence cannot be overemphasized. At times, it will be the key evidence to persuade ac-
ceptance of your theory, whether it is an insurance claim or a first-
degree arson case. It follows then that the selection of photographic equipment, films, cameras, flash equipment and associated accessories should be of the highest quality possible under budgetary
limitations.

Once equipped with this investment, it is incumbent upon the
fire investigator to train himself in the art and science of photography to the point that his photographs will yield at least a 95
percent success factor. When the investigator is in doubt regarding
exposure, he will soon learn to take a series of shots ranging from
overexposure to underexposure to guarantee a usable negative of
each pertinent shot.

In cases of a serious nature, the investigator will postpone destroying the scene until he is certain that usable negatives have
been obtained.

Photographs should be taken as the investigator proceeds with
the search. His normal approach is to proceed from the outside
of the building and work inward. He should take photographs of
the complete exterior of the building, including oblique shots of
each corner. This will show all windows, doors, entries and their
relationship to each other and to power poles, propane tanks, storage tanks, streets and access. If forced entry is discovered, close-
ups of marks should be obtained. Photographs of any debris
(fragments of windows taken as evidence) should be taken as the
debris is found, with a notation regarding the camera position,
height of camera from the ground, type of film and time of day
and date. These should be recorded as part of the operation. A
separate photographic log should be kept and become part of the
investigator's records.

If the investigator is using a tape recorder, the data can be dictated and later transcribed into a separate photographic log. Pictures should be taken of the burn pattern and each room that con-

tains what could possibly be an origin. Many photographs showing the relationship of a possible origin to the room, walls and contents should be taken. Since the cause of the fire may be unknown at this time, these photographs should not be overlooked. One of these photographs of an origin may show the cause of the fire. The investigator will find this photograph invaluable.

The camera should be an extension of the notebook. The notebook log should record the details of each photograph so that the photograph can be identified and its significance brought back to the investigator not only by the photograph itself but also by the notes.

Before any evidence is removed, at least two photographs from different angles should be taken. This will help identify not only the position of the evidence but also its condition prior to removal.

In the event of an arson fire, it is vitally important to examine all screws and connections on equipment to a piping or electrical system to discern any indications of tampering before the fire, during the fire or after the fire and prior to examination of the evidence. These should be photographed to show their condition.

Identification of Negatives

The fire investigator will take many photographs. Most of these will not be used in court. Therefore, a highly organized retrieval system must be established to retrieve photographs years after they were taken. Any alphanumeric index for identifying photographs will work efficiently. Each roll should be identified by a case number. The logical identification is to number the rolls consecutively as they are shot. Each roll has a negative number which identifies the negative. The combination of the case number, roll number and negative number completely identifies any photograph printed from the roll.

It is customary to store negatives in a negative file. Negatives should be cut in strips and placed in separate glassine envelopes with the case and roll number marked on the outside of the envelope. As an aid to quick identification, the numbers of the negatives can be added to the outside of the envelope. If these are then filed in numerical order, they can be located quickly and identified.

After the films are developed, *contact sheets* (prints the same size as the negative) of each roll can be made on 8 by 10 or 8½ by 11 single-weight glossy photographic paper. Any good commercial laboratory will do this. These contact sheets are a convenient method of showing a fire investigator the pictures included in his file. They are positive photographs and, although they are small, can be viewed under a magnifying glass to remind the fire investigator of negatives in his possession. When a report is being prepared, or a case for court, he can refer to the contact sheets and select the photographs that illustrate points he wishes to explain.

Contact sheets are inexpensive and cut the cost of maintaining a photographic file. They take the place of the 3 by 5 jumbo prints normally associated with 35 mm film. When the contact sheets are received from the laboratory, the case number and roll identification number should be written on the contact sheet. When prints are ordered, merely specify the negative number needed and send the negative strip to the photo lab. When the finished prints are received, the identification number should be promptly inscribed on the back of the photograph. This will ensure that duplicates can be quickly ordered if needed.

Photo Index

Upon receipt of the contact sheet, the investigator should immediately verify that the photographs are his and check the accuracy of his photographic log. If none was made, he should immediately make a photo index for each contact sheet to include the following information:

1. Identification number for each negative
2. Date and time of the photograph
3. Location of each photograph
4. Direction camera was facing
5. Height of camera above ground
6. Film type
7. Type of camera
8. Type of lens
9. Exposure

10. Details of what the photograph represents
11. Comments regarding lighting, any departure from normal, reason photograph was taken, etc.

By making an index as soon as the contact sheet is received, the fire investigator can take advantage of the freshness of his recollection and make notes which were not included in his log. This forces him to evaluate the photographs and decide if more need to be taken. It also gives him the opportunity to evaluate his photographic techniques and upgrade his photographic capabilities.

A SHORT COURSE IN PHOTOGRAPHY

The camera is the tool with which the fire investigator takes photographs to supplement his written report or testimony in court. A camera, regardless of price, consists of a light-tight box with a lens at one end and a film-holding device at the other. A shutter which controls the amount of light striking the film is placed between the film and lens.

The entire process starts with the film, whether it is black and white or color. A chemical coating on a base material, usually of acetate, is sensitive to light. The differences in reflection of light shining on the scene cause proportional reaction to the chemicals on the film. After exposure, the latent image on the film is developed chemically to a permanent fixed image. Depending upon the chemical characteristics of the film, negative or positive films are available for negative or positive images.

Traditionally, black-and-white films use a negative image; color films may have either a negative image or a positive image. The image depends upon the type of film purchased and required by the photographer. For slides, a positive color film is used; for prints, a negative color film is used.

Exposure

The secret of good photography is to correctly expose the film to the proper amount of light to ensure a good image. To assist the photographer, the film manufacturer provides the user with the ASA (American Standards Association) index number and instructions. The ASA number is a comparative film rating based on the film's sensitivity to light; the higher the index, the more sensitive the film. With each roll of film, an instruction sheet is included with full directions for proper use of the film. By explicitly following these instructions, good results can be obtained with the least expensive camera.

The camera is equipped with two devices which control the amount of light reaching the film. The first has already been mentioned (the shutter). The second is the aperture opening on the lens, which controls the quantity of light passing through the lens and the shutter to strike the film. This is accomplished by varying the size of the aperture opening. The larger the aperture, the more light will enter while the shutter is opened. Conversely, the smaller the aperture opening, the less light will pass through the lens to the film.

The amount of light can also be varied by leaving the shutter open for longer or shorter periods of time. It is the combined adjustment of these two variables—the aperture on the lens and the shutter speed—which most beginning photographers find confusing. This is because many settings of the two variables will allow the same quantity of light to the film.

By opening the aperture, more light is permitted to enter the camera. However, this can be restricted by limiting the time the shutter is opened, thereby limiting the quantity of light striking the film. Conversely, by using a long shutter speed and a smaller lens aperture opening, the same amount of light can be permitted to strike the film. This variation permits the photographer to take advantage of one characteristic of the lens which has been only briefly mentioned. That characteristic is *depth of field*.

The depth of field is the area in the camera field which is in focus at given aperture settings. For example, if the camera was set at $f/3.5$, the depth of field would be less than if the aperture lens was set at $f/16$. At $f/16$, near and far objects would be in focus; at $f/3.5$, they would not be in focus. Therefore, the photographer interested in showing near and far objects in focus would normally use as large a numerical aperture opening possible (a smaller physical size of the aperture opening) and a correspondingly slow shutter speed. As a general rule, the fire investigator would be wise to use the smallest aperture available and merely adjust the shutter speed.

Unfortunately, this is not always possible. The shutter speeds may be so slow that they will exceed the capability of the film to record the image or will allow blurring of the picture due to camera movement. To eliminate this problem, shutter speeds slower than $\frac{1}{100}$ second should be taken with the camera on a tripod to avoid camera movement, or flash equipment should be used. If neither of these is possible, the illumination level of the scene can be increased by using flood lights.

Taking the Picture

In order to eliminate any possible sources of error, the investigator should take photographs using the following sequence:

1. Load the camera with fresh film.
2. Set the proper ASA index on the camera light meter, camera exposure index or hand-held light meter index.
3. Carefully select the scene to be photographed.
4. Focus the camera on the principal point of interest.
5. Use the smallest aperture opening possible, consistent with the following:
   A. Hand-held shots of less than $\frac{1}{100}$ second exposure time
   B. Tripod-held shots of less than $\frac{1}{10}$ second exposure time
   C. Flash illumination
6. Use the fastest shutter speed possible consistent with the smallest aperture opening possible, or a compromise of the two.
7. Take a deep breath, let half of it out, and carefully *squeeze* the shutter release button.
8. *Immediately* advance the film to avoid double exposure.
9. After roll is exposed, *immediately* rewind film into container; remove from camera. Develop as soon as possible. In the meantime, keep film away from *heat*.

Directions for proper exposure settings can best be obtained from the manufacturer of the flash equipment, film and camera. When in doubt regarding exposure, remember that film is much cheaper than trying to recreate a scene after finding that photographs have been either overexposed or underexposed. Therefore, take several shots ranging from deliberate underexposure to overexposure to guarantee that at least one usable negative will be obtained. If you can consistently overexpose the first shot, get the second correctly exposed and the third underexposed, you can then have complete confidence in your ability and can eliminate the two extra shots.

CHAPTER 14

Evidence

E VIDENCE IS ANYTHING admitted into the court record and presented to the jury which tends to support one side of an issue being tried by the jury. Evidence is normally divided into five categories:

1. Direct evidence
2. Real evidence
3. Circumstantial evidence
4. Opinion evidence
5. Documentary evidence

The fire investigator should be concerned with all five and be familiar with the legal definitions, procedures and laws regarding evidence within his jurisdiction. Since each jurisdiction has different laws, this chapter touches only on general principles involved in the recognition, identification, preservation and presentation of evidence.

Evidence is the accumulation of material, facts, photographs, statements, records, reports, tests, expert opinions and statements of witnesses, which the fire investigator presents to the prosecutor or attorney for presentation in court. Evidence is the foundation on which a case is built. In order to be presented in court, the evidence must be admissible, competent, relevant and material.

TYPES OF EVIDENCE
Direct Evidence

Direct evidence is the testimony of a witness's personal experience of hearing, seeing, smelling, tasting or feeling directly related to the facts being presented. The lay witness, i.e. the ordinary person with no particular expertise related to fire, would be permitted to testify as to his sensory experience and express opinions in areas which come under the broad knowledge of the average person. These include, but are not limited to, identity, physical, emotional and mental condition and sobriety of a person. A lay witness may also express opinions on matters of common knowledge such as the speed of vehicles or normal temperatures and weather conditions (rainy, cloudy, clear, sunny, windy). He may not be qualified and, therefore, not permitted to offer an opinion concerning the intensity of the rain, degree of cloudiness, precise temperature, exact vehicle speed, etc.

Real Evidence

Real evidence is any physical substance relevant to the case which tends to prove the facts in question before the court. Real evidence could consist of a faulty furnace or components that caused a furnace to malfunction, such as a defective valve seat which permitted gas to escape and cause an explosion.

Real evidence could consist of the odometer reading on a vehicle, whether or not it corresponded with mileage declared in a statement admitted as direct evidence.

Real evidence is the physical material the fire investigator seeks as the scene of the fire is examined. The evidence may range from broken glass, the burn pattern at the origin of the fire, pipes, appliances and charred wood, to an entire building. The investigator may need to photograph the real evidence, since it would be impractical to bring it, i.e. the charred wall, into court. The photograph would be documentary evidence.

Circumstantial Evidence

Circumstantial evidence is indirect evidence related to the facts presented. Although the facts presented as circumstantial evidence are not the facts under direct consideration by the jury, by inference or deduction they lead directly to the facts under consideration. Circumstantial evidence may consist of any or all four types of evidence: direct evidence, opinion evidence, real evidence and documentary evidence.

The following illustrates real and direct evidence admissible as such. A witness was in a room and saw a man light a newspaper with a cigarette lighter which resulted in a fire. The eyewitness's testimony would be *direct* evidence; the cigarette lighter would be *real evidence.*

The following illustrates circumstantial evidence. A witness was standing on a street corner waiting for a bus and saw a person enter a house carrying a red 1 gallon container and emerge a few minutes later without the container. A short time later, he observed the house burst into flames. In this case, the real evidence would be the container itself, the accelerant and fingerprints (matching those of the person observed entering the house) found on the container and on door knobs. The circumstantial evidence would be what the witness observed.

Opinion Evidence

The ordinary or lay witness may testify regarding any input received through the five senses. As a rule, testimony will be limited to what has been seen, heard, smelled, felt or tasted. Under certain circumstances, if his background, experience and training qualify him, the lay witness may express an opinion. By providing sufficient background information regarding his training, education and experience, an attorney may qualify him to testify as an *expert witness.*

An *expert witness* is a term used by the court to describe a witness who, by virtue of education, training and experience, has more knowledge regarding a subject than the ordinary lay person. As such, the court grants this expert witness the privilege of expressing an opinion based upon facts admissible to the court and presented as foundation for the expert's opinion. The opinion thus obtained is termed *opinion evidence.*

The following is an example of opinion evidence. A fire chief could testify that an extraordinary amount of heavy black smoke, accompanied by an abnormal volume of bright yellow flames, was seen at a residential fire. A lay person could testify to the above facts, but no further. The fire chief, based on his education, training and experience, could express an opinion that an accelerant was used.

Assume tests were conducted on gasoline from a container found near the scene and on samples of accelerant taken from carpeting in the room where the fire originated as well as from uncontaminated samples of the same carpeting. Further assume that these samples were tested and evaluated by a forensic chemist qualified as an expert witness.

The forensic chemist could testify not only regarding the test results but could form and express an opinion by virtue of his education, training and experience. He could state that, in his opinion, the gasoline from the can and accelerant-soaked carpeting were identical and that no similar hydrocarbons were found in the control sample. The chemist's testimony would be considered *opinion evidence.*

The impression that an expert witness is always a highly educated person has been formed because these are the persons normally involved in providing opinion evidence. Such is not always the case. For example, an illiterate cowhand with twenty years experience in roping, branding and rounding up cattle may be called as an expert witness to give his opinion regarding the saddling of a horse. He could also be qualified to give an opinion concerning the conduct of a roundup regarding a stampede which resulted in personal and property damage. The basis for an expert opinion is the education, training and experience of the person voicing the opinion. The qualification is not based on education alone.

The fire investigator must present his own education, training and experience to qualify him to voice opinions when his attorney deems it necessary. In complex cases, an attorney may feel that the opinion of other experts may enhance the fire investigator's work. He may call upon other experts for opinions based upon the evidence presented by the fire investigator.

Due to the complexity of a case, it may be necessary to call in many experts for opinions on other possible causes. The fire investigator may have called upon an electrician and electrical engineer to eliminate the possibility of an electrical fire. They may be called as expert witnesses to voice their opinions. The same procedure may be followed to confirm or eliminate other possible causes of the fire.

All of these experts may testify, giving their opinion, based on the results of the fire investigator's investigation, their own investigation, tests and other evidence introduced. Each expert will consider all admissible evidence. The use of experts in addition to the fire investigator relieves him of the total responsibility for the elimination of other possible causes. Concurrence of experts may increase the credibility of the fire investigator in the eyes of the jury. If the fire investigator testified that he alone had eliminated all other possible causes, the jury may form the impression that he claims expertise in all areas–electricity, gas appliances, lightning, chemistry of spontaneous combustion, etc. By using other experts for one or more of the possible causes, the credibility of the fire investigator is enhanced and his opinion becomes more credible.

Documentary Evidence

Documentary evidence is real evidence primarily supplied in written or printed form such as insurance policies, deeds, business records, receipts, invoices, bills of lading or any other type of written document. It can also consist of audio or video tape recordings, X-rays and photographs. Documentary evidence is used to prove relationships between the accused and the crime. It is presented to prove a fact or aid the jury in reaching a decision.

Documentary evidence in a civil case tends to be extensive and may include industry standards, warranties, instruction manuals, contracts, invoices, expanded or exploded drawings, photographs, X-rays, sketches, scale drawings, sales literature and brochures, or any pertinent data which tends to prove or disprove the issue before the jury.

VALUE OF EVIDENCE

The value of evidence depends upon its admissibility. If the evidence is inadmissible, it has no value whatsoever. Therefore, the fire investigator must prevent actions which could render valid evidence inadmissible. The admissibility of evidence is ruled upon by the judge at the time it is presented. Preservation of evidence is the responsibility of the fire investigator from the time of its discovery until admittance into court. The investigator should obtain legal guidance in his jurisdiction to ensure protection and admissibility of evidence.

The next element in the evaluation of evidence is its *competency.* Proper foundation must be laid for the presentation of evidence to the court. This includes a brief history of its discovery and relationship to the case, with positive identification of evidence from its discovery until presentation in court. The identification should be in the form of a continuous chain of evidence, with possession accounted for chronologically.

The evidence, regardless of its admissibility and competence, will be inadmissible unless it directly relates to the issue before the court and bears directly upon some facet of the case. The attorney must prove the relationship between the evidence and the theory presented, or its *materiality*.

In order to prove materiality, the evidence must contribute to the decision-making process of the jury. Each piece of evidence should fit as a link in a chain to show it is material to the issue being tried before the court.

Determination of Evidentiary Value

When the fire investigator first arrives on the scene, he does not know whether the fire was of accidental, natural or incendiary origin. The presumption of the courts is that a fire is accidental until proven otherwise (the presumption of innocence). This poses a dilemma for the fire investigator. If his investigation is conducted on the presumption that the fire was accidental, he would ordinarily exercise less care than in the investigation of an arson fire. Later, he may determine that the fire was of incendiary origin and may have destroyed or lost part of the evidence. Therefore, every fire must be investigated with the greatest possible care in the preservation of evidence.

When an investigation is begun, all areas should be examined for both positive and negative evidence. The lack of open windows and doors, holes in the roof or indications of forced entry are all negative aspects of an arson case. However, these negative aspects should be recorded and photographed in a routine manner, even if no positive evidence was found. These do not constitute elimination of the possibility of arson but do add to the cumulative circumstantial evidence to be weighed at the end of the investigation. This same attitude toward positive and negative evidence should be maintained until the investigation is completed and "all the facts are in."

Since the investigator does not know which direction the case will take–arson, accidental or natural fire cause–he should continue to accumulate photographs and notes of both types of evidence. When sufficient evidence is available to indicate a definite cause, his steps can be retraced and real evidence can be obtained to disprove or substantiate his theory. The elimination of possible causes should be started with concentration on the less probable causes. A pattern will normally emerge indicating the direction of the preponderance of evidence and enable elimination of the questionable possibilities.

After the scene search has been completed, the fire investigator will have a good idea what corroborative evidence is needed in the form of statements from witnesses, owners, etc. At that time, a witness matrix can be set up as an aid to corroborate or refute the physical evidence.

PRESERVATION OF EVIDENCE

Real evidence must be either photographed, which captures the essence of the physical evidence on film, or impounded. If the evidence is photographed, the date and time of the photograph must be recorded. A series of photographs showing the relationship of the evidence to the building, with close-ups of the pertinent parts, should be made. A detailed scale drawing of the evidence should be made to indicate its relationship to the building. An overall plan of the building is usually a good reference point from which to start.

If the physical evidence can be removed, the same photographic sequence should be followed to show that the evidence has not been altered. Each item should be identified by engraved initials or indelible marks placed in an inconspicuous place so that the identity and credibility of the evidence will not be destroyed. The mark should be distinctive so that it can be recognized later.

A log should reflect when the evidence was photographed and measured, its location in relationship to the building and by whom it was removed. All details regarding the removal should be documented so that when the evidence is presented in court, the circumstances related to its removal can be explained to a jury.

In addition to the investigator's mark, the evidence should be identified with a tag, showing the case number, date of removal, special circumstances and the signature of the fire investigator and witnesses involved in the removal. It should then be removed and preserved for testing or court. If the material is a volatile substance, it should be placed in a suitable airtight container.

Samples of accelerants are normally placed in new unpainted, clean, 1 gallon paint cans. The lid should be secured and sealed. The identification should be attached to the can, not the lid, as the lid may become separated from the container. The accelerant, in its container, should then be transported to a laboratory for analysis. Some flammable materials require special handling.

When an accelerant is removed from organic material, i.e. carpeting, a control sample should be taken of the same material located in an area remote from where the accelerant sample was removed. Hydrocarbons are found everywhere. The manufacturing process may result in the presence of hydrocarbons in carpeting. Therefore, a carpet sample of an area of suspected hydrocarbons or accelerants may yield a positive result, as it would if the material was tested as it came from the factory. Hydrocarbons found in the suspected area should be in excess of those found in the *control* sample. Unless a control test is made, the opposition can argue that the hydrocarbons were the result of the manufacturing process. The testing of a control sample can refute this argument.

Chain of Evidence

In order to establish positive identification of the evidence, the chain or continuity of custodianship must be established. Each time possession of evidence changes hands, a written release must be exchanged. The person delivering the evidence should give and obtain a receipt from the recipient. This *chain of evidence* establishes who had the evidence and will be an unbroken chronological succession of possession. Any break in the chain may constitute grounds for inadmissibility of the evidence. Therefore, it cannot be stressed too strongly that this chain of evidence must be maintained from the time the investigator first observed the evidence until its presentation in court. Presentation of the evidence to the court ends the chain of evidence.

Testing the Evidence

Quite often evidence, such as suspected accelerants, valves, glass, residues or electrical appliances, must be tested. In order to ensure admissibility of the tests as evidence, the proper authorities should be notified of the date, location and persons performing the tests, as well as the proposed tests and their procedure. Written acknowledgment that notification of testing has been made should be demanded. These precautions should be taken in civil and criminal cases to prevent the opposition from barring evidence from admissibility; they will guarantee that all interested parties will have the opportunity to witness the testing.

Nondestructive Testing

The most satisfactory tests do not alter the evidence. These are termed *nondestructive* tests and do not alter the shape or form of the evidence. For example, a valve may be subjected to tests by running a fluid through the valve without destroying the valve function. However, if the pressure through the valve exceeds the design pressure of the valve, or the pressures under which the valve is normally subjected, the test may inadvertently become destructive. This must be avoided at all costs.

Each test should be carefully analyzed for its potential effect on the evidence before any testing commences. Persons conducting the test should be experienced in the test to be performed and be qualified to anticipate results or complications which could affect the evidence. If any doubt exists regarding the possible effect on the evidence, a similar piece of evidence should be subjected to the same anticipated tests to determine any destructive aspects.

All tests should be conducted under strict supervision of the investigator and should be fully documented with the time, place, type of equipment used, type of tests, persons present and their interest in the case, and the results. The specimen to be tested should be photographed before, during and after testing as evidence of the tests and proof that the evidence was not altered by the testing. Time and date records should be kept for each photograph and a written record of the test logged with all pertinent data, results and signatures.

Destructive Testing

The same general procedures outlined in nondestructive testing should be followed in *destructive testing*. Since the evidence will be destroyed as a result of the tests, it is imperative that everyone associated with the case be informed. Under no circumstances should destructive tests be conducted unless all parties have been notified in writing and have expressed their opinions regarding the matter. Then, the persons involved in the testing can weigh the advisability of the destructive tests. Quite often, rather than destroy the physical evidence from a fire, an alternative test can be conducted using an identical item to see the effects of the destructive tests on the duplicate compared with the "culprit." It may be necessary to destroy several duplicates to satisfactorily reconstruct the situation regarding the "culprit." However, this is a preferred alternative to destroying the original evidence.

If no alternative is available, complete records of the destructive testing should be kept. In some instances, high speed photography may be necessary to record details of the test. Since it is a one-time operation, every precaution should be taken to record all vital statistics of variables related to the "culprit." Complete documentation of the tests is an absolute necessity to show the justification of destroying the evidence. Photographs should be taken before, during and after the tests and the remains of the object preserved.

DOCUMENTING EVIDENCE

In every case, the fire investigator will talk to persons who, through their five senses, have knowledge of facts related to the case. The problem of documenting all of the facts related by these witnesses will now be considered.

When an eyewitness at the scene of a fire is contacted, the problem facing a fire investigator is how to extract pertinent data from the memory of the witness. This is done in a question-and-answer interview while the investigator records the answers in writing. These witnesses may include the person who first observed the fire, firemen who extinguished the fire, passersby who saw the fire and stopped, policemen responding to the emergency call who controlled traffic, neighbors or even the person who set the fire and lingered to enjoy the excitement and thrill of the moment.

All of the above persons should be interviewed to corroborate versions of the story. A witness matrix may be set up to check any discrepancies or differences. In situations where many persons must be interviewed, a tape recorder can save many hours. Before taping the interview the witness should be informed that it is being recorded. After a few minutes of exposure to the tape recorder, the witness will overcome his fear of a microphone. The recording should include the name, address and identification of the witness, the interviewer and the time and place of the interview. The sooner the interview takes place after the event, the more reliable the witness's memory. Statements taken long after the event are often incomplete and inconsistent. Therefore, the interviews should be conducted as soon as possible.

Eyewitnesses of the fire at its start should be questioned regarding the time, their observations, exact physical locations during the fire and their movements. These facts should be detailed and precisely recorded during the interview or as soon as possible after the interview. It is advisable to make a sketch or have a scale drawing available during the interview for witnesses to pinpoint their positions in relation to the fire.

The preceding pertains to informal interviews taken at the scene or shortly afterwards. These interviews will give a good understanding of the circumstances surrounding the fire from the vantage point of the eyewitnesses. The next step in converting these statements into testimony (evidence which can be used in court) is to take a formal statement under oath in a more formal setting.

Sworn Statements

A witness may be requested to give a formal, sworn statement, called a deposition, at an appropriate location where a court reporter, investigators, police officers or fire marshal are present. A body of questions must be prepared, following a chronological sequence to elicit or embellish pertinent facts in the witness's story.

After the witness is sworn, the fire investigator must ask questions to enhance or corroborate the informal interview. After the statement has been completed, the witness must read the statement and sign it to attest that it is his sworn testimony.

Interrogations

Interrogations are interviews conducted with a suspected criminal. A civil investigator should not attempt interrogations unless he is a licensed criminal investigator. Interrogations should be conducted only after properly advising the suspect of his constitutional rights, following the Miranda decision. These interrogations are best conducted without paper, pencil or tape recordings, which have a tendency to impose additional barriers to free exchange between the suspect and interrogator. After interrogation is completed and the suspect has left, the interrogator should immediately, before any distractions, record the pertinent points, attitude and response of the subject. Note if further investigation appears necessary to corroborate or refute statements made by the suspect.

After sufficient information has been garnered, the fire investigator can frame questions based on the knowledge and previous testimony of the suspect. A deposition or formal statement can then be taken with a court reporter present so that the suspect's story will be on record.

CHAPTER 15

The Expert Witness

MORE SCIENTISTS are living today than have lived in all the history of the world. Scientific knowledge has escalated so rapidly that it is impossible for any one person to become an expert in all fields encountered in fire investigation. Therefore, the court recognizes two general classifications of witness: the lay witness and the expert witness.

The average or *lay witness* may testify only to what he saw, heard, smelled, tasted or felt. These are admissible facts which the jury, not the witness, may use as a basis to draw conclusions and render a verdict.

The second type is the *expert witness*. By presentation of his education, training and experience credentials to the court, he may be classified as an *expert* on a subject considered beyond the general knowledge of the lay person. The expert witness may express an opinion, based upon established facts, on the case being considered by the jury. The sole purpose of testimony and opinion rendered by an expert witness is to *assist, not direct,* the jury in reaching a verdict.

His specialized training and experience will qualify the fire investigator as an expert witness. However, the fire investigator will frequently require the professional assistance of other experts. When confronted with complex technical problems, special skill and knowledge may be necessary to support the fire investigator in his efforts.

The reconstruction of large fires in modern business buildings and skyscrapers is a very extensive and complex project. The remains of fires are fragile and often must be disposed of quickly. Careless or ignorant action may destroy evidence. The expert can significantly enhance the results of the investigation while saving time and money by using other experts. Also, working with other experts will expose the fire investigator to new techniques and provide excellent on-the-job training.

The opinion of an expert may be a double-edged sword. There is no guarantee that his findings will confirm the theory of the fire investigator. The independent expert will seek the facts and give his opinion as an unbiased, disinterested third party. However, it is this unbiased attitude which makes the independent expert most valuable in the search for truth.

Resolving conflicts in testimony, evidence and expert opinion is the responsibility of the jury. Clear, truthful presentation of all relevant facts is the duty of the fire investigator.

An expert witness must be unimpeachable and have a reputation of unquestionable integrity. His opinion may outweigh a dozen confused eyewitnesses.

The most effective use of experts requires establishing a rapport with them so they work as a team with free exchange of information and opinions during the investigation. They should have a preliminary briefing where the problem is fully discussed and special assignments made. Additional meetings should be arranged as required.

In the interest of economy, time and results, the fire investigator may be negligent in his duty if he does not employ specialists on his "team." The expert contributes to saving time, an increase in the quality of the investigation and added knowledge for the fire investigator.

RETAINING AN EXPERT

When you come to the realization that an expert is needed, you undoubtedly have a problem which must be resolved. For example, did a solenoid valve malfunction, allowing gas to flood the area and cause an explosion? A gas expert may be consulted to determine the validity of this theory. At the time an expert is retained, it is not known whether his efforts will result in an opinion which will confirm or contradict the theory. Therefore, if the expert's determination is the key point in the case, he should be retained as early as possible and his conclusions considered before proceeding further. In the event that the expert's opinion is unfavorable, further work and expense may be avoided.

Other advantages are obtained in retaining an expert. When you appear before a jury and try to answer all of the questions regarding the circumstances of a fire, you may be overextending yourself regarding your experience, training and education and thus may lose your credibility in the eyes of the jury. For example, when you testify that you have eliminated all other possible causes, you are expressing an opinion, not a fact. Therefore, under cross-examination, the opposing attorney may ask whether you are an authority on electrical matters. Since you have already testified that you had eliminated all other possible causes of the fire, including electrical, you would, of course, be forced to respond in the affirmative. He then would pursue other possible causes, including lightning, spontaneous combustion and all other areas which may have been a cause of the fire. In each of these fields you would have to respond as being an expert. The jury may find it hard to believe that one person could become an expert in so many fields.

If you are relying upon someone else to eliminate these other possible causes, this would indicate to the jury that you realize your own limitations and have called upon help as needed. This will enhance your credibility and the credibility of the expert you retained. The action of retaining outside experts would also be favored by your client or attorney, who would not be relying upon one witness to prove the case.

At any time during the investigation when you feel the case is not progressing as it should, outside help should be obtained. For help of a general nature, secure other fire investigators to assist in evaluating your case by playing the *devil's advocate* and pointing out weaknesses. This assistance can be found from experts or colleagues in your organization, independent fire investigators, government agencies or private firms.

The investigator should maintain a list of local and national fire investigators and related specialists. A national fire investigator list would include experts who have testified on nationally publicized cases of major concern and have established themselves as recognized authorities in fire investigation. These can be found from major insurance company claims departments, the national headquarters of engineering societies and the Academy of Forensic Sciences. *The Lawyer's Desk Reference,* Fifth Edition, by

Fielow, Robb and Goodman, is a good source of nationally known expert witnesses.

Specialists, such as engineers, scientists, etc., might help indirectly with the investigation. These can be located through colleagues, superiors, other fire investigators, local engineering societies, universities, local testing labs, insurance claims departments or consulting engineering organizations. A good starting point is the engineering classification in the Yellow Pages of your telephone directory or the public library.

SELECTION OF THE EXPERT WITNESS

The expert witness should be selected on the basis of needs peculiar to your particular case. Having once defined the exact problem, the proper expert can be approached and selected on the basis of fulfilling those needs. The expert should be questioned regarding his education, training and experience, particularly in the area of concern. His selection should be based on his expertise and court experience.

His background should consist of sufficient education and experience in the problem at hand so that his expertise follows as a matter of course. It is advisable that a preliminary meeting be held to get acquainted. Your attorney or client should be present, if possible, inasmuch as you are all on the same team.

At the first meeting, it is imperative that the expert understands precisely what is expected. Request that he explain the problem back to you. This is his acid test; if he cannot explain it to you so that you can understand it, he certainly will not be able to explain it to a jury. If possible, observe him testifying in court. The expert witness must establish credibility, and your decision to retain him should be based on his ability to communicate to the jury in such a manner that they will understand and accept his opinion.

Engineers

The largest classification of experts of interest to the fire investigator is that of engineers. Following are the main subclassifications used in fire investigation. Areas of expertise are provided to indicate which type of engineer to retain for typical problems.

Agricultural Engineer

Fires associated with farms or agricultural processes may require the aid of an agricultural engineer. These engineers would likely have knowledge of fires in stored grain caused by spontaneous combustion, mechanical overheating or other failures associated with farm machinery, as well as fires involving seed, feed, fertilizer, chemical plants, etc. They are especially familiar with farm machinery and the processes associated with planting, harvesting, storing, processing and transporting agricultural products.

Automotive Engineer

The automotive engineer is highly specialized and restricts his testimony to automotive industry products (buses, trains, trucks, construction equipment, automobiles). Automotive engineers may be difficult to locate, except in the Detroit area. However, contacting the local consulting engineers council or vocational schools may lead to an expert.

Chemical Engineer

In-depth knowledge of chemical processes requires intense specialization but is invaluable in analyzing a chemical process. If possible, locate a chemical engineer within the industry who is familiar with the process involved and deals with it routinely, or contact universities or government agencies.

Civil Engineer

Civil engineering is a large classification which includes construction of buildings, sewers, roads, dams, industrial plants, etc. The expertise of the civil engineer may be required where the effects of stress, wind, heat, etc. on a structure resulted in failure, collapse or expansion of materials and caused or contributed to a fire.

Combustion Engineer

Combustion engineers are primarily concerned with fuels and combustion chambers used in everything from portable propane stoves to jet transports, nuclear submarines or gigantic steam generating plants.

Control Engineer

Control engineers work on the design of control systems. Their design objectives are to achieve automation of processes through the use of automatic or semiautomatic devices. The control engineer should be intimately familiar with the type of control being tested. The most qualified man may be one working for a competitor of the manufacturer of the "culprit" control.

Electrical Engineer

The universal application of electrical energy in today's society has resulted in great specialization among electrical engineers. Selection of a qualified electrical expert requires careful delineation of the situation involved to bring appropriate skill and expertise to bear on the problem. Qualified electrical engineers can usually be found in universities, engineering societies, industry and public agencies.

Fire Protection Engineer

Fire protection engineers are usually associated with a manufacturing company supplying fire protection devices or systems for the public and are familiar with heat detection devices, sprinkler systems, fire doors, fire suppression chemicals, fire extinguishers and similar fire fighting devices. Fire protection engineers are found in major cities. A list is available from the National Fire Protection Association.

Forensic Engineer

The forensic engineer is one who has expanded his expertise to include the application of scientific knowledge to law. His credentials must be carefully examined to ensure that he has sufficient education, training and experience relevant to the problem area. Forensic engineers are not usually classified in the Yellow Pages as such, but they can be identified by attorneys, insurance companies or engineering societies. They should have extensive experience both in court and in the investigation of similar problems.

Industrial Engineer

Industrial engineers are primarily concerned with the equipment and techniques employed in large manufacturing or processing plants. Their overall understanding of industrial operations makes them uniquely qualified to explain the interaction of complex machinery, processes and chemicals in modern plants. However, due to the wide variation in modern industrial facilities, they must be selected on the basis of relevant experience. Industrial engineers often work as independent consultants but are also found in universities or industry.

Lubrication Engineer

The lubrication engineer is usually either a mechanical engineer or a petroleum engineer who has specialized in lubrication. Therefore, education and background should be considered in selection. For example, if the case involves the overheating of a bearing, retain a mechanical engineer who has lubrication experience rather than a lubrication engineer per se.

Mechanical Engineer

Mechanical engineers are involved in the design and maintenance of the machines, tools and devices required by industry and homes to achieve mechanization and automation of literally thousands of functions and operations.

Of necessity, mechanical engineers have specialized and must be selected on the basis of relevance of experience to the problem under study. For example, some universities have established divisions such as bioengineering, power and propulsion, graphics, design, particle technology, environmental, thermodynamics, industrial transportation and fluid mechanics.

Candidates may be located through universities, Yellow Pages, insurance company claims departments, industry, the American Society of Mechanical Engineers and the Society of Automotive Engineers.

Metallurgical Engineer

Metallurgy is a highly specialized branch of mechanical engineering. It deals primarily with the internal characteristics of materials. For example, a metallurgist could be retained to explain to the jury why a valve failed because of the introduction of ammonia gas into a brass or copper line. Experience in his field is a requirement in the selection of any engineer, but particularly so in metallurgy. As with many other engineering disciplines, metallurgists have become very specialized. For example, metallurgists may specialize in the study of structural failure of steel and have limited knowledge of copper, iron or brass.

Petroleum Engineer

Most petroleum engineers work with the processing of crude oil or gas to produce marketable products. Their expertise is generally in the production, storing and distribution of gasoline, kerosene, diesel oil, jet fuel, heating oil, LP gas or other flammables. Contact the American Petroleum Institute if an expert cannot be located in your area.

Safety Engineer

Safety engineers are concerned with accident prevention. They develop and implement fire prevention programs and are qualified to testify as to possible hazardous conditions which may have existed before a fire.

The Federal Government has set up the Occupational Safety and Health Adm. (OSHA) to establish safety standards in many areas, including fire prevention. Offices have been set up in major cities where information is available. Safety engineers are usually listed in the Yellow Pages or located through the fire marshal's office.

Scientists

Among the scientists, the chemist and physicist are most often recruited to assist in fire investigations. Scientists are primarily concerned with the natural laws which govern the physical universe, while engineers are primarily concerned with the application of scientific principles for the benefit of mankind.

Chemists

The work of chemists often differs from that of chemical engineers in the application of chemical principles. The chemical engineer is usually involved in commercial applications of chemistry in industry, while chemists are usually involved in pure research to learn the composition of substances or in developing new formulas and substances.

Chemists may be needed to identify substances found at a fire site or to explain a chemical process which started a fire or caused it to spread. Chemists work throughout the United States, especially in testing laboratories, industry, consulting firms, universities and governmental agencies.

Physicists

Physics has expanded into a wide range of specialties. Before selecting a physicist, the fire investigator must confirm the relevancy of his expertise. Some are involved in research unrelated to fire investigation while others, working with organizations such as the National Bureau of Standards, may be directly engaged in fire research.

Physicists are found in universities, research labs, governmental agencies and consulting firms.

Specialists

Not to be overlooked are the craftsmen or tradesmen who are experts in the installation and repair of electrical or mechanical equipment. These would include electricians, furnace men, refinery men, construction workers, boiler makers, pipe fitters or others who, by virtue of their education, training and background, are qualified to make a determination regarding the cause of a fire.

Although there are many skilled tradesmen, those selected to assist the fire investigator must also be articulate since they may be called to testify in court. Local attorneys and trade schools may be the best sources of recommendations for qualified specialists, but others such as service companies, construction firms and trade schools or unions may also be helpful.

Table XXXII is designed to assist the fire investigator in pinpointing the expert assistance most likely to be required to successfully determine the cause of a fire.

THE ADVERSARY SYSTEM AND THE EXPERT WITNESS

Our judicial system operates on the adversary principle. The burden of proof lies with the prosecution or the plaintiff, as the law presumes the accused innocent until found guilty by a jury of his peers. The defense attorney is the advocate or protector at law

TABLE XXXII

EXPERT ASSISTANCE

Cause of Fire	Engineer	Scientist	Craftsman
Electrical	Electrical	Physicist	Electrician Manufacturer of electrical component
Control malfunction . . .	Control Mechanical Electrical	Physicist	Maintenance man Machine operator Manufacturer of failed component
Explosions	Chemical Mechanical Electrical	Chemist	Operator of plant Maintenance man Blaster Demolition expert
Gas fires	Mechanical Petroleum (gas experience)	Physicist Chemist	Gas appliance man Maintenance man
Spontaneous combustion	Agricultural Chemical	Chemist	Farmer

of the accused and is ethically, morally and legally bound to provide the defendant with the best possible defense under the law.

The ethics of confidentiality in the client-attorney relationship prevent the attorney from disclosing confidential information unless it is in the client's best interest. If one accused of arson admits to his attorney in detail how he perpetrated the crime, the attorney has a moral, ethical and legal obligation not to disclose the facts of his client's confession.

When an attorney knows precisely how the fire was started, his best defense may be to establish doubt or confusion in the jury. To pursue this course, the defense may hire an expert.

In court, the defense attorney may pose hypothetical situations as possible causes. The expert, in response to such questions, may establish that the prosecution has not proven the cause *beyond reasonable doubt*.

The fire investigator must consider all information acquired as the result of an investigation or conference with attorneys as confidential until it has been presented in court. The defense attorney might reveal confidential information or confessions to an expert, knowing it could not be admitted in court since the expert is bound by the rules of confidentiality. The expert is only required to answer direct questions in court and is not guilty of perjury by withholding information of a confidential nature.

However, there is a risk in permitting the expert witness to have knowledge of the defendant's confession. The prosecution also has the privilege of cross-examination and the use of hypothetical questions. In this way, the prosecution may evoke answers from the expert which could lead to the conviction of the defendant by his own expert witness.

To the lay person, it may appear that we have a very biased and ineffective judicial system where the guilty can confess to an attorney and still go free. Such has been the case. However, the law is aimed at protecting the innocent, and the rule of confidentiality has long been recognized as essential in our system. It is the responsibility of the prosecution to skillfully use legal procedures and prove the guilt of the defendant *beyond reasonable doubt*.

Standards of Foundation

No law exists which regulates evidence or sets requirements for what an engineer or expert must know or have done before he can form an opinion. Experts may take the witness stand who have not visited the fire scene nor examined actual evidence and testify from photographs or reports of others. They may respond to hypothetical questions and give their opinion, with or without restriction as to *reasonable scientific certainty*. Judges, under present rules, have no alternative but to caution the jury to give the weight due the opinion, based on the qualifications of the expert.

Reasonable Scientific Certainty

Reasonable scientific certainty is analogous to a doctor giving medical testimony, when asked for an opinion, qualified by the statement *reasonable medical certainty*. No legal qualifications exist for either of these restrictions on an expert's opinions. The restrictions placed on an expert witness are derived from the canon of ethics of their profession.

For example, the following statement is included in the Code of Ethics of the American Society of Civil Engineers:

> It shall be considered unprofessional and inconsistent with honorable and dignified conduct and contrary to the public interest for any member of the American Society of Civil Engineers: . . .

10. To act in any manner derogatory to the honor, integrity or dignity of the engineering profession.

In the *Guide to Professional Practice,* under the *Code of Ethics,* this is interpreted as follows:

He shall express an opinion only when it is founded on adequate knowledge and honest conviction while he is serving as a witness before a court, commission, or other tribunal.

This is further exemplified in the National Society of Professional Engineering Code of Ethics in Section V:

The engineer will express an opinion of an engineering subject only when founded on adequate knowledge and honest conviction.

These codes of ethics are also reflected in the International Association of Arson Investigators and the National Association of Fire Investigators, both of which have established a Code of Ethics. The International Association of Arson Investigator's code is a little more to the point: "I will bear in mind always that I am a truth-seeker, not a case-maker; that it is more important to protect the innocent than to convict the guilty."

The National Association of Fire Investigators has a similar code: "I will at all times strive to be a fact-finder, seeking for the truth and attempting to determine the true and accurate cause of all fires and explosions or other occurrences which I investigate. . . ."

In summary, the extent of your investigation is entirely within the realm of your conscience. When you appear as an expert witness, you will be expected to give an opinion. This opinion should be based upon adequate foundation of evidence and honest conviction. When you feel adequate evidence is available upon which to base your opinion, submit it to the following tests. Ask yourself, if you were retained by the opposition, where you would attack. Are there weak spots in the case? If so, more tests and depositions should be taken. The facts have a way of being revealed at trials, and it is far better and less expensive to discover these before trial.

Therefore, when you encounter problems outside your area of expertise, retain other experts. When you accept any expert's opinion, make doubly certain that it is valid. If the expert's theory collapses in court, yours does likewise. If in doubt, obtain a copy of the expert's report and duplicate the tests so that you will have a better understanding of the principles involved. This will confirm your judgment of the expert's opinion and enforce your own.

CHAPTER 16

Reports and Court Appearance

THE GOAL OF ANY fire investigation is to determine the origin and cause of the fire. To achieve this objective, accurate records must be kept throughout the investigation. Many photographs will be taken and many pieces of physical evidence collected. Tests may be conducted. Witnesses will be interviewed and statements or depositions taken. All of these activities must be completed and documented in a professional manner to facilitate preparation of the final report or court appearance.

RECORD KEEPING AND REPORT WRITING
File Organization

Every investigation should be given a file number and all associated evidence, documents and photographs identified with that number as soon as possible. The contents of the file may be referenced and augmented frequently. To facilitate these activities, the file contents should be organized logically. The file should be divided into sections such as notes, letters, depositions, statements, evidence records, photographs, photograph indexes and test reports. Within each section, the material should be arranged in chronological order.

For example, while inspecting the fire scene, notes and photographs will be taken. Your observations should be filed under *Notes* and the record of photographs should be filed in the *Photograph Index* section.

It is important that each page of every document or every handwritten note and photograph is clearly labeled with the identifying case number, date and identification of the person who initiated the item. It must be possible to identify each in case it becomes separated from the file. It should be possible to trace each item back to its origin.

Reviewing the File

It is always good practice to reserve judgment until "all the facts are in." Locating the one and true origin is critical to a successful determination of the cause of a fire. Periodically during the investigation, review the evidence accumulated to date to see if alternative origins or causes are possible. Reevaluate where you have been searching for the cause of the fire to assure that you have not overlooked another possible origin.

During the evaluation period a number of theories will be rejected. Each theory should be documented as to the reason for its rejection, and related evidence should be preserved. This will ensure that in the event of trial, the opposition cannot claim that damaging evidence was destroyed, implying that you are a "casemaker." Zealous attorneys are quick to use any opportunity to show reasonable doubt (which may have been created by you when you destroyed negative evidence).

The Point of Diminishing Returns

The investigator must realize that he can rarely obtain "all the facts." Perhaps only experience can teach you to recognize the exact *point of diminishing returns,* i.e. the point where additional efforts may yield little or no further evidence. Because of extensive destruction in many fires, much of the evidence will have been destroyed or altered and rendered useless as evidence in court. Once eyewitnesses have been interviewed and their responses checked on the Witness Matrix, it is highly unlikely that their stories will change. The phrase "all the evidence is in" must be modified to read "all the evidence available is in."

Once the point of diminishing returns has been passed, it is prudent to seek advice to confirm your conviction that the evidence supports your theory and that no other theory could be expounded which would refute yours. This can only be done by eliminating all other possible causes.

Elimination of All Other Causes

At this time, you should critically review the efforts you have made to eliminate other causes and determine whether the testimony of experts in the field of these other possible causes will be required. If outside expertise is required, now is the time to obtain it, not after the judge has ruled that your testimony will not be admitted regarding the chemical causation of a fire.

Report Preparation

A concise but comprehensive report should be the goal of the fire investigator. The report should cover all of the information required in NFPA Form 902, although it is not necessary to use that form. Since Form 902 has no provision for illustrations or photographs, these should be attached as a separate appendix if Form 902 is followed.

Before preparing a report, you should consider your reader's preferences and needs. Unless you are preparing the report for your own file and no one else will read it, the person who will read it should be consulted as to his needs and requirements. A brief consultation with the recipient of the report would acquaint you with his needs. If you are preparing the report for an attorney or prosecutor, a sample form of his brief can be obtained and can be used as a guide in the preparation of your report.

Before writing the report, you should discuss your findings with those who authorized your investigation as to whether a written report is desired. In cases where a trial appears inevitable, it is often the attorney's decision to forego a written report as long as the essential facts are presented to him through verbal communication. If the attorney is in possession of a written report, the opposing attorney can request it for use in preparing his case. The preparation of a written report is a tactical decision which only the attorney can make. Through the rules of disclosure and discovery, the opposition can obtain the information by deposition from the investigator if he is named as a witness to be called at the trial.

As soon as possible after completion of the investigation, the formal report should be written. If no formal report was requested by the people authorizing the investigation, it is still advisable for the investigator to prepare an outline or rough draft in the event that a formal report is requested at a later date. This should be done while the results of the investigation are fresh in his mind.

In some cases, the time lapse between the investigation and a potential court trial may be months or years. Preparing a report just before trial may be difficult, and information overlooked at the time of the investigation may no longer be available. The report should be a complete chronological record of all activities involved in the investigation. It should begin with a statement of the time and authorization of the investigation. All individuals involved should be clearly identified, with their names, addresses and phone numbers and the time and place of their activities noted.

Include all of the negative as well as the positive aspects of the investigation. If a witness appears to be uncertain or reluctant to give information, this fact should be reported. Your attorney will want to know the negative as well as positive evidence and aspects of the case.

Develop your conclusions carefully. Present your opinion in the most descriptive and simple language possible. The statement of your opinion is the most significant result of your entire investigative efforts. It should be written to convey your opinion as forcefully as possible without equivocation. Avoid the use of technical terms which are likely to confuse your reader. Your goal should be to *inform,* not to impress your reader.

All reports should be neatly typewritten and double spaced for fast, easy reading. This also enables the reader to make notes between the lines of the report. The case identification number and description, date of preparation, author's name and address and that of the recipient must be included. The report may be in the form of a letter or any logical format. An index and explanation of attached photographs and documents should be included.

Out-of-Court Settlements

The formal investigation report is the key instrument in preparation for trial. However, a comprehensive, well-written report may also be the key instrument in forcing settlement of civil cases without the necessity of a trial. A trial is necessary only when elements of doubt exist which can only be resolved by a jury. Therefore, the report should resolve all possible issues of doubt. A thorough investigation and report are the best way to achieve out-of-court settlements. Because most property destroyed by fire is covered by insurance, insurance companies enter the fire investigation scene when claims are made to recover losses incurred. Most insurance claims are settled out of court if the investigation provides convincing evidence of the cause of the fire. Insurance companies who have paid claims as a result of fire damage can recover part or all of these claims through *subrogation*.

Subrogation

Subrogation is the transfer of one party's interest to another. In the case of fire insurance, subrogation rights are transferred from the insured to the insurer or insurance company once the insured has been paid. The insurance company then has the right to recover the monies paid by them if they can prove another party was responsible for the loss. This responsible party can then be subrogated against by the insurance company who has already paid the claim, and their loss can be recovered from the subrogated company.

For example, assume a man's house has accidentally been destroyed by fire. He made a claim against his fire insurance company. They paid the claim and then retained an investigator to determine the cause of the fire. It was determined that a faulty valve in the furnace permitted gas to escape, causing a minor explosion, which resulted in the fire that destroyed the building. They filed a claim against the manufacturer of the furnace and were in turn paid by the furnace manufacturer's insurance company. The manufacturer of the furnace then sued the manufacturer of the valve and recovered their loss, again through subrogation.

The investigator's report became the instrument through which the insurance companies were able to subrogate and place the responsibility for the loss on the responsible party. Thus it can be seen that a thorough investigation and a well-written report are invaluable tools in eliminating the time and expense of trials and providing a means for the just disposition of insurance claims.

COURT PREPARATION
Pretrial Preparation

One of the ultimate purposes of the investigator's work is to present his opinion in court as an expert witness. In many cases, however, this may never happen. When the investigation clearly defines the cause and responsibility for the fire, the opposing parties may choose to settle their differences in an out-of-court settlement. A defendant (suspected arsonist) may plead guilty, or the evidence may clear a suspect so that no indictment is handed down, or no significant facts may be uncovered on which to form an opinion.

In some situations, a fire investigation may reveal evidence indicating arson. This must be reported to the proper authorities, which would include the fire chief or district attorney.

The district attorney must evaluate the evidence and determine whether it is sufficient to present before a grand jury so that they may return an indictment with reasonable assurance of a conviction.

In cases where no charges are brought against the suspected arsonist, or a grand jury does not return an indictment, the insurance company covering the fire loss may be persuaded that the evidence of fraud is sufficient to deny the insurance claim. The owner of the property must then bring a civil suit against the insurance company and prove the fire was of accidental origin in order to collect his insurance. Publicity of such cases, depriving the arsonist of his profit or bringing him to justice, may act as a deterrent to fraud arson fires.

Where sufficient evidence pointing to arson can be shown, the authorities will appeal to a grand jury for an indictment. Upon issuance of an indictment, the case will be tried in criminal court.

Investigators from governmental agencies, such as fire marshals, are required to assist the prosecutor in obtaining an indictment and conviction in criminal cases. Independent experts may be retained as witnesses for the prosecution or defense in either criminal or civil cases. Because of the possible infringement of legal rights, all information concerning the cases is considered highly confidential. It is the duty of the fire investigator to report first to the person who authorized or assigned him to the case and not reveal information to other than the proper authorities.

If you are assisting the prosecution, you will be expected to counsel with the district attorney and review pertinent points of your investigation, particularly the photographs and physical evidence. After your report has been prepared, you should select a *devil's advocate* to review the investigation in preparation of your review by the district attorney. The *devil's advocate* can be any person whom you respect as a qualified investigator who will give an unbiased evaluation of your work and recommend further action where necessary. You and your *devil's advocate* should be satisfied that the case is as strong as possible and that you have passed the point of diminishing returns.

You should prepare a list of the various types of tests, models and demonstrations you anticipate displaying to the jury. You should review and work together with the district attorney in preparing your court presentation and demonstrative evidence. Photographs to be used as evidence should be ordered well in advance of the trial date. You should familiarize yourself with every detail (negative as well as positive) of these photographs. As you review the case, you should prepare a list of the principal points, physical evidence, tests, experiments and photographs upon which you based your opinion. This will save you from thumbing through the file to give the basis for your opinion.

Depositions

A deposition is a formal pretrial hearing where the potential witness is asked questions under oath. The purpose of the deposition is to find out what the witness's testimony will be. In the deposition, both the plaintiff and defendants are represented by attorneys. A court reporter is present to record the proceedings.

When an expert witness or fire investigator is deposed, he may be asked to bring his notes, papers, documents, photographs or all materials relevant to the investigation. If the subpoena of deposition is formally rendered, it may state *"Duces Tecum,"* a Latin term meaning to bring your papers or documents. If the subpoena does not state *"Duces Tecum,"* the fire investigator should confer with his attorney and be advised whether he should bring his file.

It is perfectly proper and advisable to rely on notes when being deposed as well as when testifying. If notes are not used at the deposition and trial, the investigator may make conflicting statements which can be challenged by the opposing attorney, resulting in destruction of the investigator's credibility. It is far better to rely on notes rather than memory so that all answers are precise and accurate. Whether in a review, deposition or testimony in court, the investigator must be careful to be completely factual and truthfully accurate at all times.

Sometimes new evidence will be discovered after your deposition has been taken. This may cause you to revise your opinion. In reviewing the files for court, a copy of your deposition should be reexamined. Any discrepancies noted between your latest opinion and your opinion as stated in the deposition should be brought to the attention of the attorney. This will allow the attorney the opportunity to point out the discovery of new evidence. An opinion at the time of court testimony, of course, is based upon all of the evidence available at that time. This includes evidence discovered between the time of the deposition and the time of trial. Since this period may extend to several years in civil cases, the use of notes and the review of the file, including depositions, is vitally important.

Exhibits

Exhibits consist of physical evidence, physical demonstrations which illustrate scientific principles or reenactments of events which resulted in the fire. They include still photographs, motion pictures and video tape presentations and may take any form which will aid the jury in understanding the principles involved and help them arrive at a verdict, whether it be a criminal or civil case.

As long as the presentation is relevant, material and pertinent, practically any technique which will aid the jury will be admissible.

Fires are a highly complex physical-chemical relationship, and quite often a practical demonstration of a principle will assist the jury in understanding the problem. Most courts frown upon dangerous or theatrical presentations which can startle, endanger or disturb the jury. For this reason, most demonstrations must be performed in the judge's chambers prior to court presentation if they are out of the ordinary.

Demonstrations

It is impossible to reconstruct the original fire, but it is quite often feasible and desirable to reconstruct the origin and demonstrate the cause of the fire. Should the cause and reenactment of the origin of the fire be too dangerous to be presented in court, it may be possible to film the actual presentation by burning abandoned buildings and then presenting the film.

Whenever a test or demonstration is anticipated, the fire investigator must guarantee that it will perform as prescribed. Nothing is more embarrassing or damaging to a case than a demonstration which fails. Therefore, before any tests or demonstrations are attempted in the judge's chambers or court, the fire investigator must be familiar with all the variables. The tests should be conducted many times before demonstrating them before the judge. Quite often, the demonstration must be repeated in order to satisfy everyone involved. Therefore, the investigator must provide himself with an adequate supply of any expendable items which will be consumed in the demonstration.

Models

Many model makers specialize in the preparation of scale models for court use. They are listed in the Yellow Pages of the telephone directory under *Model Makers*. When it is impossible or impractical to bring actual items of evidence into the courtroom for the observation of the jury, it may be possible to construct a scale model.

If you anticipate difficulty in making a verbal explanation of how a mechanism works, it may be advisable to present a model or mock-up to demonstrate the principles involved. The scale should be appropriate to the courtroom so that all members of the jury may see the model and realize its significance. The model can usually speak for itself as a graphic aid for your dissertation.

In a fire where buildings were destroyed while under construction, it is possible that the architect may have a model of the proposed work available. Great caution is necessary regarding the use of models in the reenactment of fires. Except under unusual circumstances, to burn a small model is not a true representation of the actual fire. There is a problem of scale. For example, assume that a building was constructed on a scale of ¼ inch = 1 foot. Next assume that a 30 mph wind was blowing at the time of the fire. In this situation, a 30 mph wind would have to be scaled down to $^{11}\!/_{12}$ feet per second, or 0.62 miles per hour. Unfortunately, it is impossible to prevent the flame from burning at its normal rate; the flame and the chemical reaction cannot be slowed down to the equivalent of $^{1}\!/_{48}$ its original flame spread rate, which would be the scaled-down version of its normal flame spread rate.

An expert who testifies that he has burned a model of a building to demonstrate the rate of fire spread should be challenged to show that he has accounted for all the reductions in burn rate, wind velocities, vertical convective current rates and other variables which must be accounted for in reducing from full-scale fires to reduced scale.

COURT APPEARANCE

Trials, whether civil or criminal, are held in courtrooms where the judge presides and the two adversaries meet to present their cases to the judge or jury.

Court Formalities

The court consists of the judge and jury. In criminal cases, the jury must always consist of twelve people. In civil cases, the jury may be reduced to six. If agreeable to both parties, a judge may sit alone to hear the evidence and hand down a decision.

The judge is in charge of the court proceedings. His word is law within the confines of his court. It is his duty to pass judgment upon the legal aspects of any conflicts arising between the attorneys and the court. His decision is final, but the loser has the right to appeal if he feels that the judge made an error in matters of law.

It is the duty of the jury to pass judgment upon the evidence presented in the case by both sides. The jury is guided by the judge on points of law, but the judge has little jurisdiction over what the jury must decide based upon the points of fact or evidence. After each attorney has completed his case, it is the jury's duty to arrive at a verdict, based upon the evidence presented in the courtroom, in compliance with instructions given them by the judge.

A bailiff is always present for swearing in witnesses and marking evidence; he is also responsible for the physical custody of evidence during the trial. At times, these duties are shared by a clerk or court reporter who records the transcript of the trial.

All proceedings and dialogue by the attorneys, witnesses or judges are recorded verbatim by the court reporter. During heated discussions, several participants may try to speak at once, creating an impossible situation for the court reporter. He cannot simultaneously record all statements. The fire investigator should be considerate of the reporter's problems and not speak when someone else is speaking. He should speak clearly and distinctly and assist the court reporter in spelling technical terms.

In a criminal case, the prosecutor represents the state and a public defender or private attorney will represent the defendant. The burden of proof lies with the state. The state must prove beyond reasonable doubt that the accused party is guilty of the charges against him. In the case of arson, the presumption of innocence extends to the fire as well, i.e. all fires are presumed accidental until proven incendiary.

In civil cases, the plaintiff's attorney represents the person who filed suit, and under normal conditions, the state will not be involved.

Federal Rules

Sweeping changes have been made in the Federal Rules of Evidence pertaining to expert witnesses, particularly in the manner, method and means of eliciting information from an expert witness.

The Federal Rules are usually used as a guideline to establish individual State Rules of Evidence. Under Federal Rule 705, the investigator or expert, after being admitted as an expert witness by virtue of his training, education and experience, can state his opinion without any supportive evidence, i.e. foundation. Although opposing attorneys will raise objections, the judge will usually admit the opinion into evidence and advise the jury to give the opinion the weight it deserves, based upon the supportive evidence provided as well as the expert's qualifications.

PRIVILEGED COMMUNICATION. An attorney, as an officer of the court, has a duty to fulfill to the court in providing all information pertinent to a case. He can assist the investigator or expert in identifying those items of his investigation which are *privileged communications* and need not be shown at depositions or trials. Privileged communication may include letters between the attorney and the expert witness pertaining to strategy, tactics, payment of fees, hearsay evidence, inadmissible evidence, recorded telephone conversations between the attorney and the expert witness and recorded telephone conversations between the expert witness and other persons which are part of the file. The attorney must decide which of these are classified as privileged communication and need not be presented in court or at your deposition.

You should feel no compunction about removing these materials from your file. In fact, you will be doing the court a service. The criteria for determining whether the material should remain in your file lies in the answer to the question, Was this information used in forming my opinion? If the answer is yes, you should advise your attorney and leave the decision to him regarding the disposition of the documents in question. Any materials in your possession at the time of your testimony, either in deposition or trial, can be treated as evidence and entered as such by the opposing attorney. Much time in court can be wasted poring over valueless documents. By removing these from your files, you will save the court the time needed to review these valueless documents.

The Expert Witness in Court

The expert witness is called to give an opinion based upon facts introduced as evidence. His testimony depends upon his education, training, experience and ability to convey his opinion in a convincing manner to the jury. This characteristic of the expert witness is referred to as his *credibility*.

Credibility

Webster's New Collegiate Dictionary defines *credible* as "offering reasonable grounds for being believed." The opinion of the most eminent authority will not be believed if his credentials are not established in the minds of his audience. Further, his opinion will be of no value if it is not conveyed in understandable language. If the investigator's opinion is expressed in technical terms understood only by other experts, his opinion will not be communicated to the jury.

To establish credibility, the investigator must state his opinion in terms that a jury can understand and must present sufficient evidence as foundation to support the opinion. The goal of the opposing attorney is to destroy credibility if he cannot destroy the fire theory.

Qualifications Sheet

As an expert witness, you will have to be qualified by the court on the basis of your education, training and experience. Your counsel will question you regarding these facts and will need this data prior to trial to prepare his questions. Therefore, sometime before the trial, you should submit to him a list of your qualifications. He will want background information with as much detail as possible pertaining to the area of expertise in which you will be testifying.

For example, if you are an electrical engineer and are testifying on the temperature of electrical arcs, the attorney will want to know your related experience in the field of designing electrodes, circuit breakers, switches or any arc-creating device. The fact that you had spent twenty years as a field electrician may or may not be pertinent.

The following is a brief outline of suggested information to be included in your qualification sheet.

Education
1. High school, date of graduation
2. College, date of graduation, degrees, honors, fields
3. Postgraduate work, dates, degrees and courses

Training
1. On-the-job training
2. Seminars
3. Workshops
4. Professional meetings attended

Experience
1. Present employment
2. Position with the firm
3. Length of time in that position

4. Previous employment
5. Experience relating directly to testimony

Professional Affiliations
1. Membership in organizations and societies, with offices held
2. Registration, licensing and certification

Writings
1. Articles, books and publications
2. Tests or experiments (related to your area of expertise and testimony)

If your field of expertise is broad, it is probably best to list only your experience, specialized training and publication credits relating to the area in which you are testifying. Seminars, postgraduate work and professional associations should be listed even if they are not directly related. This will show that you are keeping up with the state of the art in fire investigation as related to your area of expertise. List previous court appearances as a witness, particularly as an expert witness.

The reason for being so explicit in listing education, training and experience is that the credibility of an expert is based upon these three factors. Reading the qualification records in court will give the jury a basis for comparing the qualifications of opposing experts. This is no time for the fire investigator to be modest. Include all relevant and significant achievements and let the attorney decide what is pertinent, relevant and material to your qualifications.

The investigator must appear in court as a disinterested witness whose qualifications, partiality and credibility are unquestioned. It is the attorney's responsibility to see that the court is provided with this information. By questioning the investigator as a witness, he can bring out the facts as he deems appropriate to best serve his client. By cross-examination, an opposing attorney can also question a witness to further clarify points in the testimony. Skilled attorneys can sometimes confuse a witness, cause him to contradict himself and lose credibility. The expert, therefore, must carefully analyze each question and answer it slowly and carefully, portraying the truth clearly but only as necessary to each question.

The expert witness should not volunteer any information beyond a direct answer to the question.

Impeachment

Impeachment is a legal way to challenge the credibility or validity of the testimony of a witness. This is usually accomplished by showing that a statement made prior to the court appearance has been contradicted by a statement made under oath during the trial. For purposes of impeachment, the original declaration will be read into the records and the contradictory statement repeated from the records so that the jury can compare the contradictions. Contradictory statements are not uncommon and arise from several sources. The most common is from a misunderstanding of the question. In such a situation, the witness must immediately admit the misunderstanding and ask permission to correct his testimony.

Another problem arises with the development of new evidence and the time interval between the two conflicting statements. If the introduction of new evidence changed the expert's opinion, this must be explained to show that they are both honest opinions based upon "evidence available at that point in time." If there are contradictions in two statements, the witness must admit the error and request an opportunity to correct the statement.

This type of admission will damage credibility, and the jury may question other theories which may have been proposed. The opposing attorney will do everything possible to reinforce the doubt implanted in the minds of the jury by contradictory statements.

Testimony

A fire investigator must anticipate that he will be called upon to testify in court. It is of little significance whether he testifies for the plaintiff or the defense. His testimony is that of an unbiased, impartial witness. The ethical code of the fire investigator states that he is a truth-seeker, not a case-maker. His testimony will be based upon honest conviction and sound evidence. When called upon to testify, he will be providing direct testimony for one side or the other. On rare occasions, he may be called as an *amicus curiae* (friend of the court). When the judge feels that an impartial outside expert could help the jury in resolving two conflicting expert's opinions, he may exercise his prerogative and retain an expert witness as a "friend of the court."

Direct Testimony

Under our adversary system, the ethical position of an attorney is entirely different from that of a fire investigator. While the fire investigator is impartial and unbiased, the attorney has a moral and ethical obligation to defend his client to the best of his ability within the framework of the law. The accused is presumed innocent and is entitled to the best possible defense. In legal cases, the fire is presumed to be an accidental fire until proven otherwise by the prosecution or plaintiff.

The investigator should bear in mind that the attorney is an adversary for his client and will call the fire investigator as an expert witness only if it will aid his client.

If the attorney feels it may not help his client, he may not call for the investigator's opinion. It is not the investigator's responsibility to carry his opinion to others, including the opposition, for to do so would be violating the ethics of confidentiality. The opposition has the same opportunity to retain an investigator to examine the same evidence and uncover the same facts, which can be presented to a jury.

An investigator may be retained by an attorney for different reasons. In some cases, he will be retained to act as a consultant to aid the attorney in understanding the problems involved. The results of this investigation may become a *work product* for the attorney in preparing his case. None of this work is discoverable, i.e. it is not available to the opposition. The expert is not named as a witness to be called at trial. An investigator can also be retained as an *expert witness* whom the attorney plans to call as a witness. In this case, he will conduct an investigation and all his opinions and data are available to the opposition before trial. When an attorney retains a fire investigator, he is not certain that the investigator's findings will aid his client's case. Therefore, in some cases, an expert may be retained with the understanding that he will not testify if his findings are unfavorable.

USE OF NOTES. Concern for accuracy and truth in the investigator's testimony in court cannot be stressed too strongly. By no means should he attempt to memorize the complete file. The most expected is to recall key dates and events. Rely upon notes to refresh the memory and specific details of the investigation. The investigator's credibility may be destroyed in cross-examination or by contradictory witnesses.

ERRORS IN TESTIMONY. "To err is human." Despite thorough preparation, errors in testimony do occur. If you realize at the time that you have made an error, then correct it immediately. Turn to the judge and tell him, "Your Honor, I would like to correct my testimony." Then proceed to do so immediately. However, if you discover an error after you have left the stand, call it to your attorney's attention. He will then advise the court that you wish to correct your testimony and call you to the stand to elicit the correct testimony. The only explanation necessary is "to tell the truth" and admit that you have made an error. Under no circumstances should you answer questions on which you are not sure. Simply state that you cannot recall. It is an injustice to your client, the accused and to both sides to start guessing at answers. It will result in utter destruction of your credibility.

TESTIFYING FOR THE OPPOSITION. According to current statutes, if an attorney has named an expert as a potential witness but in the course of the trial decides that his testimony would not be in his client's best interest, he may decide not to call him as a witness. Under present rules, the opposing attorney then has the right to call the expert as *his* witness. He may assume that the testimony would favor his client since the expert was not called.

However, attorneys are reluctant to place any witness on the stand unless they are reasonably certain of what the witness will say. The opposing attorney, therefore, will request an interview with the expert witness before placing him on the stand. Under present law, the expert witness has the right to refuse to talk to the opposing attorney. This creates a dilemma.

The unbiased fire investigator is interested in presenting the truth to assist the jury in making a decision. The expert witness, having been retained by his attorney, has a moral and ethical obligation. An expert witness obtains part of his fee for his confidentiality and the exchange of privileged information. An expert receiving a fee for his services has a moral and ethical obligation to protect the privileged communications. Therefore, it would normally be unethical for an expert witness to discuss the case with the opposing attorney prior to testifying.

An expert's opinion will remain the same, regardless of whether he testifies for the plaintiff or the defense. However, by the nature of the questioning, the opposing attorney can emphasize points which will favor his client. Since an attorney in cross-examination can only ask questions on subjects discussed under direct examination, pertinent data may not be brought forth unless both attorneys use their prerogatives of calling the expert as a witness. Thus, an investigator may have the experience of appearing as a witness for both sides of a case.

Cross-examination

When the investigator is called upon to give testimony in court as an expert witness, it is assumed that his opinion is unbiased. The expert's testimony may be contrary to the interest of the opposing attorney. Therefore, during cross-examination he will attempt to destroy the expert's credibility by attacking the fallible aspects of mankind, memory and judgment. He will challenge the expert's opinion. He may imply that details have been omitted or that notes have been interpreted incorrectly. One of the best methods to prepare for cross-examination is to imagine that you have been retained by the opposition. Make a list of the weak areas of the case and immediately proceed to strengthen them. If outside experts are needed to support your theory, they should be retained and their opinions recorded as foundation for your opinion. Close attention to details, particularly the elimination of other possible causes, will prepare you to refute other theories presented. When questioned regarding the strength of your opinion, you can point out the care and number of tests which you have made prior to forming your opinion.

The prosecution faces many difficult situations in an arson case. If the accused is guilty and has confided in his attorney regarding details on how the fire was set, the defense attorney will know more about the fire than the prosecutor and his expert.

In cross-examination, the defense attorney will use this knowledge to determine the extent of the investigation and the details developed through inference. If the theory or details of the way the fire started are incorrect as presented by the prosecution, the defense attorney, armed with his superior knowledge, may refute them by testimony of the accused, who was there at the time and had intimate knowledge of how the fire actually started. The burden of proof rests with the prosecution. The opposing attorney does not have to prove his client's innocence. All that is required is proof that the theory is not consistent with the facts and therefore should not be considered. By using the truth, he can destroy the theory even though his client is guilty. While this may seem to be incongruous with justice, it can and often does occur. Although the main facts may be well established by the prosecution, the stumbling blocks are the small details which offer an opening wedge for the defense. It may be contradictions in details that will give the defense the opportunity to destroy the credibility of the fire investigator and thereby win the case. Therefore, close attention to details in all cases is mandatory.

CROSS-EXAMINING THE OPPOSING EXPERT. When the subject matter of a case is very technical, an attorney may ask the fire investigator to listen to testimony and assist in preparing a line of questioning of the opposition's expert witness. In some cases, the fire investigator may also be able to assist the attorney in preparing questions to be presented in his own testimony.

The attorney should be advised to ask the opposing investigator the number of hours spent on the case and the time and date he arrived at his conclusion. The number of hours spent on the case is important to the jury, as it reflects the depth of the investigation. The attorney should be advised to probe into the methods used to eliminate all other possible causes. Such questioning may uncover a shaky foundation for the exclusion of other possible causes and destroy his credibility. The time and date when the expert arrived at his opinion is important. If he names a time preceding the receipt of vital evidence, it would indicate prejudice or that he has formed an opinion without sufficient evidence. The investigator should always record the time and date that he has arrived at a conclusion, in anticipation of such a question.

SUMMARY

Preparation for writing the formal report and possible court appearance begins when the assignment to investigate a fire is received. When the investigation begins, there is no way to foresee all that lies ahead. Fire investigations may end in litigation, so the fire investigator must prepare for that eventuality.

The investigator must always be mindful of the canon of the International Association of Arson Investigator's Code of Ethics: "I am a truth-seeker, not a case-maker. . . ." The truth-seeker characteristically maintains an impartial, unbiased attitude and considers all relative evidence in forming his opinion as to the cause of the fire, whereas the case-maker typically preserves only evidence which supports his case and ignores evidence which tends to disprove it. The fire investigator should not take the role of adversary and should be careful to avoid any action which would create that impression.

Index

A

Accelerants, 45, 93, 95, 131, 134, 311, 325
Acetylene-oxygen, 341
Adsorption, soil, 186
Aerial photography, 250, 368-370
  calculation of lens required, 369
Aerosol cans, 355-361
  illustrations, 356, 359, 360
Air
  circulation, 181
    infiltration, 181-182
    table of air changes, 183
  combustion, 220
  gap, 264
  primary, 216
  secondary, 216
Aluminum wiring, 166-169 (*see also* Wires, aluminum)
America Burning, 4, 32, 351
Ampacity, 156
Ampere, 144
Animals, as cause of fire, 87
Aperture, function, 396
Appliance controls, 114, 257-269, 275-276, 279
Appliance malfunctions, 89, 257-290
Arson, 90-91, 137
  conviction record, 8, 119
    tables, 121, 122
  definition, 122
  delay mechanisms (*see* Delay mechanisms)
  development of criminal case, 139-141
  elimination of causes, 140-141
  evidence of, 131, 141
    building contents, 135-136
    forced entry, 97, 130-131
    multiple low points, 106
    ventilated fire, 135
  fires

    accidental, 123, 129-130
    criminal, 123
    factors to be considered, 128-129
    fraud, 8, 123-124, 140
  investigation of, 126-138
  laws, 122, 124-126
  motives, 125-126, 140
  motor vehicle, 141, 324-326
Arson investigator, 16, 126
Arsonist's Cookbook, The, 8
ASTM Standard Fire Curve, 54
  illustration, 72
Atoms, 45
Auto-ignition, 329
  spontaneous combustion, 68
Autoignition temperature, 62
  table, 60-61

B

Backdraft diverter, 277
Basic Incident Report, 9, 10
Bibliography, 32-33
BLEVE, 222-223, 238
Blow-back, 76-77
"Boiling Liquid-Expanding Vapor Explosion" (*see* BLEVE)
Boiling point, 57-58
  table, 60-61
Bomb, pipe, 252
Brake drums, overheated, 315
Brazing, 343
Building, contents of, 75
  fire load, 100-101
  in arson, 135-136
Building construction, 53, 56, 75
Building settlement, 87
Burden of proof, 419
  beyond reasonable doubt, 125, 420, 425
  in civil cases, 125
  in criminal cases, 125
  reasonable scientific certainty, 421